Combustion Noise

Anna Schwarz · Johannes Janicka
Editors

Combustion Noise

Editors

Anna Schwarz
Technische Universität Darmstadt
Institute for Energy and
Powerplant Technology
Petersenstr. 30
64287 Darmstadt
Germany
schwarz@ekt.tu-darmstadt.de

Prof. Dr.-Ing. Johannes Janicka
Technische Universität Darmstadt
Institute for Energy and
Powerplant Technology
Petersenstr. 30
64287 Darmstadt
Germany

ISBN 978-3-642-42610-0 ISBN 978-3-642-02038-4 (eBook)
DOI 10.1007/978-3-642-02038-4
Springer Dordrecht Heidelberg London New York

Cover design: eStudio Calamar S.L.

Springer is part of Springer Science+Business Media (www.springer.com)

Foreword

November, 2008 *Anna Schwarz, Johannes Janicka*

In the last thirty years noise emission has developed into a topic of increasing importance to society and economy. In fields such as air, road and rail traffic, the control of noise emissions and development of associated noise-reduction technologies is a central requirement for social acceptance and economical competitiveness. The noise emission of combustion systems is a major part of the task of noise reduction. The following aspects motivate research:

- Modern combustion chambers in technical combustion systems with low pollution exhausts are 5 - 8 dB louder compared to their predecessors. In the operational state the noise pressure levels achieved can even be 10-15 dB louder.
- High capacity torches in the chemical industry are usually placed at ground level because of the reasons of noise emissions instead of being placed at a height suitable for safety and security.
- For airplanes the combustion emissions become a more and more important topic. The combustion instability and noise issues are one major obstacle for the introduction of green technologies as lean fuel combustion and premixed burners in aero-engines. The direct and indirect contribution of combustion noise to the overall core noise is still under discussion. However, it is clear that the core noise besides the fan tone will become an important noise source in future aero-engine designs. To further reduce the jet noise, geared ultra high bypass ratio fans are driven by only a few highly loaded turbine stages.
- The development of layout tools for "quiet" technologies requires calculation methods for predicting of sound power in new technical products.

The research Initiative "Combustion Noise" was founded in the year 2002 and was funded by the German Research Foundation (DFG) for six years. The objective of this project was the development of methods to identify basic principles required to reduce the noise in combustion systems.
This book reflects the main findings of the research Initiative "Combustion Noise"

over the last six years. The chapters summarize the main scientific results of the sub-projects of this Initiative.

The editing of this book required the cooperation of all authors. This excellent collaboration is greatly appreciated. The financial support of the German Research Foundation (DFG) is gratefully acknowledged.

Contents

1 **Numerical RANS/URANS simulation of combustion noise** 1
Bernd Mühlbauer, Berthold Noll, Roland Ewert, Oliver Kornow and Manfred Aigner
1.1 Introduction 2
1.2 Theoretical Background 2
1.2.1 RANS/URANS approach 3
1.2.2 Boundary conditions 4
1.2.3 RPM-CN approach 4
1.3 Results and Analysis 9
1.3.1 Indirect combustion noise 9
1.3.2 Direct combustion noise 19
1.4 Conclusions 26
References 28

2 **Measurement and Simulation of Combustion Noise emitted from Swirl Burners** 33
C. Bender, F. Zhang, P. Habisreuther, H. Büchner and H. Bockhorn
2.1 Introduction 33
2.2 Theoretical Background 35
2.2.1 Experimental Setup 35
2.2.2 Numerical Methods 37
2.3 Results and Analysis 41
2.3.1 Experiment 41
2.3.2 Numerical Simulation 49
2.4 Conclusions 59
References 60

3 **Modeling of noise sources in combustion processes via Large-Eddy Simulation** 63
Anna Schwarz, Felix Flemming, Martin Freitag and Johannes Janicka
3.1 Introduction 63

3.2 Theoretical Background ... 64
3.2.1 Non-Premixed Flames ... 65
3.2.2 Partially Premixed Flames ... 67
3.2.3 Premixed Flames ... 68
3.2.4 LES/CAA Hybrid Approach ... 70
3.3 Results and Analysis ... 73
3.3.1 Open, Non-Premixed Jet Flames ... 73
3.3.2 Model Combustor (Partially Premixed Flames) ... 77
3.3.3 Tecflam Burner (Premixed Flames) ... 81
3.3.4 LES/CAA Coupling ... 83
3.4 Summary and Conclusions ... 85
References ... 86

4 Modelling of the Sound Radiation from Flames by means of Acoustic Equivalent Sources ... 89
Rafael Piscoya, Haike Brick, Martin Ochmann and Peter Költzsch
4.1 Introduction ... 89
4.2 Theoretical Background ... 90
4.2.1 Hybrid Approach ... 90
4.2.2 Equivalent Source Method (ESM) ... 92
4.2.3 Boundary Element Method (BEM) ... 93
4.2.4 Numerical Simulation of the Flames ... 95
4.3 Results and Analysis ... 95
4.3.1 Numerical Aspects of the Hybrid Method ... 95
4.3.2 Location of the Control Surface ... 102
4.3.3 Inclusion of Ground Effects ... 102
4.3.4 Measurement of the Flame ... 104
4.3.5 Results of the Simulation and Comparison with the Measurement ... 105
4.3.6 Sound Propagation in a Non-Homogeneous Medium ... 107
4.4 Conclusions ... 119
References ... 120

5 Investigation of the Correlation of Entropy Waves and Acoustic Emission in Combustion Chambers ... 125
Friedrich Bake, André Fischer, Nancy Kings and Ingo Röhle
5.1 Introduction ... 126
5.2 Theoretical Background, Test Specification and Data Analysis ... 126
5.2.1 Test Specification and Data Analysis ... 129
5.3 Results and Discussion ... 134
5.3.1 Entropy Wave Generator Test Rig (EWG) ... 134
5.3.2 Combustor Test Rig ... 137
5.4 Conclusions ... 141
References ... 142

6 Influence of boundary conditions on the noise emission of turbulent premixed swirl flames 147
Fabian Weyermann, Christoph Hirsch and Thomas Sattelmayer
6.1 Introduction 148
6.2 Theory and Methods 149
6.2.1 Calculation of the acoustic power spectrum 149
6.2.2 Modeling of the spectral heat-release 152
6.2.3 Coherence volume 156
6.2.4 Acoustic power spectrum of an unconfined flame 157
6.2.5 Simulation of confined flames 157
6.2.6 Experimental setup and measurement techniques 161
6.3 Results and Analysis 166
6.3.1 Validation of the noise-model for unconfined flames 166
6.3.2 Adiabatic flames 168
6.3.3 Unconfined flames, modeling based on CFD-data 168
6.3.4 Sound emission from a complex combustion system into the environment 169
6.4 Conclusions 170
References 171

7 Theoretical and Numerical Analysis of Broadband Combustion Noise 175
Thanh Phong Bui and Wolfgang Schröder
7.1 Introduction 175
7.2 Aeroacoustic theories to compute combustion generated noise 178
7.2.1 Acoustic analogies based on a scalar wave equation 178
7.2.2 Acoustic perturbation equations for reacting flows (APE-RF) 179
7.2.3 Summary of the APE-RF formulation 182
7.2.4 Source term formulations 183
7.2.5 Rayleigh's criterion for acoustic wave amplification 186
7.3 Hybrid CFD/APE-RF method to simulate combustion noise 187
7.3.1 CFD/CAA interface conditions 187
7.3.2 Numerical methods used in the CAA 196
7.4 Results and Analysis: Application of the APE-RF system to open turbulent flames 197
7.4.1 H3 Flame: A non-premixed open turbulent flame 197
7.4.2 Premixd Methane Flame 199
7.4.3 DLR-A Flame: A non-premixed open turbulent flame . . . 204
7.5 Summary and Conclusions 209
7.6 Acknowledgments 211
References 211

8 Investigations Regarding the Simulation of Wall Noise Interaction and Noise Propagation in Swirled Combustion Chamber Flows 217
Christoph Richter, Łukasz Panek, Verina Krause and Frank Thiele
8.1 Introduction 217
8.2 Theoretical background 220
8.2.1 Mathematical models 220
8.2.2 Numerical Method 221
8.2.3 Intensity-based analysis of the result 225
8.3 Results and Discussion 227
8.3.1 The entropy wave generator (EWG) model experiment 227
8.3.2 Swing-off response of a premixed swirl combustor flow 231
8.3.3 10 kW model combustion system with 17 mm exit nozzle 233
8.4 Conclusion 235
References 237

9 Direct Numerical Simulations of turbulent flames to analyze flame/acoustic interactions 239
G. Fru, H. Shalaby, A. Laverdant, C. Zistl, G. Janiga and D. Thévenin
9.1 Introduction 239
9.2 Theoretical Background, Numerical methods and procedure 242
9.2.1 DNS code family *Parcomb* 243
9.2.2 DNS code family π^3 245
9.2.3 Steps involved by the computational procedure 246
9.2.4 Modeling of chemical reactions 247
9.3 Flame/acoustics interactions investigated with DNS 253
9.3.1 Local Rayleigh's criterion 254
9.3.2 Examples of results and discussion 255
9.4 Post-processing challenge: AnaFlame 261
9.4.1 Introduction 261
9.4.2 Content of the post-processing toolbox AnaFlame 262
9.5 Conclusions and perspectives 263
References 266

10 Localization of Sound Sources in Combustion Chambers 269
Christian Pfeifer, Jonas P. Moeck, C. Oliver Paschereit and Lars Enghardt
10.1 Introduction 269
10.2 Theoretical Background 271
10.2.1 Theory of Nearfield Acoustic Holography 271
10.2.2 Sound Pressure Field in the Combustion Chamber without Mean Flow 272
10.2.3 Modal Composition of *G* 273
10.2.4 Sound Pressure Field in the Combustion Chamber with Mean Flow 275
10.2.5 Reflection at the Combustion Chamber Outlet 276
10.2.6 Reconstruction of Sound Sources 277

10.3 Results and Analysis 280
10.3.1 Optimization of the Sensor Arrangement 281
10.3.2 Effect of Noise on the Reconstruction Accuracy 284
10.3.3 Reconstruction of Sound Sources not located on assumed Source Distribution 285
10.3.4 Effect of Reflection at the Combustion Chambers Outlet . 287
10.3.5 Effect of Mean Flow 288
10.4 Conclusion 289
References 290

Preface

November, 2008 *Anna Schwarz, Christoph Richter and Johannes Janicka*

The reduction of noise emissions is a topic of increasing relevance to the public and the economy. While the research effort has previously concentrated on aerodynamic noise formation by airplanes and other vehicles, the research Initiative "Combustion Noise" promoted the development of methods and design criteria to minimize the noise formation by combustion sources. The research was focused on the generation, propagation and radiation of the combustion noise. With the help of suitable experiments the group developed a better understanding of the underlying processes. Key simulation techniques were the Large Eddy Simulation (LES) and well established industrial application URANS-methods (Unsteady Reynolds Averaged Navier Stokes) to compute the noise sources related to the combustion process. Non-intrusive measurements were performed using laser diagnostics, yielding a better insight into the phenomena and providing data for validation. Direct sources of noise emissions due to the time dependent changes of the heat release and the indirect sources from entropy fluctuations were investigated and included in the modeling process.
The research Initiative "Combustion Noise" was split into 10 sub-projects, each presenting their main results within a separate chapter. Thus, the chapter structure corresponds to the project number given in the following. The organization of the project as distributed research collaboration allowed the contribution of various german experts in combustion as well as in aeroacoustics with experiments, analytical and numerical contributions. Therefore, this book provides a comprehensive collection of works with relation to combustion noise. In the following the involved institutions and their projects will be summarized in short:

1. Institute of Combustion Technology, Stuttgart. (Prof. Dr.-Ing. M. Aigner).
   The first sub-project, "Numerical URANS Simulations of combustion noise", focused on the development of URANS methods. The heat release fluctuations due to the periodical combustion oscillations were described through the various modeling methods. The turbulent broadband was captured through different

modeling approaches. These methods were applied to simulations of indirect entropy noise and direct combustion noise.

2. Institute for Chemical Technology and Engler-Bunte Institute, Division of Combustion Technology, University of Karlsruhe (Prof. Dr.-Ing. H. Bockhorn).
   The objective of the second sub-project "Measurements and simulation of noise emitted from swirl-burners with different Burner Exit Geometries" was an experimental identification of sound intensity of turbulent, enclosed and swirled flames, which should provide a basis for physical modeling of quantitative noise formation. The simulations of experimentally investigated configurations with a compressible LES approach were also conducted.
3. Institute for Energy and Powerplant Technology, Technical University of Darmstadt (Prof. Dr.-Ing. J. Janicka)
   The third sub-project, "Modeling of noise sources in combustion processes via Large-Eddy Simulation", was focused on a further development of the Large Eddy Simulation. The resulting noise sources provide a basis for the noise propagation modeling methods like Linearized Euler-Equations (LEE), Acoustic Perturbation Equations (APE), Acoustical Equivalent Source Methods (ESM) and Boundary Element Methods (BEM). The development of subgrid scale modeling under influence of acoustic perturbation was another aspect of this project.
4. Department of Mathematics, Physics and Chemistry, Technical University of Applied Sciences Berlin (Prof. Dr.-Ing. habil., Dipl.-Math. M. Ochmann) and Institute of Acoustics and Voice Communication, Technical University of Dresden (Prof. Dr.-Ing. habil. P. Költzsch).
   The main objective of the fourth sub-project, "Modeling of the sound radiation by means of the equivalent source method", was an advancement of the simulation of noise formation and radiation due to the acoustic Equivalent Source Methods (ESM) and Boundary Elements Methods (BEM). The aims of these investigations were the coupling between LES-ESM/BEM, the simulation of noise field based on incompressible and compressible LES, the validation of simulation methods of the far fields of flames and the adaption of simulation processes for sound radiation of enclosed flames.
5. Institute of Propulsion Technology, DLR Berlin (Dr.-Ing. I. Röhle).
   The "Investigation of the correlation of entropy waves and acoustic emissions in combustion chambers" was the focus of the fifth project. Based on a model combustion chamber with variable length and variable cross section of the combustion chamber outlet nozzle, the contribution of the direct noise and the entropy sound to the total noise in a combustion chamber was investigated. In a model experiment featuring electrical heating to generate non-isentropic perturbations in a spatially varying average flow field a reference test case was set up. Comprehensive experimental data was provided for the validation of numerical methods with respect to entropy noise.
6. Institute of Thermodynamics, Technical University München (Prof. Dr.-Ing. T. Sattelmayer).
   The acoustic ambient conditions are very significant for simulations of combustion sources. The sixth sub-project,"Influence of boundary conditions on the

noise emissions of turbulent premixed swirl flames", concentrated on contactless methods for local and high dynamic heat release and velocity measurements. Premixed and diffusion flames were the subject of the study. The focus of the enhanced understanding of combustion induced noise in this work was the formation and development of simpler usable models.
7. Institute of Aerodynamics, RWTH Aachen University (Prof. Dr.-Ing. W. Schröder). The development of noise propagation simulation methods was the aim of the seventh sub-project, "Simulation of combustion noise in the near field of premixed and diffusion flames". The acoustic perturbation equations were developed for the simulation of sound propagation in strong inhomogeneous fields based on the LES of the instantaneous source field.
8. Institute of Fluid Mechanics and Engineering Acoustics, Berlin Institute of Technology (Prof. Dr.-Ing. F. Thiele).
The technical combustion systems are usually enclosed, so that the interaction of sound, wall and the mean flow are very significant. In the eighth sub-project, "Investigations regarding the simulation of wall noise interaction and noise propagation in swirled combustion chamber flows", a computational aeroacoustics method was applied for the simulation of entropy modes and their sound generation.
9. Institute of Fluid Dynamics and Thermodynamics, Otto-von-Guericke University Magdeburg (Prof. Dr.-Ing. D. Thevenin).
The ninth sub-project, "Direct numerical simulation of the interaction between flame and acoustic waves", delivered information about the interaction of sound waves and combustion due to the direct numerical simulation of partial adjustment ranges in premixed and diffusion flames. This information was used for detailing the inner structures of flames and for improving and validating the models used in the LES and CAA simulations.
10. Hermann-Föttinger Institute of Fluid Mechanics and Engineering Acoustics, Technical University of Berlin (Prof. Dr.-Ing. C. O. Paschereit) and Institute of Propulsion Technology, DLR Berlin (Dr. rer. nat. L. Enghardt)
The tenth sub-project, "Acoustical near field holography in combustion chambers", dealt with the indirect determination/identification of sound sources from acoustic pressure measurements. The focus of this work was the reconstruction of sound sources using a Green's function representation of the sound pressure field for the investigated combustion chamber geometries. The method was developed for the combustion chamber from sub-project five and delivered additional validation data for theoretical-numerical sub-projects.

The cooperation of experts in different areas allowed the development of new methods and understandings which gain input from different fields. Besides the contribution of each sub project, this led to an additional impact of the present results. The major advances in describing and understanding combustion noise through the project are summarized as following:

- The understanding of the mechanisms of noise generation was addressed using suitable experiments and simulations in which a wide variety was developed in the current project. Several improvements in combustion noise modeling are presented in chapter 1 for URANS based method as well as in chapter 2 and 3 for LES based methods. A direct numerical simulation was applied in chapter 9. Acoustic propagation and far-field methods for combustion noise are presented in chapter 4, 7 and 8. A large experimental parameter variation for premixed and non-premixed flames is provided in chapter 2, therein LES simulation was applied to identify large scale structures as noise source. Further experiments including advanced instrumentation were provided in chapter 6. Finally, experimental investigations of the indirect entropy noise generation are presented in chapter 5.

- Another topic was the modeling of the reflections from connected duct systems. The numerical simulations in chapter 1 as well as 8 underline the importance of impedance modeling for the simulation of combustion noise. Due to the correct consideration of the reflection from up- and downstream duct sections, the prediction of peak frequencies becomes possible and the swing off was adjusted with the correct impedance applied.
- The contribution of experts from the different fields allowed the development of a variety of hybrid methods for the prediction of the noise propagation and radiation which combine the specific numerical methods. Combustion noise required quite different numerical modeling assumptions for the combustion process and the noise propagation and radiation. The idea to couple two methods which were developed for each of these objectives was obvious. Consequently, there is a large variety of hybrid approaches.
  A hybrid RANS-CAA approach was applied using an extension of the random particle method (RPM) in chapter 1. A boundary element method and an equivalent source method which were both based on the Helmholtz equation were applied in chapter 4 to obtain the far-field characteristics of an open flame based on LES (chapter 3). The extension to a dual reciprocity boundary element method which is capable of handling temperature gradients in the field was discussed in chapter 4. Several parameters influencing the accuracy of the acoustic prediction were also investigated. The acoustic perturbation equations described in chapter 7 were applied using the unsteady sources from LES described in chapter 3 as well as from a direct numerical simulation described in chapter 9. Last but not least the application of a hybrid approach based on URANS-CAA coupling is presented in chapter 8.
- An outstanding result of the project was the development of a theoretical model for the prediction of the sound spectrum. Based on a dimensional analytical consideration a general model for the prediction of the sound power spectrum of a flame was developed in chapter 6. The model parameters were influenced by innovative measurements like simultaneous temporally resolved PIV/LIF measurements. These measurements form a breakthrough in the understanding of flame dynamics and allow the adjustment of the model parameters. The model was shown to be in a good agreement with a wide variety of open and thermal enclosed flames (comp. chapter 6).

- The modeling of indirect combustion noise sources was addressed within the project as a potential noise source in a realistic application of gas turbines and aero-engines. The first experimental evidence of the generation of indirect noise was achieved in the model combustion chamber described in chapter 5. The original model of the combustion chamber was investigated numerically in chapter 3. A more realistic redesign of this combustion chamber was used in chapter 5 in order to provide further evidence of the source mechanism together with comprehensive benchmark data for code validation. However, due to the complexity of the system it was not applicable for a first validation of numerical codes.
  For this reason a simplified experiment was developed which has soon become a reference for code validation with respect to entropy noise. It was described in chapter 5 as well. Based on this simplified experiment with controlled electrical heating a validation of numerical methods regarding the indirect source mechanism was carried out for an URANS-method in chapter 1 and for a CAA-method in chapter 8.
- The interaction between the noise generation and the combustion was investigated numerically as an important source of noise as well as combustion instabilities. The basic source mechanisms for the combustion-acoustic feedback were investigated numerically by a direct numerical simulation in chapter 9 and the contribution of the species to the resulting amplification or damping of sound waves due to the interaction with the flame front was uncovered.
- Finally, for the purpose of modeling and identifying of noise sources, identification techniques were developed in the current project. For identification in a numerical simulation the source terms of different acoustic analogies were applied in chapter 1 as well as chapter 2, whereas the method used in chapter 8 was based on the acoustic intensity and identifies radiating sources. The methods presented in chapter 1 and chapter 8 were feasible to identify the flow in the nozzle of the model experiment and therewith the indirect noise as major source of sound.
  An extended theoretical and numerical analysis of the source terms in reacting flows was provided in chapter 7, in which the source terms were also analyzed in chapter 3. The material derivative of the density was found to be the major source of noise in an incompressible simulation. The analysis is shown in chapter 7. Compared to the numerical simulation with full time resolved field data, the location of the source in experiments is even more challenging. This topic was addressed by a holographic method in chapter 10. It provides a mathematical study of the reproduction of the internal sound field with a minimum number of microphones as they are available for a realistic experimental instrumentation. The fast development of this method is astounding on the background that it joined the project in the last two years.

Chapter 1
Numerical RANS/URANS simulation of combustion noise

Bernd Mühlbauer, Berthold Noll, Roland Ewert, Oliver Kornow and Manfred Aigner

Abstract In the present work, numerical simulation tools for two different combustion noise source mechanisms are presented. The generation and propagation of entropy noise is computed directly using a compressible CFD approach in combination with appropriate acoustic boundary conditions. The Entropy Wave Generator (EWG) experiment is taken for validation of the proposed approach and for evaluating the acoustic sources of entropy noise. Simulation results of pressure fluctuations and their spectra for a defined standard test configuration as well as for different operating points of the EWG agree very well with the respective experimental data. Furthermore, a new numerical approach called RPM-CN approach was developed to predict broadband combustion noise. This highly efficient hybrid CFD/CAA approach can rely on a reactive RANS simulation. The RPM method is used to reconstruct stochastic broadband combustion noise sources in the time domain based on statistical turbulence quantities. Subsequently, the propagation of the combustion noise is computed by solving the acoustic perturbation equations (APE-4). The accuracy of the RPM-CN approach will be demonstrated by a good agreement of the simulation results with acoustic measurements of the DLR-A flame. The high efficiency and therefore low computational costs enable the usage of this numerical approach in the design process.

Bernd Mühlbauer · Berthold Noll · Manfred Aigner
German Aerospace Center (DLR), Institute of Combustion Technology, Pfaffenwaldring 38-40, 70569 Stuttgart, Germany, e-mail: bernd.muehlbauer@dlr.de

Roland Ewert · Oliver Kornow
German Aerospace Center (DLR), Institute of Aerodynamics and Flow Technology, Technical Acoustics, Lilienthalplatz 7, 38108 Braunschweig, Germany

A. Schwarz, J. Janicka (eds.), *Combustion Noise*, DOI 10.1007/978-3-642-02038-4_1,

1.1 Introduction

Combustion systems, for example of aero-engines, emit direct and indirect combustion noise [49]. Direct combustion noise is due to periodic combustion oscillations or stochastic fluctuations of the heat release [17, 18]. Indirect combustion noise or entropy noise is generated mainly by temperature non-uniformities, which are convected out of the combustion chamber and then accelerated [18, 37, 35] for example downstream the gas-turbine combustion chambers in the first turbine stage.

The approaches of the computational combustion acoustics (CCA) are very similar to those of the computational aeroacoustics (CAA) [14]. However, the numerical simulation of combustion noise is more complex since turbulence, acoustics and chemical reaction coincide. Like in CAA [52] direct computational methods (DCM) are considered as the more accurate technique for CCA [14]. DCM solve the complete, fully coupled compressible Navier-Stokes equations and resolve the unsteady reactive flow and the acoustic field at the same time. However, due to the application of direct numerical simulations (DNS) or large eddy simulations (LES) the prediction of combustion noise is expensive. Thus, DNS/LES methods especially for industrial users are very costly. However, the strongly varying length- and time scales of turbulence, acoustics and chemistry as well as the need of high order discretization schemes to avoid dissipation and dispersion of acoustic waves initiated the development of hybrid CFD/CAA-approaches, which split the combustion noise simulation into a simulation of the turbulent reactive flow and a subsequent simulation of the acoustic processes in the time domain.

Therefore, the applications of direct Reynolds Averaged Navier-Stokes (RANS) approaches or hybrid approaches which are based on RANS computations are a promising and computational efficient method for combustion noise prediction. Since there is not much known about the applicability of much more efficient RANS/URANS approaches the present work was devoted to the application of RANS/URANS methods as the underlying CFD method for the numerical computation of direct and indirect combustion noise. For the indirect combustion noise a compressible URANS method was applied and the sound generation and propagation was treated inherently. In contrast to this, the simulation of direct combustion noise was done by a newly developed multi scale CFD/CAA approach.

1.2 Theoretical Background

In this section, theoretical background to the applied simulation techniques is given. First, the employed RANS/URANS modeling approach and acoustic boundary conditions for CFD application in the time-domain are briefly discussed. Subsequently, the applied hybrid CFD/CAA approach for the prediction of turbulent combustion noise is depicted.

1.2.1 RANS/URANS approach

1.2.1.1 Statistical averaging of the transport equations

The governing equations of fluid flows can be cast in a general form which, for instance, in tensor notation reads as

$$\frac{\partial(\rho\phi)}{\partial t} + \frac{\partial(\rho u_j \phi)}{\partial x_j} = \frac{\partial}{\partial x_j}\left(\Gamma \frac{\partial \phi}{\partial x_j}\right) + S_\phi . \tag{1.1}$$

The URANS (Unsteady Reynolds Averaged Navier-Stokes) transport equations of turbulent flows can be derived applying Ensemble-averaging. Hereby, the stochastic fluctuations of turbulence are separated from the deterministic oscillations [7, 39]. The flow variable ϕ is decomposed into a time-dependent but deterministically changing value $\phi_0(t)$ and a superimposed stochastic temporal fluctuation $\phi'(t)$, Fig. 1.1

$$\phi(t) = \phi_0(t) + \phi'(t), \tag{1.2}$$

whereas

$$\phi_0(\tau) = \frac{1}{N}\sum_{i=0}^{N} \phi(\tau + iT). \tag{1.3}$$

Here T is a representative time-interval of the deterministic changes of ϕ.

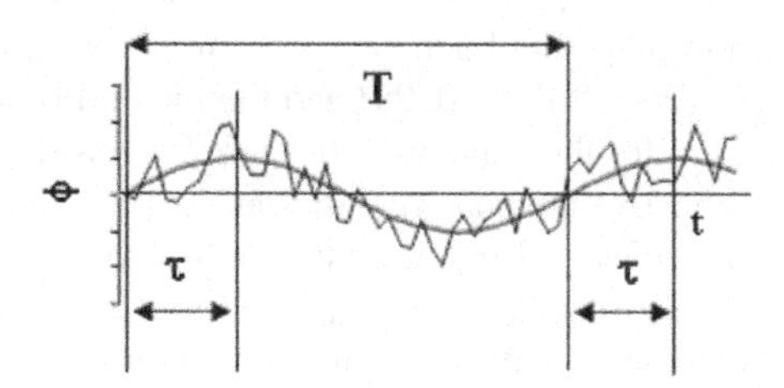

Fig. 1.1 Periodic oscillation with superimposed turbulent fluctuation [39]

Introducing this decomposition in the transport equations and taking the Ensemble-average of these equations over a proper time interval results in the following Reynolds-averaged transport equations for unsteady turbulent flows [7, 39] with constant density

$$\frac{\partial(\rho\phi_0)}{\partial t} + \frac{\partial\left(\rho u_{j,0}\phi_0 + \rho\langle u_{j,0}\phi'\rangle_0 + \rho\left\langle u'_j\phi_0\right\rangle_0 + \rho\left\langle u'_j\phi'\right\rangle_0\right)}{\partial x_j}$$
$$= \frac{\partial}{\partial x_j}\left(\Gamma\frac{\partial\phi_0}{\partial x_j}\right) + \langle S_\phi\rangle, \tag{1.4}$$

with $\langle . \rangle$ denoting Ensemble-averages where $\langle u_{j,0}\phi' \rangle_0 = 0$ and $\left\langle u'_j\phi_0 \right\rangle_0 = 0$. Favre-averaging of the equations of variable density flows leads to similar equations with ρ denoting the Ensemble-average of the density.

1.2.1.2 Turbulence modeling

A large number of different closures for the unknown correlations $\left\langle u'_j\phi' \right\rangle$ exist [43]. They can be classified depending on the statistical order of the closure and the number of additional model equations required. In this work the standard k-ε model, a two equation closure model of first order and a Reynolds Stress Model (RSM), a seven equation model of second order have been applied.

1.2.2 Boundary conditions

In order to predict thermoacoustic phenomena like combustion noise applying CFD methods the accurate modeling of the acoustic behavior at the boundaries is indispensable. In the case of computing combustion noise of open flames the surrounding boundary conditions have to allow acoustic waves to leave the computational domain without any reflections. To prevent reflections of physical and numerical waves different approaches have been developed in the past. For the implementation of non-reflecting boundary conditions common approaches are the Perfectly Matched Layers (PML) [30, 29] and the method based on the characteristics [41, 54]. However, for the simulation of thermoacoustic phenomena in enclosed geometries non-reflective boundary conditions are often not appropriate. Here, instead the acoustic impedance of the downstream and the upstream geometry, i.e. the inlet and outlet boundaries, has to be captured by the boundary conditions. In general, only time-domain impedance boundary conditions which are able to capture the frequency-depending impedance behavior of acoustic waves must be used [55, 31].

1.2.3 RPM-CN approach

The RPM-CN approach is a hybrid CFD/CAA approach for the prediction of turbulent combustion noise. Based on mean statistical turbulence quantities of a preceding reactive RANS simulation the turbulent combustion noise sources are stochastically reconstructed by the RPM-CN method. This section briefly discusses the underlying RPM method and the applied discretization and afterwards the realization of turbulent combustion noise sources is derived.

1.2.3.1 RPM method

The Random Particle-Mesh (RPM) method was introduced by Ewert [22, 19, 21, 20] as a stochastic method to generate unsteady turbulent fields with locally prescribed one- and two-point statistics, with special emphasizes on its application in CAA. The RPM method is an Eulerian-Lagrangian stochastic method, which generates a statistically steady fluctuating sound source $Q(x,t)$ in the Eulerian frame used in CAA methods. The cross-covariance $\mathscr{R}$ generated by the method as applied in this work is Gaussian in space and exponential in time, i.e.

$$\begin{aligned}\mathscr{R}(x,r,\tau) &= \langle Q(x,t)Q(x+r,t+\tau)\rangle \\ &= \hat{R}\exp\left\{-\frac{|\tau|}{\tau_s}-\frac{\pi(r-u_c\tau)^2}{4l_s^2}\right\},\end{aligned} \tag{1.5}$$

where the brackets denote ensemble average. The parameters l_s and τ_s define respectively integral length- and time scales and $\hat{R}$ denotes the variance of the correlated quantity for vanishing separation space r and time τ. Taylor's hypothesis is taken into account by the convection velocity u_c. For inhomogeneous turbulence u_c, l_s, τ_s, and $\hat{R}$ depend on position x.

The fluctuating quantity Q is obtained by spatially filtering a spatial white-noise field $\mathscr{U}$, the latter of which is generated with a specific stochastic partial differential equations. The filtering reads

$$Q(x,t) = \int_{V_S^n} \hat{A}(x)\mathscr{G}\left(|x-x'|,l_s(x)\right)\mathscr{U}\left(x',t\right)\mathrm{d}^n x'. \tag{1.6}$$

In this expression $\mathscr{G}$ is a spatial filter kernel, n indicates the dimension of the problem, and V_S^n is the considered source region. The amplitude function $\hat{A}(x)$ realizes a local target variance of the fluctuating quantity Q, its scaling is discussed below. The spatial white noise field is generated by a Langevin equation [43]

$$\frac{D_0}{Dt}\mathscr{U} = -\frac{1}{\tau_s}\mathscr{U} + \sqrt{\frac{2}{\rho_0^c\tau_s}}\xi(x,t), \tag{1.7}$$

formulated in a Lagrangian frame. Here $D_0/Dt = \partial/\partial t + u_0^c\cdot\nabla$ denotes a substantial time derivative. The steady velocity field $u_0^c(x)$ determines the finally achieved convection velocity u_c. For a constant velocity u_0^c the convection velocity becomes $u_c = u_0^c$. However, in general an arbitrary local convection velocity can be achieved. The density ρ_0^c is defined such that $\nabla\cdot\left(\rho_0^c u_0^c\right) = 0$ is satisfied. The quantity $\xi(x,t)$ is Gaussian distributed spatio-temporal white-noise, i.e. it satisfies

$$\langle\xi(x,t)\rangle = 0, \tag{1.8}$$

$$\langle\xi(x,t)\xi(x+r,t+\tau)\rangle = \delta(\tau)\delta(r), \tag{1.9}$$

where δ denotes the Dirac δ-function.

Based on the definition for Q, Eq. (1.6) its cross-covariance in the fully written form becomes [38]

$$\begin{aligned}\mathscr{R}(x,r,\tau) &= \langle Q(x,t)Q(x+r,t+\tau)\rangle \\ &= \frac{\hat{A}(x)\hat{A}(x+r)}{\rho_0^c(x)} l_s^{\frac{n}{2}}(x) \\ &\exp\left(-\frac{|\tau|}{\tau_s} - \frac{\pi|r-u_0^c\tau|^2}{4l_s^2(x)}\right),\end{aligned} \tag{1.10}$$

To achieve the appropriate variance $\mathscr{R}(x,0,0) = \hat{R}$ of Eq. (1.5) the parameter $\hat{A}$ must be chosen as

$$\hat{A}(x) = \sqrt{\frac{\rho_0^c(x)\hat{R}(x)}{l_s^n(x)}}. \tag{1.11}$$

1.2.3.2 Numerical discretization of the RPM method

The numerically method used here to discretize the filtered stochastic partial differential equation as presented in the previous section represents the convecting white noise field by convecting particles, which carry random values. The random values are Gaussian deviates with a variance proportional to the inverse of the particle density. A bundle of streamlines spans the resolved source domain over the field u_0. In this work the convection field is identified with the time-averaged mean-flow $\widetilde{u}$ from RANS. Random particles are seeded with a constant clock rate at the foremost upstream position on each streamline. The particles drift along the streamline until being finally removed downstream. The spatial filtering is conducted sequentially. In the first step the random values are filtered along the streamline. Next, the values are weighted and distributed in direction normal to the streamline onto the CAA mesh.

To discretize the Langevin equation, Eq. (1.7), the random values carried by each particle are not kept constant but rather change over time according to the discrete equation

$$r_i^{n+1} = \alpha r_i^n + \beta s_i^n. \tag{1.12}$$

Here r_i^{n+1} and r_i^n denote the random value of a particle at time-level $n+1$ and n, respectively. The quantity s_i^n is a Gaussian deviate with same variance as r_i. This procedure realizes an exponential decay [6]. The constant α follows, by discretizing the Langevin equation Eq. (1.7). It is related to the time-scale τ_s via

$$\alpha = 1 - \frac{\Delta t}{\tau_s}, \tag{1.13}$$

where Δt denotes the time-increment between levels $n+1$ and n. To preserve the root-mean square value of r_i over time, β must be chosen as $\beta = \sqrt{\frac{2\Delta t}{\tau_s}}$.

1.2.3.3 Realization of turbulent combustion noise sources

For the acoustic combustion noise simulations the acoustic perturbation equation system (APE-4) introduced by Ewert & Schröder [23] is applied. This modification of the linearized Euler equations (LEE) reads

$$\frac{\partial \mathbf{u}'}{\partial t} + \nabla\left(\mathbf{u}_0 \cdot \mathbf{u}'\right) + \nabla\left(\frac{p'}{\rho_0}\right) = \mathbf{q}_m \tag{1.14}$$

$$\frac{\partial p'}{\partial t} + c_0^2 \nabla \cdot \left(\rho_0 \mathbf{u}' + \mathbf{u}_0 \frac{p'}{c_0^2}\right) = c_0^2 q_c \tag{1.15}$$

with the right hand side source terms

$$\mathbf{q}_m = -\left(\omega \times \mathbf{u}\right)' + T'\nabla s_0 - s'\nabla T_0 - \left(\nabla \frac{(u')^2}{2}\right)' + \left(\frac{\nabla \cdot \underline{\tau}}{\rho}\right)' \tag{1.16}$$

$$q_c = -\nabla \cdot \left(\rho' \mathbf{u}'\right)' + \frac{\rho_0}{c_p}\frac{D_0 s'}{Dt}. \tag{1.17}$$

Here, all non-entropy sources can be neglected since sources of aerodynamic sound are usually negligible in the relatively low speed flows prevailing in flames [14] which means that the combustion noise sources are dominant [8, 17, 28, 48, 47]. Thus, neglecting all non-entropy source terms delivers the following sources of the APE-4 system which include among others all acoustic sources due to chemical reaction.

$$\mathbf{q}_m = T'\nabla s_0 - s'\nabla T_0 \tag{1.18}$$

$$q_c = \frac{\rho_0}{c_p}\frac{D_0 s'}{Dt} \tag{1.19}$$

First the assumption was made that the sources of the momentum equations $\mathbf{q}_m$ are negligible and the substantial time derivative of the entropy fluctuation $D_0 s'/Dt$ is the dominant acoustic combustion noise source [12]. Secondly it was assumed that the entropy fluctuation is a strong function of the temperature fluctuation only [53], which is true for constant pressure.

$$ds = \frac{c_p}{T_0} dT \tag{1.20}$$

$$s' = \frac{c_p}{T_0} T' \tag{1.21}$$

Thus, the following acoustic source term is obtained.

$$q_c = \frac{\rho_0}{c_p}\frac{D_0}{Dt}\left(\frac{c_p}{T_0} T'\right) \tag{1.22}$$

The main advantage of modeling the entropy fluctuation by the temperature fluctuation is the fact that the prediction of temperature statistics is well established and can be validated against measurements. The prefactor ρ_0/c_p of the source term is a result of the RANS simulation and the substantial time derivative is modeled by the RPM method which will be derived in the following.

By reconstructing the substantial time derivative $D_0\left(c_p T'/T_0\right)/Dt$ with RPM, a scaling similar to that used for the Tam & Auriault [50] RPM realization for jet noise prediction reported in [9, 20, 38], yields a variance of the source

$$\hat{R} = \frac{\frac{c_p^2}{T_0^2}\widetilde{T''^2}}{c^2\tau_T^2}. \tag{1.23}$$

Use of Eq. (1.23) in conjunction with Eq. (1.11) delivers the amplitude $\hat{A}$, where n indicates the dimension of the problem.

$$\hat{A} = \frac{c_p\sqrt{\rho_0^c\widetilde{T''^2}}}{c\tau_T l_T^{n/2} T_0} \tag{1.24}$$

For the final use with the APE-4 system the RPM generated source has to be complemented with the prefactor ρ_0/c_p to realize Eq. (1.22). The modeled correlation function has three characteristic parameters which are the length scale l_T and time scale τ_T of the temperature fluctuation as well as the temperature variance $\widetilde{T''^2}$. The modeling of these variables is presented in the following. Note that the turbulent temperature length and time scales might differ from the one used e.g. for the Tam & Auriault jet noise model [50]. This means that different calibration constants have to be applied. To clearly distinguish the temperature scales, the related parameters are labeled l_T, τ_T, c_{Tl}, and $c_{T\tau}$, instead of l_s, τ_s, c_l, and c_τ in the following. This convention already has been used in Eq. (1.24), where the length and time scale l_T and τ_T instead of l_s and τ_s occur.

The temperature length l_T and the time scale τ_T are linked to the turbulence model scales via

$$l_T = c_{Tl}\frac{k^{3/2}}{\varepsilon} \tag{1.25}$$

$$\tau_T = c_{T\tau}\frac{k}{\varepsilon}. \tag{1.26}$$

Simultaneous to the RANS calculation a transport equation for the temperature variance is solved [26].

$$
\begin{aligned}
\frac{\partial \bar{\rho}\widetilde{T''^2}}{\partial t} + \frac{\partial \bar{\rho}\tilde{u}_j\widetilde{T''^2}}{\partial x_j} = \frac{\partial}{\partial x_j}\left\{\left(\overline{\mu} + \frac{\mu_t}{Pr_t}\right)\frac{\partial \rho \widetilde{T''^2}}{\partial x_j}\right\} \\
+ c_{prod}\frac{\mu_t}{Pr_t}\left(\frac{\partial \widetilde{\overline{T}}}{\partial x_j}\right)^2 - c_{diss}\overline{\rho}\frac{\overline{\varepsilon}}{\tilde{k}}\widetilde{T''^2}
\end{aligned}
\tag{1.27}
$$

The constants of the transport equation are $c_{prod} = 2.0$ and $c_{diss} = 2.0$.

At least the suitability of the RPM method to model combustion noise sources will be discussed. The correlations of the scalar source Eq. (1.22) will show a correlation given by Eq. (1.6) with τ_s and l_s replaced by τ_T and l_T, i.e. the combustion sound sources are assumed to have Gaussian spatial correlations. Note that currently no data is available about the typical shape of the correlation function associated to the combustion source term used in the present work. However, acoustically the meaning of the correlation length scale of a turbulent sound source is to take into account coherent and incoherent source areas. If the turbulent length scale is smaller than the acoustic wave-lengths the coherent sound sources can be deemed to be compact, which effectively means that acoustically only the local correlation length scale, respectively the local correlation volume, plays a role but not the special shape of the correlation function. The assumption of compactness is satisfied for combustion noise problems with good accuracy, hence a Gaussian spatial correlation of the combustion source might be appropriate, irrespective of its actual shape.

1.3 Results and Analysis

1.3.1 Indirect combustion noise

1.3.1.1 Experimental configuration

The Entropy Wave Generator (EWG) is a test rig for non-reactive flows. The main part of the rig is basically an accelerated tube flow that allows the generation of entropy modes of perturbation [15] by an electrical heating module upstream of the nozzle. The detailed setup is described in Bake et al. [1]. A sketch of the design is shown in Fig. 5.1 in chapter 5.2.1.1.

The air flow, which is supplied by a compressor, is lead into a settling chamber with a honeycomb flow-straightener before it enters the tube section via a bell-mouth intake. The inner diameter of the tube is 30 mm. The heating module consists of four ring sections with ten platinum wires each stretched through the cross section. The wires have a diameter of 25 μm. In the current setup, the wires can be heated with an electrical power up to 200 W. The length of the heating module in stream-wise direction is equal to 32 mm.

The tube section following the heating module has a length of 92.5 mm. Further downstream the flow is accelerated through the convergent part of a convergent-divergent nozzle and then decelerated in the divergent section of the nozzle. The nozzle throat has a diameter of 7.5 mm. The following 1020 mm long tube section has a diameter of 40 mm. The acoustic pressure excited by the accelerated entropy modes in the convergent-divergent nozzle is measured by wall-flushed microphones at four different axial positions for acoustic analysis in the tube. Following this measurement section, the flow enters a tube of 1000 mm in length and 44 mm in diameter, which is connected with a smooth transition from circular cross section to square cross section over a length of 250 mm. Due to this change of cross sectional area partial reflection takes place and thus acoustic waves propagate back into the measurement section. Finally, flow and acoustic waves leave the experimental apparatus through a nearly anechoic termination. A picture of the Entropy Wave Generator is displayed in Fig. 5.2 in chapter 5.2.1.1.

The standard configuration is defined by a mass flow of 42 kg/h at ambient conditions to the EWG. The wires of the heating module were electrically heated once every second with a time duration of 0.10 s to induce an energy pulse excitation. The supply of energy induces a temperature perturbation of 9 K. In the following the experimental data of the standard configuration are discussed.

The transistor-transistor logic signal (TTL) that controls the pulse excitation as well as the temperature measured downstream close to the source area are plotted in Fig. 1.2. Although the TTL signal shows a stepwise increase the air temperature is increasing with a time delay and a certain gradient. Hwang et al. [32] showed that even rapid heating causes a delayed and retarded increase of the wire temperature what explains the dynamic behavior of the air temperature.

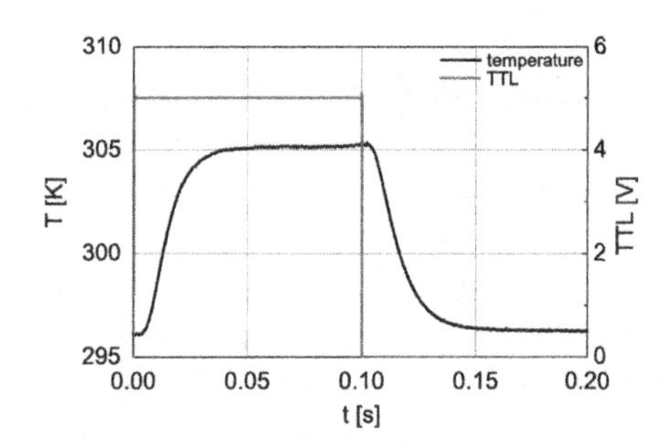

Fig. 1.2 Temperature and TTL signal [1]

The phase averaged signal of the microphone located 1.15 m downstream of the nozzle throat is depicted in Fig. 1.3. The supply of energy to the air flow generates a local area with increased temperature and decreased density. This entropy mode convects with flow speed towards the nozzle. During the passing of the entropy mode through the convergent-divergent nozzle noise is generated. This so called entropy noise propagates downstream towards the microphone positions with the speed of sound and causes a pressure signal. The pressure signal of the entropy noise is detected with a time delay after the energy is delivered to the system. The time delay is the sum of the time the entropy mode needs to convect to the nozzle with the flow speed and the time of the generated acoustic wave propagates to the microphone positions predicted at the speed of sound. The measured time delay of 0.01 s is almost consistent with the estimated time delay of 0.0095 s. Signal S1 in Fig. 1.3 shows a positive pressure signal generated by the accelerated increasing entropy in the nozzle. The decrease of energy supply to the flow field analogously causes a negative pressure signal S2. The oscillations found in the pressure graph following the positive and negative pressure signals S1 and S2 are due to the partial reflection which arise in the transition from circular to quadratic cross section. This causes superposition of downstream and reflected upstream propagating pressure waves.

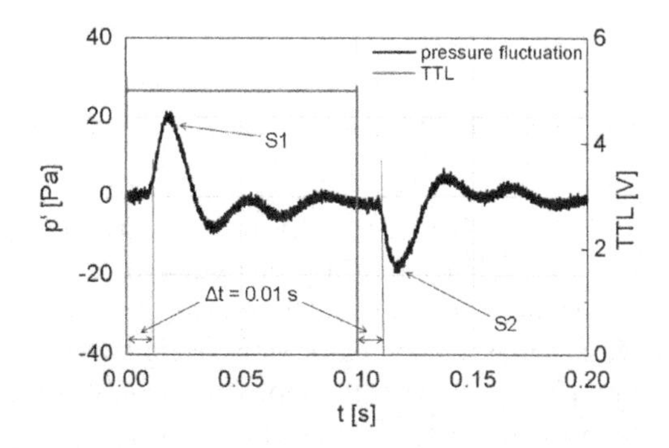

Fig. 1.3 Pressure fluctuations and TTL signal [1]

1.3.1.2 Numerical setup

The commercial software package ANSYS CFX 11.0 was used to conduct the simulations. The numerical simulation of the EWG experiments can be performed in a axisymmetric frame. However, in order to meet the requirements of ANSYS CFX the numerical simulations were performed on a three dimensional unstructured grid. The rotational symmetric geometry was represented by a segment of 10°. This restricts the number of nodes to 57 000. The mesh is appropriate to adequately resolve acoustic perturbation frequencies up to $f = 2800\,\text{Hz}$ and hydrodynamic perturbation frequencies up to $f = 99\,\text{Hz}$ based on a sufficient accuracy above 50 points per wavelength (PPW). For the calculations, the average speed of sound and the average axial velocity of the standard configuration downstream the heating module were considered.

For a precise numerical simulation of acoustic phenomena like entropy noise, the acoustical behavior at the boundaries has to be taken into account appropriately. In the present case, acoustic reflections occur at the smooth transition from circular cross section to square cross section of the experimental setup. In the present work, the outlet boundary condition was located at the beginning of the smooth transition component. The effects of fully reflective, non-reflective and partially reflective pressure outlet boundary conditions were investigated. A sketch of the computational domain is depicted in Fig. 1.4.

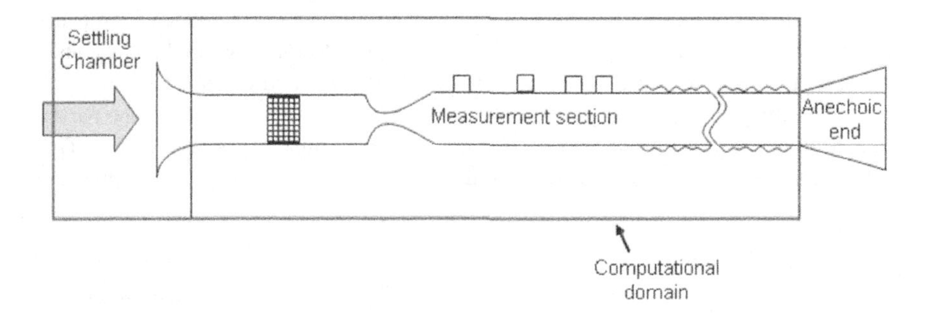

Fig. 1.4 Sketch of the computational domain of the Entropy Wave Generator (EWG)

In contradiction to fully reflective boundary conditions, non-reflective or partially reflective boundary conditions allow acoustic induced waves to leave the computational domain. The boundary condition used in the present work can be classified as a formulation of acoustic non-reflective characteristic boundary condition [54, 41, 3]. It is to emphasize here, that the non-reflective boundary condition used in this case, in contrast to the method described by Poinsot & Lele [41], was derived for pressure based solvers [54]. The realized reflection coefficient $r(\omega)$ of the boundary condition formulation is a function of the angular frequency ω and of the relaxation coefficient $K = \sigma\left(1 - Ma^2\right) c/L$ [46],

$$r(\omega) = \frac{1}{\sqrt{1 + \left(\frac{2\omega}{K}\right)^2}} \tag{1.28}$$

where Ma is the Mach number of the mean flow, c the speed of sound, and L the size of the domain. An alternative relation was given by Polifke et al. [42]. By modifying the coupling parameter σ the reflection at the boundary condition can be adjusted and thus a partially reflective boundary condition can be realized.

At the inlet of the computational domain a classical fully reflective mass flow inlet boundary condition was applied. A constant mass flow of 42 kg/h with a temperature of 300 K was set for the standard configuration. The side planes of the computational domain were modeled by symmetry boundary conditions. This simplification can be justified because of negligible swirling velocity component. The walls of the EWG were modeled with no-slip wall boundary conditions.

A three dimensional compressible URANS approach was applied and turbulence was considered applying the standard k-ε model. This can be justified by the only slight impact of turbulence. In the case considered here according to preliminary investigations the dissipation and dispersion of the entropy mode is very low. In addition, the influence of the turbulence on the propagation of the entropy noise downstream the nozzle is negligible. For the inclusion of the non-reflective boundary conditions, a software extension of ANSYS CFX 11.0 which was developed at the DLR-Institute of Combustion Technology, Stuttgart was applied [54]. The spatial discretization was performed using the 'high order resolution scheme', which is essentially second order accurate and bounded. A second order backward Euler transient scheme was applied for the time discretization. Transient simulations were performed with a time step size of $50\,\mu s$. The application of this time step size implies a cut-off frequency of $f = 400\,\text{Hz}$ when considering 50 discretization points per period (PPP) as sufficient.

The heated energy was added to the flow at the cell layers of the computational grid that correspond to the location of the heating wires of the experimental setup. The energy transferred from the wires to the air flow depends on the actual wire temperature. So in accordance to measurements [1] the energy supply inserted to the air flow was modeled in setting a linear increase and an exponential decrease of the energy source. The source term in the energy equation is implemented using a CFX User Fortran routine.

At the same positions where the microphones are located in the experimental setup the pressure fluctuations were monitored in the numerical simulation. In a first step a RANS simulation was carried out to achieve a steady flow field solution of the EWG. Then, based on the steady simulation, a subsequent URANS calculation was performed to simulate the time-dependent source of heat, the convection of the induced entropy mode and the generation and propagation of the entropy noise.

1.3.1.3 Standard configuration

The Mach number distribution and the temperature distribution in the convergent-divergent nozzle achieved by the RANS simulation are displayed in Fig. 1.5. The air is supplied at the inlet and flows through the tube section upstream the nozzle, which contains the heating module with an average axial velocity of 12.4 m/s and a temperature of 300 K. Downstream the heating module the air flow is accelerated in the convergent-divergent nozzle to a maximal Mach number of 1.32. Due to the high acceleration of the air flow in the nozzle the temperature is decreasing to 222 K. Subsequently, the air flow decelerates in the downstream tube section. The acceleration leads to a low pressure area within the nozzle. In the following all displayed and discussed acoustic pressure fluctuations correspond to the microphone located 1.15 m downstream the nozzle throat.

Fig. 1.6a illustrates a comparison with the experiment of the simulated pressure fluctuations, applying the fully reflective outlet boundary condition. The first pressure signal fits well with measurement but all following pressure fluctuations are falsified due to unphysical reflections that occur at the outlet boundary. Downstream propagating pressure fluctuations generated by accelerated entropy gradients are fully reflected at the outlet boundary condition. Thus, they propagate back into the measurement section, superpose with downstream propagating pressure waves and cause an oscillating pressure signal. This effect also explains the underestimated

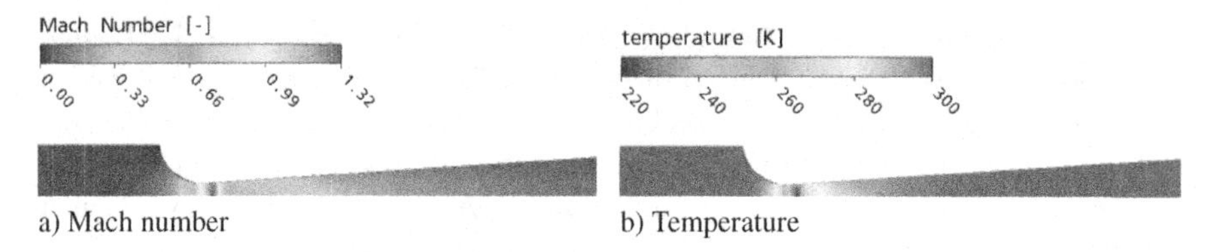

a) Mach number b) Temperature

Fig. 1.5 Mach number and temperature distribution of the steady flow field

amplitude of the first pressure signal. Thus, these results show that the fully reflective boundary condition is inappropriate to simulate thermoacoustic phenomena in the present case. Fig. 1.6b depicts the pressure spectra of the simulated pressure fluctuations and the experimental data. The spectra were obtained by a Discrete Fourier Transform (DFT) of the pressure signals. The pressure spectrum of the microphone measurement indicates the energy containing waves in a frequency range below 100 Hz. The energy containing frequencies of the measurements can be found in the simulation, too. In comparison to the measured spectrum the computed spectrum depicts an over prediction of the amplitude for the energy containing frequency of 5 Hz.

Fig. 1.7a displays the simulated pressure fluctuations if a non-reflective boundary condition at the outlet is applied. Both incident pressure signals agree well with the measurements. Due to the absence of reflected pressure waves the magnitude of the computed pressure amplitude is maximal and no oscillations are found in the calculated pressure signal. Thus, the application of the non-reflective outlet boundary condition is a way to derive the maximum amplitude of entropy noise which is generated by a certain entropy mode within a certain configuration. The pressure spectrum of the simulated pressure fluctuations displayed in Fig. 1.7b emphasizes discrepancies with respect to the measured spectrum. In the frequency range below 40 Hz the amplitudes are under predicted by the simulation. On the contrary the amplitudes of the frequencies of 45 and 125 Hz are over predicted.

The comparison with the experimental data reveals the necessity of applying a boundary condition which reproduces properly the acoustic behavior at the change of cross sectional area of the test rig. The adjustment of the coupling parameter σ in the boundary condition formulation results in the realization of a certain reflection according to Eq. (1.28). In the present work the coupling parameter was adjusted in a way that the calculated pressure response showed close accordance to the measured one. Fig. 1.8a shows simulated pressure fluctuations applying a coupling parameter of σ=1.8. The superposition of downstream and reflected upstream propagating pressure waves results in pressure fluctuations that are in close agreement with measurements. Both computed energy containing pressure signals indicate accordance in the shape as well as in the amplitude to the measurements. Only the oscillations after the two pressure signals show minor discrepancies. Furthermore, the simulated pressure spectrum applying the partially reflective boundary condition displayed in Fig. 1.8b is in very good agreement with the experimental spectrum. Both spectra depict identical dominant frequencies. Additionally, the pressure amplitudes of the dominant frequencies are nearly identical. Only small deviations can be identified.

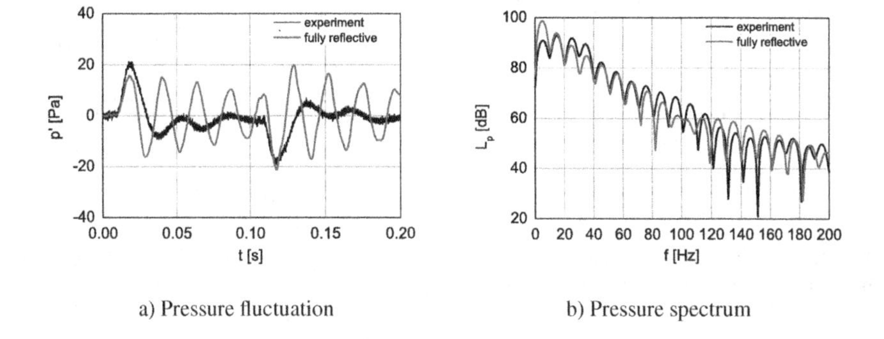

a) Pressure fluctuation b) Pressure spectrum

Fig. 1.6 Pressure fluctuation and pressure spectrum simulated with the fully reflective outlet boundary condition in comparison to measurements.

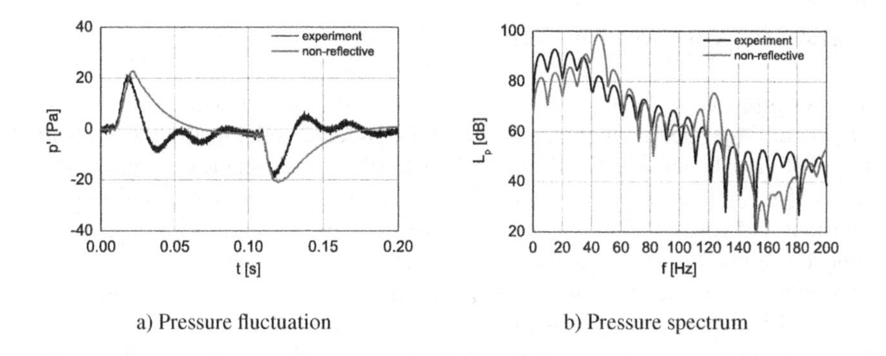

a) Pressure fluctuation b) Pressure spectrum

Fig. 1.7 Pressure fluctuation and pressure spectrum simulated with the non-reflective outlet boundary condition in comparison to measurements.

The remaining minor discrepancies between simulation and measurement are probably due to the simplified modeling of the acoustic boundary condition. The smooth transition of change of cross sectional area of the EWG acts as an impedance. This means, acoustic waves interacting with the geometry are reflected with a certain reflection coefficient and phase shift, both function of the frequency. In the present work the employed partially reflective boundary condition neglects the phase shift and assumes a constant reflection coefficient for all frequencies. Additionally, there is a numerically caused frequency dependency of the reflection coefficient according to Eq. (1.28) and a small numerically induced phase shift, which do not comply

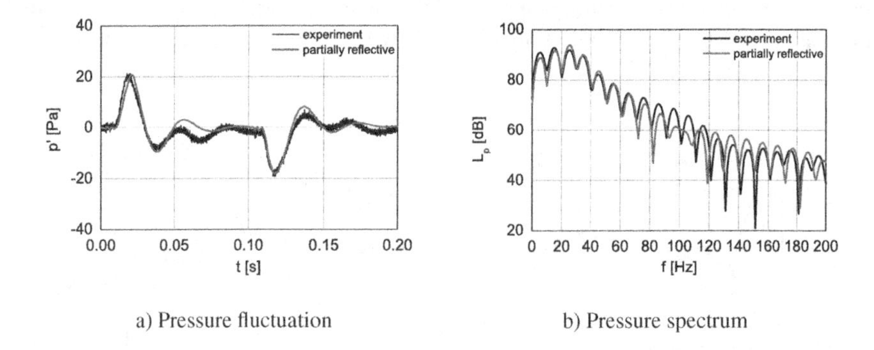

a) Pressure fluctuation

b) Pressure spectrum

Fig. 1.8 Pressure fluctuation and pressure spectrum simulated with the partially reflective outlet boundary condition in comparison to measurements.

with the real impedance. Thus, the exact acoustic behavior at the boundary could be captured by the application of proper time-domain CFD impedance boundary conditions that consider the measured reflection coefficient and the phase shift both as a function of the frequency [55, 31].

1.3.1.4 Different operating points

The performance of the applied numerical approach is further demonstrated by simulating different operating points of the EWG. Here, the inlet mass flow was varied to investigate the influence of the mean flow Mach number on the generated entropy noise. The supplied energy source was adjusted to keep the temperature increase constant and equal to 9 K. All simulations were carried out applying the partially reflective outlet boundary condition with a coupling parameter σ=1.8.

Fig. 1.9 illustrates the measured and calculated entropy noise in terms of the maximal pressure fluctuation as a function of the Mach number in the nozzle throat. The maximum in the pressure fluctuations of the experimental data increases up to a nozzle Mach number of 0.8 and then slightly decreases for higher nozzle Mach numbers. Fig. 1.9 reveals that the numerical results are in very good agreement and within the precision of the measurements. They describe the same behavior namely the nearly linear increase up to nozzle Mach number of 0.8 and the subsequent decrease for higher nozzle Mach numbers.

In order to find an explanation for the decrease of entropy noise at higher nozzle Mach numbers an investigation based on numerical simulations was performed. At the end of the computational domain a non-reflective boundary condition was applied. Furthermore, to avoid reflections at the transition to the plenum upstream of the heating module the plenum was truncated and the inlet was modeled applying a non-reflective boundary condition. For this configuration the results of the computations show the decline of generated entropy noise for higher nozzle Mach numbers, too. Thus, reflections do not cause the pressure fluctuation decrease. Additional numerical investigations to find an explanation for the decline of the generated entropy noise for higher nozzle Mach numbers are discussed in the next section.

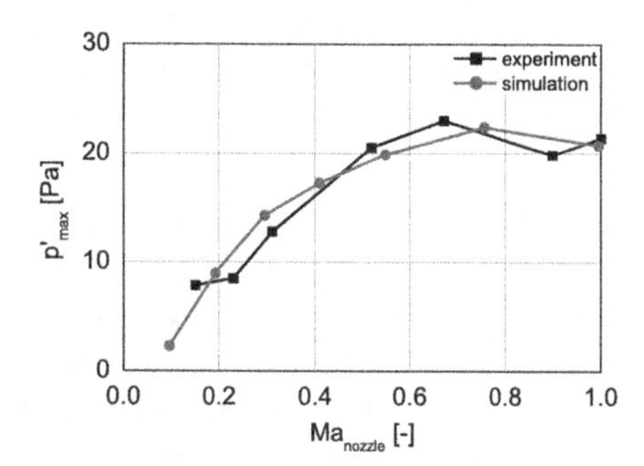

Fig. 1.9 Maximal pressure fluctuation over the nozzle Mach number

1.3.1.5 Acoustic sources

Dowling [17] derives the following inhomogeneous wave equation with source terms that are due to thermoacoustics.

$$
\begin{aligned}
\frac{1}{c_0^2}\frac{\partial^2 p}{\partial t^2} - \nabla^2 p = & -\frac{\partial}{\partial t}\left(\frac{\alpha \rho_0}{c_p \rho}\left(\sum_{n=1}^{N} \left.\frac{\partial h}{\partial Y_n}\right|_{\rho,p,Y_m} \rho \frac{DY_n}{Dt} + \nabla \cdot q - \frac{\partial u_i}{\partial x_i}\tau_{ij}\right)\right) \\
& + \frac{\partial^2}{\partial x_i \partial x_j}(\rho u_i u_j - \tau_{ij}) \\
& + \frac{1}{c_0^2}\frac{\partial}{\partial t}\left(\left(1 - \frac{\rho_0 c_0^2}{\rho c^2}\right)\frac{Dp}{Dt} - \frac{(p-p_0)}{\rho}\frac{D\rho}{Dt}\right) \\
& + \frac{\partial^2}{\partial x_i \partial t}(u_i \rho_e)
\end{aligned}
\tag{1.29}
$$

The 'excess density' ρ_e is defined as

$$
\rho_e = \rho - \rho_0 - \frac{p - p_0}{c_0^2}, \tag{1.30}
$$

where the suffix 0 denotes a suited reference value. The right hand side source terms of Eq. (1.29) describe all thermoacoustic sources. The first term represents the acoustic source due to irreversible flow processes. Lighthill's jet noise is considered in the second term. The third term implies acoustic sources generated by flow unsteadiness with varying mean density and sound speed from the ambient fluid. The last term is the one that describes the generation of acoustic waves by the change of momentum due to density inhomogeneities. This entropy noise source term is of dipole character in contrast to the direct combustion noise source terms which are monopole sources.

For a comprehensive investigation of acoustic phenomena the analysis of the acoustic sources is indispensable. To achieve a reduction in noise level the acoustic source mechanism has to be understood and the acoustic sources located. Dowling [17] derived an inhomogeneous wave equation (1.29) including the source term

$$q = \frac{\partial^2}{\partial x_i \partial t}(u_i \rho_e) \tag{1.31}$$

describing the acoustics generated by the acceleration of density inhomogeneities. Accordingly, in the present work, numerical simulations of the EWG were conducted and the transient acoustic sources of entropy noise were computed.

The acoustic sources according to Eq. 1.31 were evaluated by numerical simulations in order to find an explanation for the decline in generated entropy noise for higher nozzle Mach numbers in the EWG (Fig. 1.9). In order to investigate the strength of the acoustic sources a volume integral Q_V of the scalar acoustic source term q, Eq. (1.31) over the convergent-divergent nozzle domain V_N was computed.

$$Q_V = \int_{V_N} q\, dV \tag{1.32}$$

A cutout of the computational domain with the nozzle domain highlighted is displayed in Fig. 1.10. The volume integral as a function of the time was calculated for the same flow conditions as covered in the 'Different Operating Points' section and is displayed in Fig. 1.11. The graphs indicate a monotonic increase of the strength of the acoustic source with increasing nozzle Mach number. This result thus evidences that the decline of the generated entropy noise for higher nozzle Mach numbers is not caused by the strength of the acoustic source.

Besides the amount of the acoustic sources the transient simulations provide the distribution of the acoustic sources in the acceleration/deceleration area. Fig. 1.12 illustrates representative instantaneous distributions of the acoustic sources of entropy noise for varying nozzle Mach number. Here, the qualitatively distributions of the timely varying acoustic sources at the moment of the strongest source intensity are shown. It should be noted that different scales are used for the displayed acoustic source distributions.

For all subsonic simulations the acoustic sources have a similar distribution. The acoustic sources are located in the convergent and partly in the divergent part of the

Fig. 1.10 Nozzle domain for volume integration.

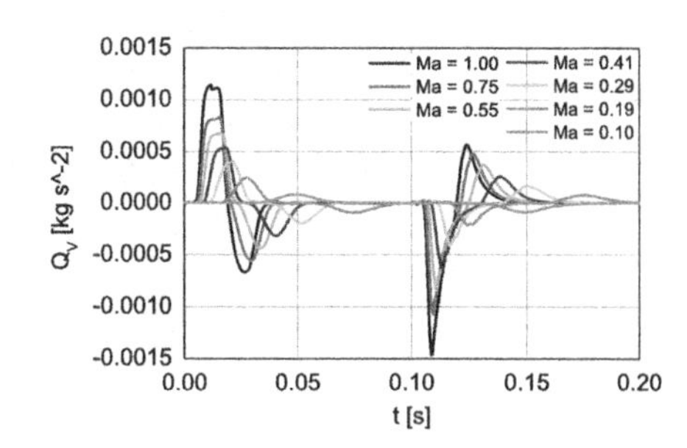

Fig. 1.11 Volume integral of the acoustic source over the convergent-divergent nozzle domain

nozzle. Positive acoustic sources are located nearby the axis and negative acoustic sources close to the wall. Furthermore, the area including positive sources is moving slightly downstream with increasing nozzle Mach number. In contrast to the subsonic configurations the source distribution for a nozzle Mach number of 1.0 is significantly different. The acoustic sources are mainly located in the divergent part of the nozzle. Positive acoustic sources are located upstream of the shock location (see Fig. 1.5). Negative acoustic sources are located downstream of the shock position.

This acoustic source distribution in the supersonic flow configuration causes the decline of the maximal pressure fluctuation for high nozzle Mach numbers, Fig. 1.9: Positive and negative acoustic sources generate positive and negative pressure fluctuations, that propagate downstream and superimpose. Due to this superposition an attenuation of the acoustic waves occurs.

1.3.2 Direct combustion noise

1.3.2.1 Experimental configuration

The DLR-A flame, a benchmark flame of the "International workshop on measurement and computation of turbulent non-premixed flames" [2] is used to validate the applied acoustic simulation approach. The configuration was a non-premixed turbulent N_2 diluted CH_4-H_2 jet flame. The fuel with a composition of 22.1 % CH_4, 33.2 % H_2 and 44.7 % N_2 was injected through a 0.35 m long straight stainless steel tube of diameter $D = 0.008$ m. The mean jet exit velocity corresponded to $u_{jet} = 42.15$ m/s. The Reynolds number with respect to D has been $Re = 15\,200$. The tube was surrounded by a contoured nozzle supplying co-flowing dry air at an exit velocity of 0.3 m/s at a temperature of 292 K. The outer nozzle had a diameter of 0.14 m. The temperature of both jets was 300 K. The configurations have been experimentially investigated by Bergmann et al. [4], Meier et al. [36] and Schneider et al. [44]. The spectral sound emmisions were measured by Singh et al. [47].

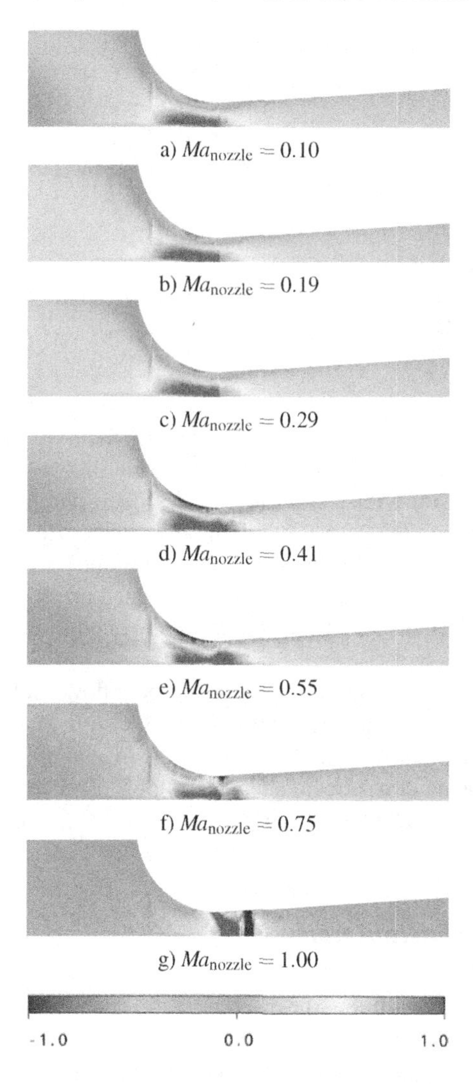

Fig. 1.12 Instantaneous acoustic source distribution for varying nozzle Mach number. The scaling is normalized with the respective maximum acoustic source.

1.3.2.2 CFD setup

A fully three dimensional compressible steady-state RANS simulation was performed to calculate the reactive flow field. The turbulence was considered by the Baseline Reynolds Stress Model (BSL-RSM). The chemical reaction was simulated applying the Burning Velocity Model (BVM) [40, 56]. For the spacial discretization the so called 'high order resolution scheme' which switches between first and second order upwind was used and the time was discretized with a second order Euler backward differencing scheme. Furthermore, buoyancy was taken into account whereas heat radiation was neglected. The CFD simulations were carried out applying the commercial software package ANSYS CFX-11.0.

The computational domain was discretized with a cylindrical unstructured hexahedron grid comprising 370 000 grid nodes. The grid has dimensions of 94 D in axial and 113 D in radial direction and is strongly refined in the combustion zone. The fuel and the co-flow were admitted to the simulation domain by mass flow inlet boundary conditions whereas the mass flow and the temperature were imposed according to the experimental setup. The free stream boundary conditions of the computational domain are approximated by static pressure opening boundary conditions.

1.3.2.3 CAA setup

The acoustic processes were simulated solving the APE-4 equations, which are integrated by the DLR-CAA code PIANO [16]. The code applies the fourth-order DRP scheme of Tam & Webb [51] in space and a LDDRK method [30] in time on block structured meshes. The stochastic reconstruction of the broadband combustion noise sources is implemented in a development version of the PIANO code. Currently, the implementation of the stochastic source term reconstruction is restricted to two dimensional simulations.

The computational CAA domain is discretized by a two dimensional plane block structured grid of 34 000 grid nodes. The mesh coverages 90 D in axial and 110 D in radial direction. The mesh is appropriate to solve frequencies up to $f = 11\,000\,\mathrm{Hz}$ based on 7 points per wavelength (PPW). The chosen time step size allows for a resolution frequency $f = 71\,000\,\mathrm{Hz}$ based on 7 points per period (PPP). Acoustic non-reflective radiation boundary conditions by Tam & Webb [51] surround the computational domain.

1.3.2.4 CFD results

The simulated axial profiles on the center line of the axial velocity, the mixture fraction, the temperature and the temperature rms of the DLR-A flame are compared to experimental data in Fig. 1.13. The reference diameter is the nozzle diameter $D = 0.008\,\mathrm{m}$. The results of the DLR-A flame including radial profiles have been already published and discussed in Ref. [38]. The computed axial velocity shows a good agreement with the experimental data for $x/D < 20$ and $x/D > 60$. In the axial range $20 < x/D < 60$ the computed profile shows deviations compared to the measurements. Detailed investigations evidenced that these deviations are not dependent on the boundary conditions, on the grid or on the quality of the combustion model. Different turbulence models have been tested and it has been found, that the BSL-RSM model in this case provided the best results. The mixture fraction was calculated using Bilger's definition [5]. The axial profile of the mixture fraction is in good agreement with the the experimental data. However, discrepancies are also found in the axial range of $20 < x/D < 60$. Accordingly, the simulated temperature profile agrees well with measurements. Along the center line the simulated maximal temperature is over predicted by 46 K. The remaining over prediction of the temperature can be attributed to the fact that heat radiation was not considered in the calculations. The transport equation of the temperature variance in combination with the RANS simulation delivers satisfactory results. The maximum value of the temperature rms agrees well with the experimental data.

It is known that many RANS turbulence models are not fully appropriate for the accurate prediction of free round jets. However, the applied RANS BLS-RSM model approach delivers acceptable results at low computational costs. Deviations of the temperature rms, which is a key input variable for the reconstruction of the combustion noise sources, compared to experimental data can be caused by turbulence modeling and combustion modeling. In the following, it will be shown that the dominant acoustic sources are in the range $x/D < 20$ where the RANS simulation delivers good agreement compared to measurements.

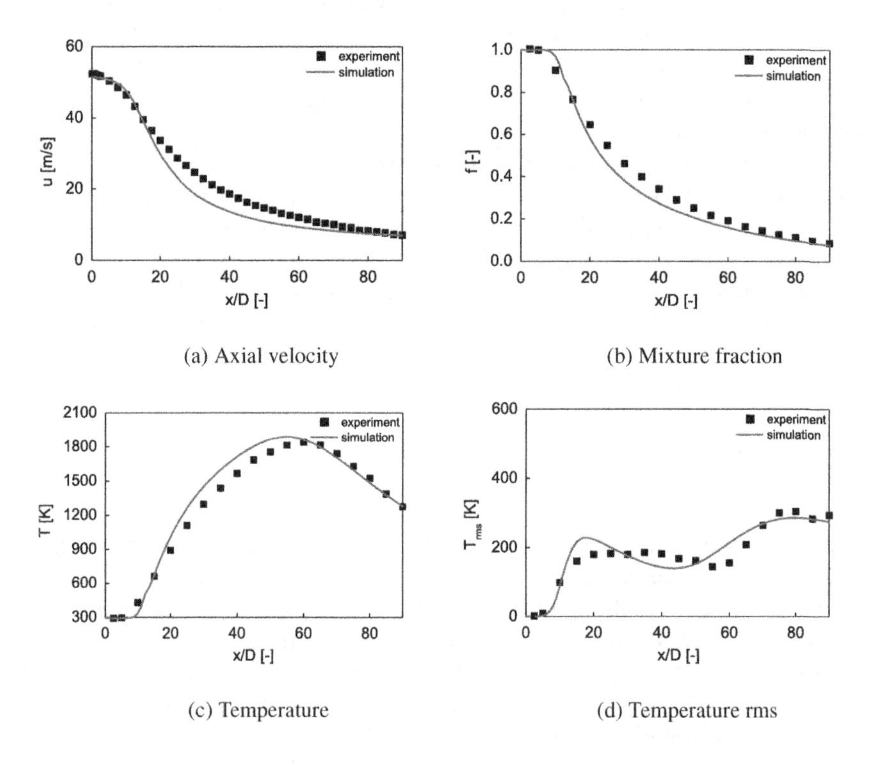

(a) Axial velocity

(b) Mixture fraction

(c) Temperature

(d) Temperature rms

Fig. 1.13 Axial profiles at $y/D = 0$ of the axial velocity, the mixture fraction, the temperature and the temperature rms. Comparison of the experimental data (■) and simulations (—) of the DLR-A flame.

1.3.2.5 CAA results

Due to the absence of available data concerning the typical shape of the correlation function associated to the used combustion source term as already mentioned before, in the present work the constants c_{Tl} and $c_{T\tau}$, Eqs. (1.25,1.26) were defined in accordance to the Tam & Auriault [50] RPM realization for jet noise prediction [9, 20, 38]. Hence, the present two dimensional numerical results have been obtained applying the constants $c_{Tl} = 0.273$, $c_{T\tau} = 0.233$.

In a first step an evaluation was performed regarding the capability of the RPM method to realize the defined two-point correlations and to reconstruct the resolution of the target source variance. Fig. 1.14 shows the statistical analysis of a fluctuating combustion noise source from RPM for a constant convection velocity in x-direction u_c, constant length scale l_T and constant time scale τ_T. R_{12} denotes the cross-correlation between two sampling points inside the generic test field. The cross-correlation between two sampling points i for relative time shift τ is evaluated from the unsteady time samples $Q_i(x_i t)$ by integrating over sampling interval ΔT, i.e.

$$R_{12}(\tau) = \overline{Q_1(x_1,t)Q_2(x_2,t+\tau)} = \frac{1}{\Delta T}\int_{T_0}^{T_0+\Delta T} Q_1(x_1,t)Q_2(x_2,t+\tau)\mathrm{d}t. \tag{1.33}$$

All $i = 12$ points used are distributed with equidistant spacing along the x-axis. Starting from the farthest left cross-correlation curve, the sequence of correlation curves refer to the cross-correlations between point combinations 1-2 up to 1-12.

Fig. 1.14 shows that two-point correlations can be reconstructed very well with the RPM method. That is, for a given combination of two sampling points, the related cross-correlation curves have Gaussian shape. The realized two-point correlation of the RPM method shows a good agreement with the analytical correlation according to Eq. (1.5). Furthermore, the maximally achieved correlation decays with farther downstream chosen second sampling point. The envelope to all curves has exponential behavior and only small deviations compared to the analytical envelope $\exp(-\tau/\tau_T)$ can be determined. The horizontal location of the cross-correlation maximum is successively shifted to higher shift times τ for the cross-correlation with farther downstream chosen second sampling point. The constant time difference of $\Delta\tau = 0.025$ between the maxima indicates the achievement of a constant convection velocity for the equidistantly distributed sampling points. Based on the constant time difference as indicated in Fig. 1.14 and the equidistant spacing between the sampling points the defined constant convection velocity in x-direction can be exactly determined.

An important feature that has to be accomplished by the stochastic model is the accurate resolution of the source variance $\hat{R}$ topology, Eq. (1.23). To evidence this capability of the model, Fig. 1.15 juxtaposes the RANS target solution for the source variance to the according results from the stochastic models. For this purpose, 100000 time levels of the stochastically generated sources are sampled and

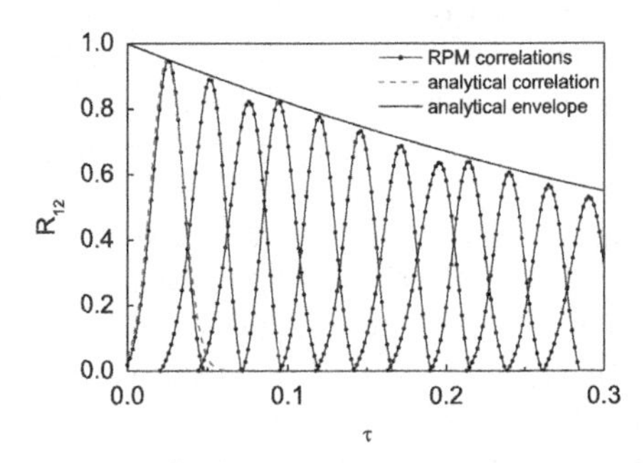

Fig. 1.14 Two-point space-time correlations of the RPM sources, analytical correlation and analytical envelope

averaged. A satisfactory realization of the source variance for the energy topology as well as the absolute magnitudes is found.

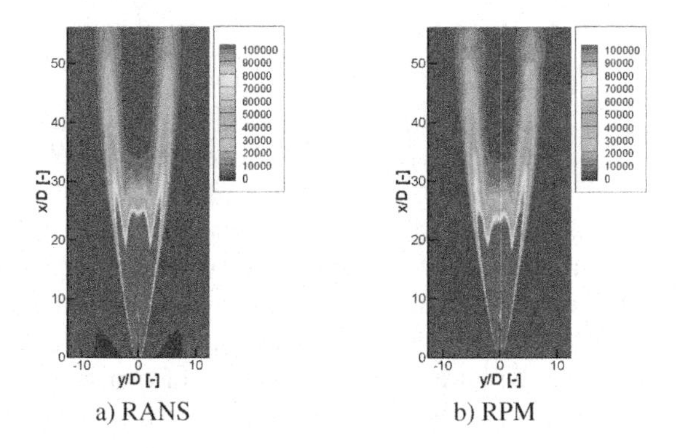

Fig. 1.15 Source variance distribution of the RANS simulation and results from the stochastic reconstruction by RPM.

An instantaneous acoustic perturbation pressure distribution of the CAA simulation of the DLR-A flame is indicated in Fig. 1.16. As can be seen here the acoustic radiation is not omni directional even though the main combustion noise sources have monopole character. The perturbation pressure distribution indicates a cone of silence like in heated jets, which is due to acoustic wave refraction by density and velocity gradients of the mean flow. The so called cone of silence caused by refraction effects has been proved experimentally [27, 45] and numerically [13, 34, 10]. Omnidirectional radiation could be observed for acoustic simulations based on an isothermal mean flow excluding density and velocity gradients which are not presented here.

The sound pressure level spectrum of the computed acoustics of the DLR-A flame is compared to the experimental data from Singh et al. [47] in Fig. 1.17. The Strouhal number Sr is based on the nozzle Diameter D and the mean jet exit

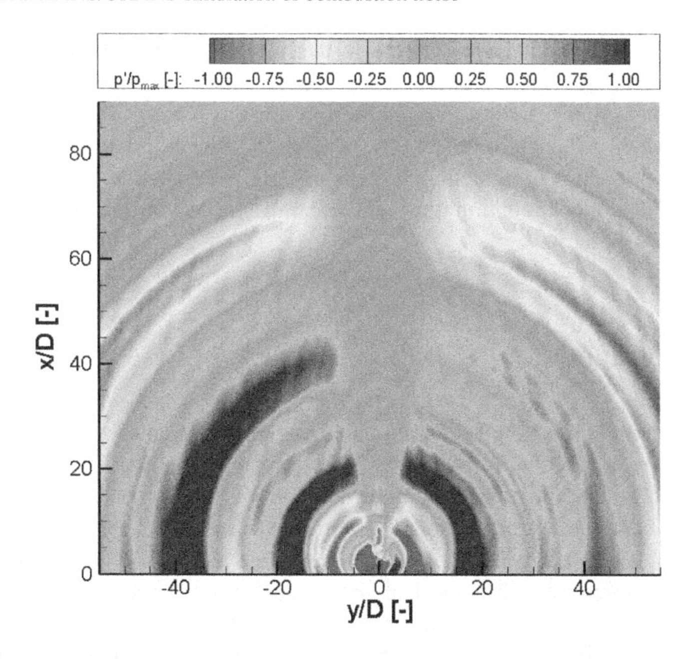

Fig. 1.16 Instantaneous distribution of the calculated perturbation pressure of the DLR-A flame.

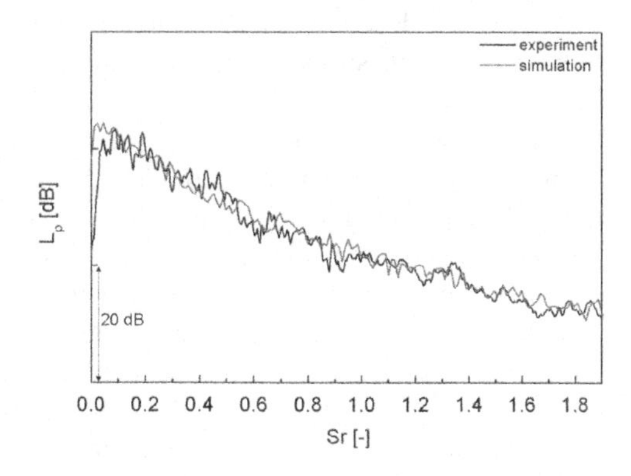

Fig. 1.17 Computed and measured sound pressure level spectra of the DLR-A flame at $x/D = 0$ and $y/D = 25$ in the Strouhal range $0 < Sr < 1.9$ which corresponds to a frequency range $0\,\text{Hz} < f < 10000\,\text{Hz}$. The Strouhal number Sr is based on the nozzle Diameter D and the mean jet exit velocity u_{jet} of the DLR-A flame.

velocity u_{jet} of the DLR-A flame. The Strouhal range $0 < Sr < 1.9$ corresponds to a frequency range $0\,\text{Hz} < f < 10000\,\text{Hz}$. The shape of the computed sound pressure level spectrum is in very good agreement with the measured one for the entire Strouhal range. Even though the amplitude of the computed spectrum is in the range of the experimental data the sound pressure level spectrum is not compared absolutely since the source term modeling is not scaled until now. Nevertheless, it should be noted that the deviations of the computed sound pressure level spectrum predicted by this highly efficient numerical approach is not higher than deviation of the spectrum predicted by high fidelity methods or statistical methods for this test case [33, 25, 11].

The CFD simulations have been performed in parallel processing on two IBM Power5 processors. Thus, the cost of a simulation was 11 cpu hours. The subsequent acoustic simulation including the combustion noise reconstruction was performed sequentially on a 2.0 GHz processor spending 50 cpu hours. Consequently, the RPM-CN approach is computational economic and compared to hybrid LES/CAA approaches the long simulation times and the transient source data exchange can be avoided.

1.4 Conclusions

Numerical simulations of the generation and convection of entropy modes and associated entropy noise of a test rig called Entropy Wave Generator (EWG) have been presented. The three dimensional compressible URANS approach was successfully validated by simulating a standard configuration of the EWG assuming a constant reflection coefficient at the end of the computational domain. Pressure fluctuations and pressure spectra of the numerical simulation are in very good agreement with the experiments. Simulations of different operating points with varying Mach number agree very well with measurements, too. Furthermore, a numerical method was adapted to investigate the acoustic sources of entropy noise. Using this method revealed that the decrease of entropy noise for high Mach numbers in the EWG is due to the fact that the spacial structure of the entropy noise sources is changed at high Mach numbers.

The reliability of the new RPM-CN approach, a highly efficient hybrid CFD/CAA approach, for the prediction of broadband combustion noise was demonstrated. The RPM-CN approach relies on the RPM method for the stochastic reconstruction of combustion noise sources in the time domain as a function of statistical turbulence quantities which can be delivered by reactive RANS simulations. Subsequently, the combustion noise propagation is computed by integrating the acoustic perturbation equations (APE-4). In a test case CFD results of the reactive flow computations of the DLR-A flame were compared to measurements and discussed. Based on this calculation a combustion noise prediction was done. The capability of the RPM-CN method to reproduce the defined two-point correlations was clearly demonstrated by a statistical analysis. Furthermore, the computed sound pressure level spectra were compared to measurements. The predicted decay of the spectra show a very good agreement with the acoustic measurements. Although a open jet flame was used for the validation, the presented numerical approach is appropriate for the prediction of combustion noise in complex geometries like confined combustion chambers in aero-engines. In combination with very low computational costs the application of the highly efficient RPM-CN approach can be taken advantage of in the design process.

Within the DFG Research Unit FOR 486 this sub-project was devoted to the evaluation and development of calculation models relying on Reynolds Averaged Navier-Stokes (RANS) turbulence modeling. Thus, highly efficient simulation models for the numerical prediction of combustion noise were considered here. Concerning, entropy noise the work was strongly linked with the experimental work of sub-project 5. Furthermore, theoretical work of sub-project 8 was complementary to the work of the present project. The investigations concerning the simulation of turbulent combustion noise was in conjunction with the work in sub-project 6 and with the work relying on LES computations of sub-project 3 and sub-project 7.

Acknowledgements The authors gratefully acknowledge the financial support by the German Research Foundation (DFG) through the Research Unit FOR 486 "Combustion Noise". We like to thank Dr.-Ing. Friedrich Bake and Dr.-Ing. Ingo Röhle of the DLR Institute of Propulsion Technology, Berlin, Germany, for providing excellent measurements and Dipl.-Ing. Axel Widenhorn of the DLR Institute of Combustion Technology, Stuttgart, Germany, for very helpful discussions concerning the application of non-reflecting boundary conditions.

References

[1] Bake F, Kings N, Röhle I (2008) Fundamental mechanism of entropy noise in aero-engines: experimental investigation. Journal of Engineering for Gas Turbines and Power 130:011,202–1 – 011,202–6
[2] Barlow R (1996-2004) Proceedings of the TNF Workshops - Sandia National Laboratories, www.ca.sandia.gov/TNF. Livermore, CA
[3] Baum M, Poinsot T, Thevenin D (1994) Accurate boundary conditions for multicomponent reactive flows. Journal of Computational Physics 116(2):247–261
[4] Bergmann V, Meier W, Wolff D, Stricker W (1998) Application of spontaneous Raman and Rayleigh Scattering and 2D LIF for the characterization of a turbulent CH4/H2/N2 jet diffusion flame. Applied Physics 66(4):489–502
[5] Bilger RW (1988) The structure of turbulent non-premixed flames. In: Twenty-Second Symposium (International) on Combustion, The Combustion Institute, Pittsburgh, pp 475–488
[6] Billson M, Eriksson LE, Davidson L (2003) Jet noise prediction using stochastic turbulence modeling. In: AIAA 2003-3282, Hilton Head, South Carolina
[7] Bird R, Stewart W, Lightfoot E (1960) Transport phenomena. John Wiley & Sons
[8] Bragg S (1963) Combustion noise. J Inst Fuel 36:12–16
[9] Brinkmann B (2007) Numerische Simulation des hochfrequenten Effekts von Düsenrandmodifikationen. Master's thesis, Abteilung Technische Akustik, Institut für Aerodynamik und Strömungstechnik, Deutsches Zentrum für Luft- und Raumfahrt e.V. (DLR)
[10] Bui T, Schröder W (2008) Non-isotropic acoustics of open turbulent flames. In: International conference on jets wakes and separeted flows, Berlin, Germany
[11] Bui T, Ihme M, Meinke M, Schröder W, Pitsch H (2007) Numerical investigation of combustion noise and sound source mechanisms in a non-premixed flame using LES and APE-RF. In: AIAA 2007-3406, Rome, Italy
[12] Bui T, Schröder W, Meinke M (2008) Numerical analysis of the acoustic field of reacting flows via acoustic perturbation equations. Computers & Fluids 37(9):1157–1169
[13] Candel S (1977) Numerical solution of conservation equations arising in linear wave theory: application to aeroacoustics. Journal of Fluid Mechanics 83(3):465–493
[14] Candel S, Durox D, Ducruix S, Birbaud AL, Noiray N, Schuller T (2009) Flame dynamics and combustion noise: progress and challenges. International Journal of Aeroacoustics 8(1):1–56
[15] Chu BT, Kovasznay L (1958) Non-linear interactions in a viscous heat-conducting compressible gas. Journal of Fluid Mechanics 3:494–514
[16] Delfs J, Bauer M, Ewert R, Grogger H, Lummer M, Lauke T (2007) Numerical simulation of aerodynamic noise with DLR's aeroacoustic code PIANO - PIANO manual version 5.2. Braunschweig

[17] Dowling A (1992) Thermoacoustics and Instabilities - In Crighton et al.: Modern methods in analytical acoustics. Springer-Verlag
[18] Dowling A (1996) Acoustics of unstable flows - In Theoretical and Applied Mechanics. T. Tatsumi, E. Watanabe, and T. Kambe eds., Amsterdam
[19] Ewert R (2006) Slat noise trend predictions using CAA with stochastic sources from a random particle mesh method (RPM). In: AIAA 2006-2667, Cambridge, MA
[20] Ewert R (2007) RPM - the fast Random Particle-Mesh method to realize unsteady turbulent sound sources and velocity fields for CAA applications. In: AIAA 2007-3506, Rome, Italy
[21] Ewert R (2008) Broadband slat noise prediction based on CAA and stochasic sound sources from a fast random particle-mesh (RPM) method. Computers & Fluids 37:369–387
[22] Ewert R, Edmunds R (2005) CAA slat noise studies applying stochastic sound sources based on solenoidal digital filters. In: AIAA 2005-2862, Monterey, CA
[23] Ewert R, Schröder W (2003) Acoustic perturbation equations based on flow decomposition via source filtering. Journal of Computational Physics 188:365–398
[24] Ewert R, Kornow O, Tester B, Powles C, Delfs J, Rose W (2008) Spectral broadening of jet engine turbine tones. In: AIAA 2008-2940
[25] Flemming F (2006) On the simulation of noise emissions by turbulent non-premixed flames. PhD thesis, TU Darmstadt
[26] Gerlinger P (2005) Numerische Verbrennungssimulation - Effiziente numerische Simulation turbulenter Verbrennung. Springer-Verlag
[27] Grande E (1967) Refraction of sound by jet flow and jet temperature. In: NASA CR-840
[28] Hassan H (1974) Scaling of combustion generated noise. J Fluid Mech 66(3):445–453
[29] Hesthaven J (1998) On the analysis and construction of perfectly matched layers for the linearized Euler equations. Journal of Computational Physics 142:129–247
[30] Hu F, Hussaini M, Manthey J (1996) Low-dissipation and low-dispersion Runge-Kutta schemes for computational acoustics. Journal of Computational Physics 124:177–191
[31] Huber A, Polifke W (2008) Impact of fuel supply impedance on combustion stability of gas turbines. In: ASME Turbo Expo, GT2008-51193, Berlin, Germany
[32] Hwang I, Kim Y (2006) Measurement of thermo-acoustic waves induced by rapid heating of nickel sheet in open and confined spaces. International Journal of Heat and Mass Transfer 49(3):575–581
[33] Ihme M, Bodony D, Pitsch H (2006) Prediction of combustion-generated noise in non-premixed turbulent jet flames using large-eddy simulation. In: AIAA 2006-2614, Cambridge, Massacusetts

[34] Ihme M, Kaltenbacher M, Pitsch H (2006) Numerical simulation of flow- and combustion-induced sound using a hybrid LES/CAA approach. In: Proceeding of the Summer Programm 2006, Stanford, USA

[35] Marble F, Candel S (1977) Acoustic disturbance from gas non-uniformities convected through a nozzle. Journal of Sound and Vibration 55(2):225–243

[36] Meier W, Barlow R, Chen YL (2000) Raman/Rayleigh/LIF measurements in a turbulent CH4/H2/N2 jet diffusion flame: experimental techniques and turbulence-chemistry interaction. Combustion and Flame 126:326–343

[37] Morfey C (1973) Amplification of aerodynamic noise by convected flow inhomogeneities. Journal of Sound and Vibration 31(4):391–397

[38] Mühlbauer B, Ewert R, Kornow O, Noll B, Delfs J, Aigner M (2008) Simulation of combustion noise using CAA with stochastic sound sources from RANS. In: AIAA 2008-2944, Vancouver, Canada

[39] Noll B, Schütz H, Aigner M (2001) Numerical simulation of high-frequency flow instabilities near an airblast atomizer. In: ASME Turbo Expo, GT2001-0041, New Orleans, LA

[40] Peters N (2000) Turbulent combustion. Cambridge monographs on mechanics, Cambridge University Press

[41] Poinsot T, Lele S (1992) Boundary conditions for direct simulations of compressible viscous flows. Journal of Computational Physics 101(1):104–129

[42] Polifke W, Wall C (2002) Non-reflecting boundary conditions for acoustic transfer matrix estimation with LES. In: Proceedings of Summer Programm 2002, Stanford, USA

[43] Pope S (2000) Turbulent flows. Cambridge University Press

[44] Schneider C, Dreizler A, Janicka J (2003) Flow field measurements of stable and locally extinguishing hydrocarbon-fuelled jet flames. Combustion and Flame 135:185–190

[45] Schubert L (1972) Numerical study of sound refraction by a jet flow. I. Ray Acoustics. J Acoustical Soc Am 51(2A):439–446

[46] Selle L, Nicoud F, Poinsot T (2004) Actual impedance of nonreflecting boundary conditions: Implications for computation of resonators. AIAA Journal 42(5):958–964

[47] Singh K, Frankel S, Gore J (2004) Study of spectral noise emissions from standard turbulent nonpremixed flames. AIAA Journal 42(5):931–936

[48] Strahle W (1972) On combustion generated noise. Journal of Sound and Vibration 23(1):113–125

[49] Strahle W (1975) A review of combustion generated noise. Prog Astronaut Aeronaut 37:229–248

[50] Tam C, Auriault L (1999) Jet mixing noise from fine-scale turbulence. AIAA Journal 37(2):145–153

[51] Tam C, Webb J (1993) Dispersion-relation-preserving finite difference schemes for computational acoustics. Journal of Computational Physics 107:262–281

[52] Wagner C, Hüttl T, Sagaut P (2007) Large-eddy simulation for acoustics. Cambridge University Press, Cambridge

[53] Warnatz J, Maas U, Dibble R (2001) Verbrennung, 3rd edn. Springer-Verlag, Berlin

[54] Widenhorn A, Noll B, Aigner M (2006) Accurate boundary conditions for the numerical simulation of thermoacoustic phenomena in gas-turbine combustion chambers. In: ASME Turbo Expo, GT2006-90441, Barcelona, Spain

[55] Widenhorn A, Noll B, Aigner M (2008) Impedance boundary conditions for the numerical simulation of gas turbine combustion systems. In: ASME Turbo Expo, GT2008-50445, Berlin, Germany

[56] Zimont V (2000) Gas premixed combustion at high turbulence. Turbulent flame closure combustion model. Experimental Thermal and Fluid Science 21:179–186

Chapter 2
Measurement and Simulation of Combustion Noise emitted from Swirl Burners

C. Bender, F. Zhang, P. Habisreuther, H. Büchner and H. Bockhorn

Abstract A major uncertaincy, when designing combustors is the influence of geometrical patterns of the design on the combustion noise generated. In order to determine the mechanisms and processes that influence the noise generation of flames with underlying swirling flows, a new burner has been designed, that offers the possibility to vary geometrical parameters. Experimental data (flow field, noise emission) have been determined for this burner. In addition, Large Eddy Simulations (LES) have been performed to study the isothermal and reacting flow of the burner. The results of the measurements show a distinct rise of the sound pressure level, obtained by changing the test setup from the isothermal to the flame configuration as well as by varying geometrical parameters, which is also resembled by the LES simulation results. A physical model has been developed from experiments and verified by the LES simulation, that explains the formation of coherent flow structures and allows to separate their contribution to the overall noise emission from ordinary turbulent noise sources. The computed isothermal and reacting flow fields have been discussed through flow visualization; the computed acoustic pressure has been compared with the experiment and it showed good agreement.

2.1 Introduction

The development of modern gas turbines and jet engines is focused on the reduction of pollutant emissions and, increasingly, on the reduction of overall noise emission, including combustion noise. This requires the minimization of the noise sources, namely noise from the turbulent flow, combustion noise and noise caused by periodic instabilities and fluctuations of the ignition zone. This has to be achieved under

---

C. Bender, F. Zhang, P. Habisreuther, H. Büchner, H. Bockhorn
University of Karlsruhe, Division of Combustion Technology
Engler-Bunte-Ring 1, 76131, Karlsruhe, Germany
e-mail: feichi.zhang@vbt.uni-karlsruhe.de, christian.bender@vbt.uni-karlsruhe.de

A. Schwarz, J. Janicka (eds.), *Combustion Noise*, DOI 10.1007/978-3-642-02038-4_2,

conservation of the benefits of swirl flames, e.g. high ignition stability and broad operation ranges. Subproject 2 "Measurement and Simulation of Combustion Noise emitted from Swirl Burners with different Burner Exit Geometries" is focused on the description and characterization of fluctuations of the ignition zone due to different mixing and stabilization characteristics of swirl flames, the influence of the enclosure of the flames in combustion chambers and the description of flame noise caused by the formation and burning of coherent, periodic flow structures, often detected in swirl flows and flames [3, 4, 18, 25].
Most technical combustion systems use turbulent premixed or non-premixed swirl flames with high volumetric reaction densities. To realize good ignition stability, especially, when using high air equivalence ratio to prevent thermal NOx-emissions, the flow field is swirl stabilized. The swirl flow forms - when a critical value of the swirl number is exceeded - a central inner recirculation zone and causes a longer residence time of reactive species and enables the formation of short flames of high reaction densities with the benefits of high ignition stability due to the recirculation of hot gases [5, 23] and the achievable low-pollution combustion. Another advantage of premixed swirl-stabilized flames is the option to reduce pollutant emissions of modern combustion systems materialized in industrial, traffic, power plant and aircraft applications compared to diffusive flames by cost-efficient design solutions. Basic research investigation of swirl flames is given by [16, 21].
Besides the named properties of swirl flames, also some issues arise putting this type of burner into action. Swirl flows and flames tend to generate periodic flow instabilities, which lead to an increasing noise emission and problem of flame stability. Combustion instabilities have been investigated intensively in the recent years using pulsated inflows [1, 9]. Instabilities in swirl flames have been investigated by [10, 22, 28] and reveal the importance of coherent structures in the forward flow surrounding the inner recirculation zone. The existence and the influence of coherent structures on the noise emission of the flame and thus of the combustion system requires more physical knowledge to minimize combustion noise in modern industrial and aircraft applications already during the design process. Another disadvantage of swirl flames are their high noise levels caused by high reaction densities and amplified by the use of premixed flames in industrial furnaces.
Numerical methods using high resolution techniques for the description of turbulent flows, like large eddy simulation (LES) offer the possibility to resolve the unsteady flow structure with high accuracy. The present work uses the LES to adress the mechanisms of the reacting flow field. In this approach, the filtered unsteady Navier-Stokes equations are taken into account and only large scale turbulences are resolved dependent on the cut-off level, the unresolved small ones are modelled via a so-called sgs (subgrid scale)-model. Since the large turbulent motions carry the most energy of the flow and generally show a non-isotropic behaviour, it is adequate to model the small eddies, whereas these exhibit more universal features. Thus, the LES technique is well suited for studying instabilities in the combustor devices, since the flow field of concern is highly unsteady and dominated by turbulence motions that can be adequately resolved computationally [27].

2.2 Theoretical Background

2.2.1 Experimental Setup

The investigations were carried out using a modular double-concentric swirl-burner for premixed and non-premixed combustion of gaseous fuels, offering the options to vary independently geometric parameters (especially the angle of the burning gas injection and the burner exit geometry with a recessed pilot often used in industrial applications [12, 20]) as well as technically relevant operation parameters like thermal load $\dot{Q}_{th}$ and air equivalence ratio $\lambda = \dot{V}_{air}/(\dot{V}_{fuel} \times l_{min})$ with l_{min} as minimal, stoichiometric required air rate. Another advantage of the burner design is the possibility to generate periodic flow instabilities by variation of the burner exit geometry and definite conservation of the mean turbulent flow field [18]. Identical frequencies have been determined in the spectral distributions of the noise far sound field measured by a microphone probe and in the fully turbulent flow field detected via laser diagnostics (Laser Doppler Velocimetry). The part of noise caused by the underlying physical mechanism [25] can be quantified and avoided by proper design of the burner outlet geometry.
The influence of the gas injection angle in the case of non-premixed flames was investigated with variation of gas injection angles between 0° and 90° referring to the burner's axis (resulting in type-I and type-II-diffusion flames, see figure 2.13). The operating parameters' range of the pilot burner varied between premixed conditions - leading to sufficient main flame stabilization at the pilot burners exit - and the main flame without pilot flame, causing a fluctuation of the ignition zone in dependence of the gas injection angle of the main gas supply and a strong corresponding increase of the flame noise.
Figure 2.1 shows a sketch of the double-concentric burner, which produces two concentric swirl-flows: A central pilot burner flow and an outer main burner flow.

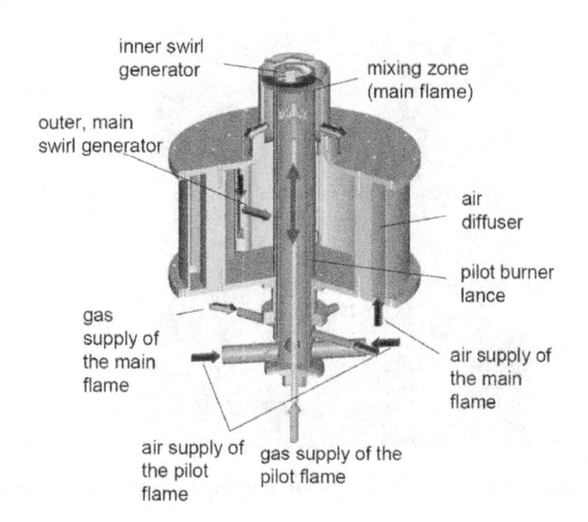

Fig. 2.1 Sketch of the double-concentric swirl burner with a central pilot burner [4]

The inner diameter of the main burner nozzle is $D_0 = 110mm$ and the outer diameter of the pilot burner is $d_0 = 70mm$. For isothermal investigations, the inner and outer flows were fed with air from a compressor, the measurements with swirl flames were carried out with premixed or non-premixed natural gas/air mixtures.
The normalized momentum flux ratio representing theoretical swirl number $S_{0,th}$ at the burner's exit is given in equation 2.1 with the tangential momentum flux $\dot{D}$, the axial momentum flux $\dot{i}_{ax,0}$ and the burner's exit radius R_0.

$$S_{0,th} = \frac{\dot{D}}{\dot{i}_{ax,0} \times R_0} \tag{2.1}$$

The swirl of the pilot flame was generated by an axial swirl generator with vanes and a theoretical swirl number $S_{0,th,pilot} = 0.79$ (equation 2.1). The main flow was swirled by tangential swirl generators with a theoretical swirl number $S_{0,th,main} =$ 0.90. The axial pilot burner position with respect to the main burner outlet is the variable geometric parameter. In figure 2.1, the pilot burner outlet is aligned to the burner outlet ($x_{pilot} = 0mm$). The position of the inner pilot burner can be set up to 40 mm behind the burner outlet ($x_{pilot} = -40mm$).
The microphone probe was positioned at a fixed measurement position with a radial distance to the center line $r/D_0 = 4.55$ and an axial distance to the burner outlet of $x/D_0 = 1.0$, after having proven the independence of the noise measurements from the probe position due to the spherical noise emission characteristics [17].
Figure 2.2 shows the experimental setup for the investigations under enclosed conditions. The combustion chamber was designed with an optical access to investigate the flow fields of enclosed isothermal swirl flows and enclosed premixed swirl flames. With the microphone probe the flow and flame noise was detected to characterize the spectral and integral distribution of the noise from isothermal flows and combustion noise of swirl flames under premixed and non-premixed conditions.

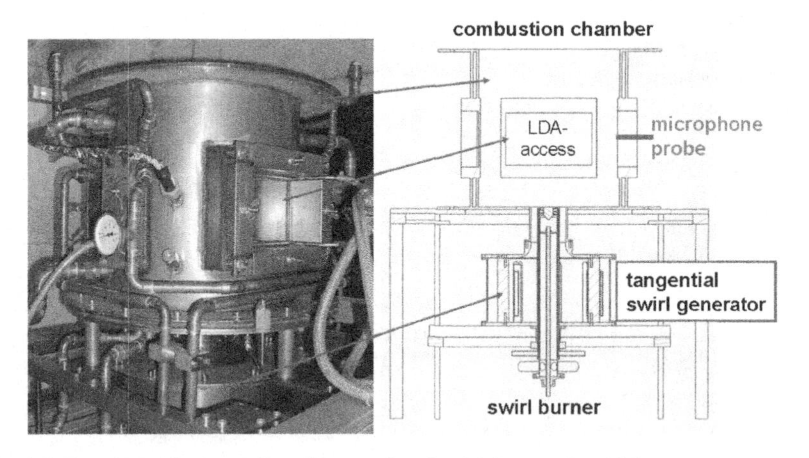

Fig. 2.2 Experimental setup to investigate enclosed swirl flows and swirl flames

2.2.2 Numerical Methods

2.2.2.1 Governing Equations

Basics of the numerical modelling are the compressible Favre-averaged Navier-Stokes equations and the energy equation; within the LES-formulation these read:

$$\frac{\partial \bar{\rho}}{\partial t} + \frac{\partial \bar{\rho}\tilde{u}_i}{\partial x_i} = 0 \tag{2.2}$$

$$\frac{\partial \bar{\rho}\tilde{u}_i}{\partial t} + \frac{\partial \bar{\rho}\tilde{u}_i\tilde{u}_j}{\partial x_j} = -\frac{\partial \bar{p}}{\partial x_i} + \frac{\partial(\tilde{\tau}_{ij} - \tau_{ij}^{sgs})}{\partial x_j}. \tag{2.3}$$

$$\frac{\partial \bar{\rho}\tilde{E}}{\partial t} + \frac{(\partial \bar{\rho}\tilde{E} + \bar{p})\tilde{u}_j}{\partial x_j} = \frac{\partial}{\partial x_j}(\tilde{u}_j + \bar{q}_j - H_i^{sgs} + \sigma_{ij}^{sgs}) \tag{2.4}$$

where overbars and tildes denote the space-filtered and Favre-averaged scales. τ_{ij} and q_j are the viscous stress tensor and heat flux, respectively. τ_{ij}^{sgs} is the subgrid scale (sgs) stress tensor which is defined as $\tau_{ij}^{sgs} = \overline{u_i u_j} - \bar{u}_i\bar{u}_j$. Using the Smagorinsky subgrid Model [31], τ_{ij}^{sgs} is modeled as

$$\tau_{ij}^{sgs} = -2\rho\nu_t\bar{S}_{ij} = -\rho\nu_t(\frac{\partial \bar{u}_i}{\partial x_j} + \frac{\partial \bar{u}_j}{\partial x_i}), \quad \nu_t = (C_S\Delta)^2|\bar{S}_{ij}|, \tag{2.5}$$

where S_{ij} is the shear strain rate and C_S a model constant, it takes the value between 0.065 and 0.24 [15] dependent on the flow cases. In our case, $C_s = 0.1$ is used. The sgs enthalpy flux H_j^{sgs} is modeled similarly using a gradient transport model and the sgs viscous work term σ_{ij}^{sgs} is neglected.

2.2.2.2 The TFC Model

The turbulent flame speed closure (TFC) model [29, 33] was used for the combustion modeling. An addtional transport equation for the progress variable θ is solved in this model, which describes chemical reaction progress. θ is linearly related to the fuel or product mass fraction and has the value 0 in the unburned mixture and 1 in the burned mixture. In the framework of LES the transport equation of θ yields the form:

$$\frac{\partial \bar{\rho}\tilde{\theta}}{\partial t} + \nabla \bullet (\bar{\rho}\tilde{\mathbf{u}}\tilde{\theta}) = \nabla \bullet (\frac{\mu_t}{Sc_t}\nabla\tilde{\theta}) + \bar{\rho}_u S_t \left|\nabla\tilde{\theta}\right|. \tag{2.6}$$

The turbulent flame speed S_t is an important parameter which covers the chemical/physical characteristics of the combustable mixture and the characteristics of the turbulent flow, it describes further more the interaction between turbulence and chemical reactions. The closure of equation (2.6) requires that S_t has to be expressed via known values. Schmid [29] has theoretically derived an expression for S_t:

$$\frac{S_t}{S_l} = 1 + \frac{u'}{S_l}\left(1 + Da^{-2}\right)^{-1/4}, \quad Da^{-1} = \frac{\tau_c}{\tau_t} = \frac{\delta_F u'}{S_l L_t} = \frac{au'}{S_l^2 L_t}. \tag{2.7}$$

u' is the velocity fluctuation and a the thermal diffusivity. L_t is the turbulent length scale quantified by $\Delta = C_S(\Delta_x \Delta_y \Delta_z)^{1/3}$; the turbulent velocity at subgrid scale is assumed as:

$$u'_\Delta = \frac{\nu_t}{\Delta} = C_S \Delta \sqrt{2\tilde{S}_{ij}\tilde{S}_{ij}}. \tag{2.8}$$

Other expressions for u'_Δ and Δ are possible and the reader is referred to [13, 15]. It has to be noted that u'_Δ is now grid dependent as well as the time resolved velocity. In the case of very small grid size $\Delta \to 0$, the filtered Navier-Stokes equations will switch to the exact unfiltered form. The turbulent diffusion terms in the filtered transport equations (equation 2.3, 2.4, 2.6) will vanish since $\mu_t \to 0$. The remaining term in S_t (equation 2.7) will be only the laminar flame speed S_l. In this case, the smallest eddies and the laminar flame front are resolved on the fine mesh and no modelling is needed. This makes a consistent transfer from LES to DNS.

S_l is the laminar burning speed and it describes the reaction solely from the mixture side. In premixed flames, the fuel is mixed with air in a certain ratio upstream of the burner, so that S_l can be given using a constant value. However, to solve flames with different air equivalence ratios λ, like the one in the current case (sec.2.2.2.4), the mixing effect of S_l has to be considered [29]. On the left side of figure 2.3, the dependency of S_l on the air equivalence ration λ is shown, the measured laminar flame speed [34] was fitted using the least-square-method via a fourth order ploynomial.

This TFC combustion model has been implemented in the commercial CFD code ANSYS CFX which uses an unstructured finite volume method (FVM) to solve the compressible Navier Stokes equations and is capable to run LES. The central difference scheme in space and 2nd order fully implicit time progress were used. Since the aim of the model is primarily to describe heat release, an integral one-step reaction $CH_4 + 2O_2 \to CO_2 + 2H_2O$ has been considered adequate for modelling methane/air

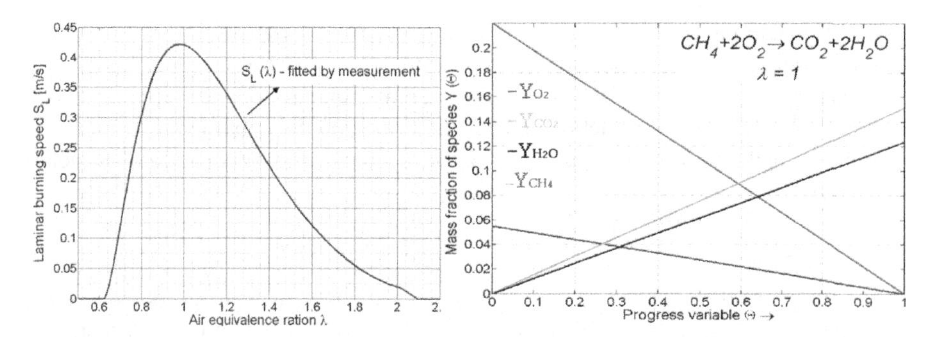

Fig. 2.3 Laminar flame speed S_L dependent on the air equalence ratio (left) and mass fractions over progress varialbe (right)

combustion. In this case, the reactants are completely burned to the stable product and all species concentrations are related to the progress variable θ via simple algebraic equations. On the right hand side of figure 2.3 the relations between the species mass fractions Y_i and the progress varible Θ for the stoichiometric mixture case is shown. The product concentrations (Y_{H_2O} and Y_{CO_2}) increase linearly with Θ, whereas the fuel concentration Y_{CH_4} straightly falls.

2.2.2.3 Lighthill's Acoustic Analogy

The Lighthill's acoustic equation for the density fluctuation results by conversion of the basic fluid mechanics equations without any approximation and linearization [24]:

$$\frac{\partial^2 \rho'}{\partial t^2} - c_0^2 \Delta \rho' = \frac{\partial^2 T_{ij}}{\partial x_i \partial x_j}, \tag{2.9}$$

$$T_{ij} = \rho v_i v_j - \tau_{ij} + \delta_{ij}(p' - c_0^2 \rho'). \tag{2.10}$$

T_{ij} is the Lighthill's stress tensor and c_0 the sonic speed. The left hand side of equation 2.9 has the form of a linear wave equation and the right hand side of this inhomogeneous wave equation can be considered as a source term presuming no viscous terms and small Mach numbers. While the turbulent reacting flow acts as a source, the acoustic waves propagate in a uniform acoustic medium. The transfer of informations from the sources to the external medium is achieved through the Lighthill stress tensor T_{ij}. In a turbulent reacting flow, the viscous stress τ_{ij} in T_{ij} is very small in comparison to the other terms and can be neglected. The velocity correlation term $\rho v_i v_j$ (Reynolds-stress term) and $p' - c_0^2 \rho'$ represent the sources caused by the turbulent flow and by the unsteady heat release. An order of magnitude analysis of the Lighthill's tensor showed [11] that the ratio of both terms scales to Ma^2, which indicates further, that for turbulent reacting flows at low Mach numbers $Ma^2 << 1$, the dominant term will be the last one on the right hand side of equation 2.10. Another version of the Lighthill's equation can be derived for the pressure fluctuation p' in the same way [24]:

$$\frac{1}{c_0^2} \frac{\partial^2 p'}{\partial t^2} - \Delta p' = \frac{\partial^2 T_{ij}^*}{\partial x_i \partial x_j} + \frac{\partial^2 W}{\partial t^2}, \tag{2.11}$$

$$T_{ij}^* = \rho v_i v_j - \tau_{ij}, \quad W = \frac{1}{c_0^2} p' - \rho'. \tag{2.12}$$

$\frac{\partial^2}{\partial x_i \partial x_j} T_{ij}^*$ and $\frac{\partial^2}{\partial t^2} W$ are then the sources responsible for the generation of pressure fluctuations.

2.2.2.4 Numerical Setup

The combustable mixture consists of fully-premixed air/natural gas. For the numerical simulation, the total thermal load of both flows is 135 kW, the important operational parameters are listed in table 2.1. $\dot{V}$ is the volume flow rate, λ the air equivalence ratio and S the theoretical swirl intensity.

Figure 2.4 shows the simulation domain and mesh of the open swirled flame with

Table 2.1 Operational parameters in the experiment

$\dot{V}_{main}$	$180m_N^3/h$	λ_{main}	1.5	S_{main}	0.9
$\dot{V}_{pilot}$	$20m_N^3/h$	λ_{pilot}	1.05	S_{pilot}	0.79

a planar pilot lance. The considered computational mesh represents the main parts of the interior flow domain of the burner and a large free domain downstream of the burner exit where reaction takes place. The axial swirl generator in the pilot flow was also included in the computational mesh. A block-structured grid with about 50 blocks and 1 mio. nodes was used. A time step of $10\mu s$ was chosen to keep the Courant-number. smaller than one. Colors of the planes denote types of the boundary conditions, which were set according to tab.2.1. The static pressure or total pressure (depending on the flow direction) and the temperature values are fixed like for the ambient air at the opening boundaries. No slip walls with constant temperature have been used for burner walls. The Reynolds-number based on the nozzle axial velocity and diameter of the main flow was Re $\approx$ 27000. The pilot lance can be retracted by 40mm as clearly shown in figure 2.5.

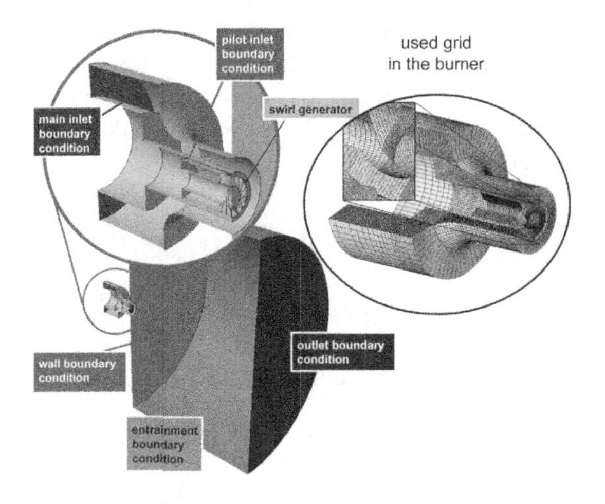

Fig. 2.4 Flow domain and mesh used in the simulation

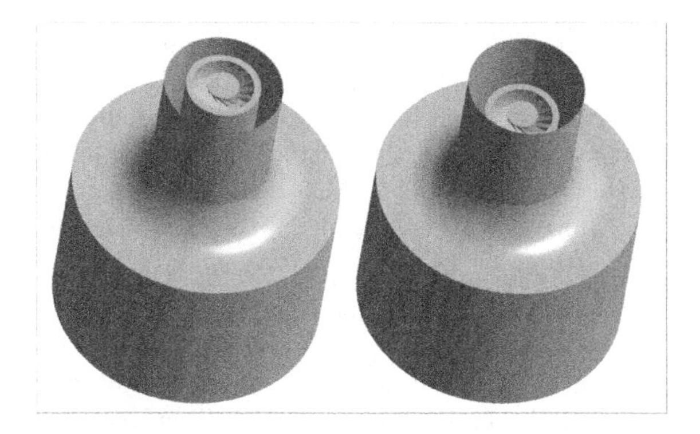

Fig. 2.5 Swirl burner with a planar (left) and a recessed (right) pilot lance

2.3 Results and Analysis

2.3.1 Experiment

2.3.1.1 Acoustic Characterization of Isothermal Swirl Flows

In the following section the results of experimental investigations of the isothermal, turbulent flow field of the new burner design are presented. The experiments were carried out under variation of the main operation and geometrical parameters like swirl intensity, flow rate and burner outlet geometry (axial position of the pilot burner). In figure 2.6 the results of the time-averaged isothermal flow field at four planes in the vicinity of the burner are plotted. The theoretical swirl number of the main and pilot flow are $S_{0,th,main} = 0.90$ and $S_{0,th,pilot} = 0.79$ and the axial position of the pilot burner is $x_{pilot} = 0mm$. The profiles of the axial velocity, starting nearby the burner outlet ($x/D_0 = 0.05$, cross) with the divided pilot and main flows, are shown. With increasing distance to the burner outlet the typical behavior of a strongly swirling flow was observed: The separation of the pilot and main flow disappears, indicating the fast mixture of the two flow fields, and at the centre line

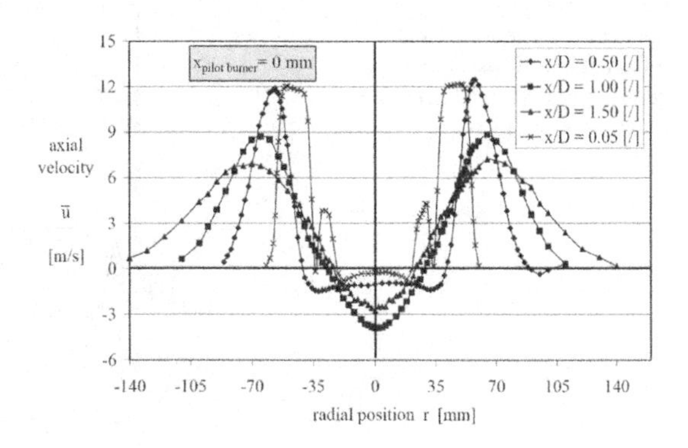

Fig. 2.6 Axial velocity profiles of the isothermal flow at four planes [4]

a inner recirculation zone with negative axial velocity (opposite to the main flow direction) can be identified.
The remaining single swirling jet spreads radially and, caused by the free flow field, entrains surrounding, non-rotating medium, while propagating further downstream ($x/D_0 = 0.50, 1.00, 1.50$). Measurements of the noise emissions of the isothermal swirl flow showed values of the sound pressure level $SPL = 20 \times log(\frac{p_{rms}}{p_{0,rms}}) = 75.7dB$ ($x_{pilot} = 0mm$, $p_{0,rms} = 2 \times 10^{-5} Pa_{rms}$) at the fixed measurement position with a radial distance to the center line $r/D_0 = 4.55$ and an axial distance to the burner outlet of $x/D_0 = 1.0$. An interesting result is the negligible influence of the swirl intensity and the strong influence of the flow rate (burner outlet velocity) on the turbulent flow noise. The value of 75 dB at a volume flow rate of $\dot{V}_{n,air} = 200m_N^3/s$ and the described measurement position is the minimal order of magnitude which cannot be undercut for the used isothermal, turbulent swirl flow.

2.3.1.2 Periodical Flow Instabilities in Isothermal Swirl Flows

To investigate the dynamic behaviour of the emitted noise more detailed, the acquired data from the microphone probe ($r/D_0 = 4.55$, $x/D_0 = 1.0$) were analysed in the frequency domain to get more information of the spectral distribution of the sound pressure ratio. Referring to the first burner configuration (see chapter 2.3.1.1) the pilot burners outlet was aligned to the burner outlet ($x_{pilot} = 0mm$) and the spectral distribution shows the typical behaviour of the turbulent disturbances in an isothermal, turbulent flow field (figure 2.7 dashed blue line; [4, 18]). In figure 2.7 the distribution of the emitted sound pressure is shown with the pilot burner drawn back 40 mm upstream the burner outlet (black line). Industrial burner applications with lean-premixed flames often use recessed components like pilot burners and bluff-bodies to prevent thermal damage of the sensitive units, examples are Siemens V64 [20] and Pratt & Whitney FT 8 [12]. The spectral distribution of the sound pressure shows a dominating peak at a frequency f = 56 Hz. In comparison to the aligned

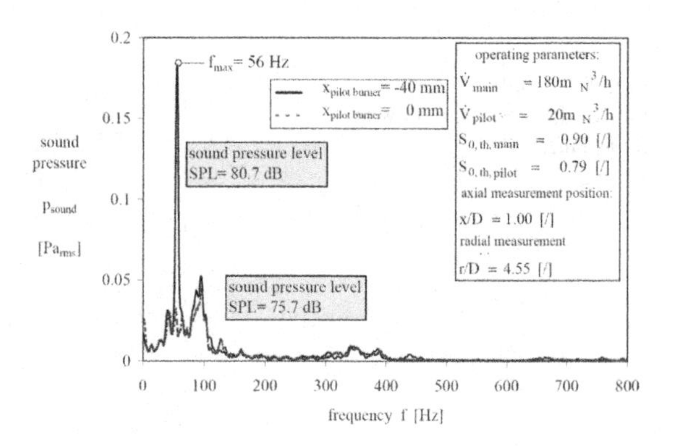

Fig. 2.7 Spectral distribution of the sound pressure of an isothermal swirl flow with different positions of the central pilot burner ($x_{pilot} = 0mm/-40mm$)

pilot burner $x_{pilot} = 0mm$ the peak increases the total SPL from 75.7 dB to a level of 80.5 dB. The influence on the acoustic field was identified as a consequence of periodic oscillations of the axial position of the inner recirculation zone in the swirl flow and the quantified increase of the sound pressure level is caused by a periodic disturbance of the isothermal flow field, shown by spectral distributed measurements of the flow field. The underlying physical mechanism is described by [10, 18] and shows a time-dependent axial movement of the recirculation zone, causing a time-dependent effective swirl number $S_{eff} = S_{eff}(t)$ and the stabilization point of the recirculation zone changes the different positions sinusoidally.
The variable burner design enables to generate periodical flow disturbances in an isothermal swirl flow. The consequence for further burner constructions is the need to prevent coherent flow structures caused e.g. by recessed pilot burner constructions, as often used in common industrial burner designs [12, 20].

2.3.1.3 Premixed Swirl Flames

Noise Sources in premixed Swirl Flames: Influence of Operating Parameters

In figure 2.8 the results of sound pressure measurements of premixed swirl flames are presented. The experiments were carried out with a constant pilot flame of a thermal load $\dot{Q}_{th,pilot} = 20kW$ and an air equivalence ratio of $\lambda_{pilot} = 1.05$ to guarantee a stable pilot flame. The pilot burner is positioned at the level of the burner outlet ($x_{pilot} = 0mm$). It is obvious, that the thermal output of the premixed main flame increases the sound pressure considerably. The reacting flow shows a sound pressure level of about 25 dB higher than the SPL of the isothermal flow (gas flow rate substituted by air). This part is called flame-induced noise and is caused by the volume expansion and density fluctuations in the flame due to the massive increase of the exhaust gas temperature. A two times enhanced thermal output of the main flame ($\dot{Q}_{th,main} = 150kW$ *and* $300kW$) causes an increase of the SPL from 103 dB to nearly 108.5 dB, whereas the influence of the air equivalence ratio λ_{main} is negli-

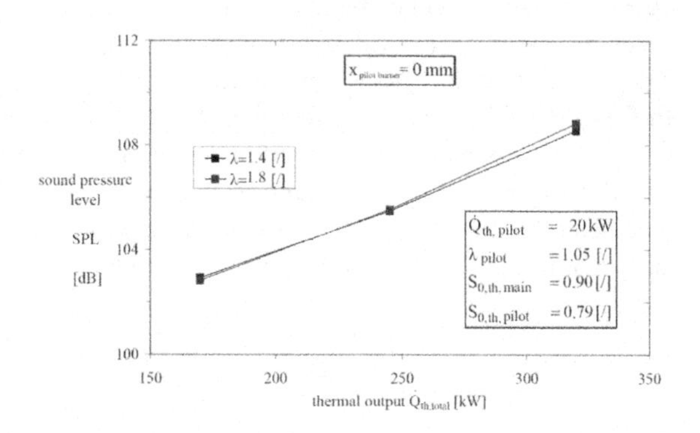

Fig. 2.8 Sound pressure level of premixed flames under variation of thermal load and air equivalence ratio main of the main flame

gible (see figure 2.8).

Noise Sources in premixed Swirl Flames: Variation of Geometric Parameters

Next, the influence on the noise of premixed swirl flames of the geometric parameters was investigated. For isothermal flows the influence was predominant (see chapter 2.3.1.2) and the question to be answered was how do premixed flames react on such disturbances of the flow field. In figure 2.9 the dependency of the burner outlet geometry on the sound pressure of premixed flames is presented. The spectral distribution is almost identical for all frequencies except f = 101 Hz. The recessed pilot burner generates dynamic, periodic structures in the flow field. It should be pointed out that the existence is not limited to isothermal flows. Corresponding to that fact such structures influence the investigated flames and their formation and combustion lead to an increase of the SPL of 2.3 dB (+30%).
Comparing reacting and isothermal cases shows that in the isothermal flow the fre-

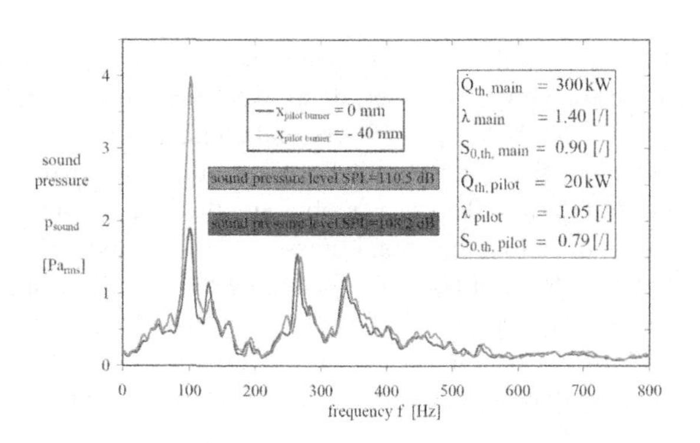

Fig. 2.9 Spectral distribution of the sound pressure of premixed flames with different burner exit geometries (different positions of the pilot burner $x_{pilot} = 0mm,\ and\ -40mm$)

quency of the periodic disturbance is lower. The frequency distribution of flames covers a wider frequency band. The Strouhal number Str ($= f_{peak} \times D_0/u_0$) for both cases reveals the same values (0.74 (premixed flame) and 0.72 (isothermal flow)). This proves that the effect is originated from the same physical phenomenon [18]. In figure 2.10 the results for identical combustion parameters with variation of the burners exit geometry are plotted. The varied pilot burner position influences the sound pressure level of the flame in the same way as in the isothermal case. The recessed pilot burner $x_{pilot} = -40mm$ generates coherent structures in the flow field and, therefore, the SPL increases in the presented case about 2.5 dB (+33.5%).

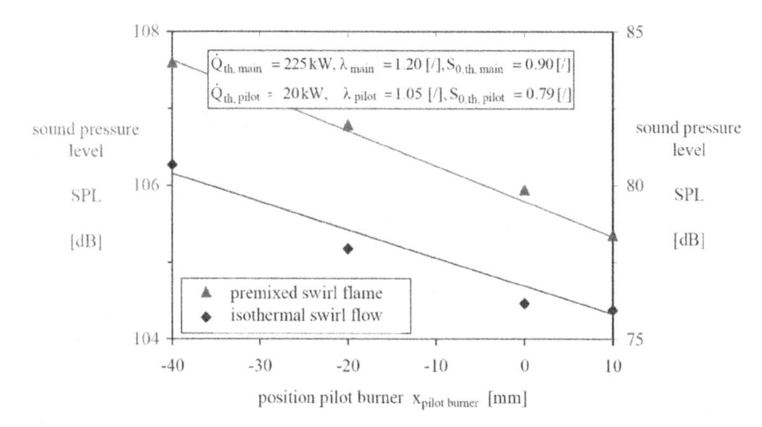

Fig. 2.10 Influence of the pilot burner position x_{pilot} on the sound pressure level of premixed swirl flames

2.3.1.4 Noise Sources in Non-Premixed Swirl Flames

In the second part of the characterization and quantification of noise from reacting flows non-premixed flames were studied and results were compared to investigations of premixed flames, which were carried out to get a classification of all technically relevant type of flames. The difference to the above described premixed flames is the location and the way of gas injection. In the case of premixed flames a mixture of fuel and air leaves the burner exit, whereas the air equivalence ratio is constant across the burner outlet area. For non-premixed flames the burner outflow consists of pure air, the gas is injected directly into the flame zone. This leads to a mixture of fuel, air and hot combustion products in the flame zone with the effect of fuel-rich and fuel-lean regions and the effect that the flames are often longer than premixed flames of same operating parameters, caused by a time delay for mixture formation before ignition occurs. Consequently, a lower volumetric reaction density is characteristic for non-premixed flames, which causes disadvantages like higher, local temperatures (NOx-formation) and soot formation.

Fig. 2.11 Sketch of the gas injection for non-premixed flames

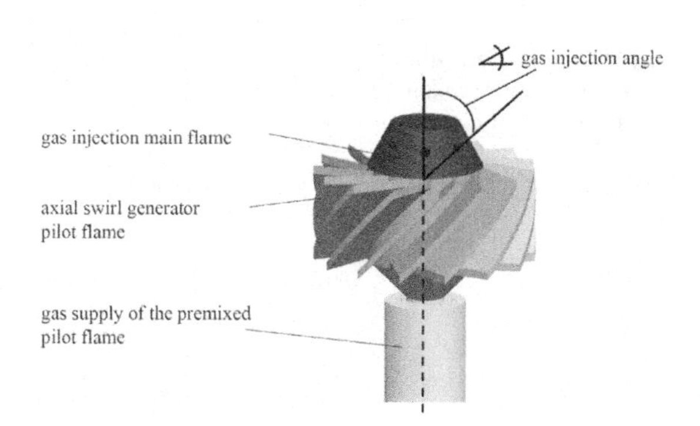

Variations of Geometric Parameters

The investigations of non-premixed flames were carried out with the same double-concentric swirl burner configuration, mentioned before. The gas injection was located at the central bluff-body of the pilot burner and the parameter to influence the flame characteristics (type-I- / type-II-flame) is the gas injection angle to generate technically relevant flames for different applications like jet engines or industrial furnaces. The angle (see figure 2.11) is defined between the direction of the gas flow and the burner axis. Figure 2.12 shows the realization of different types of non-premixed flames by varying the gas injection angles in a range of 0° to 90° in steps of 15°. Figure 2.13 shows the results for variation of gas injection angles between

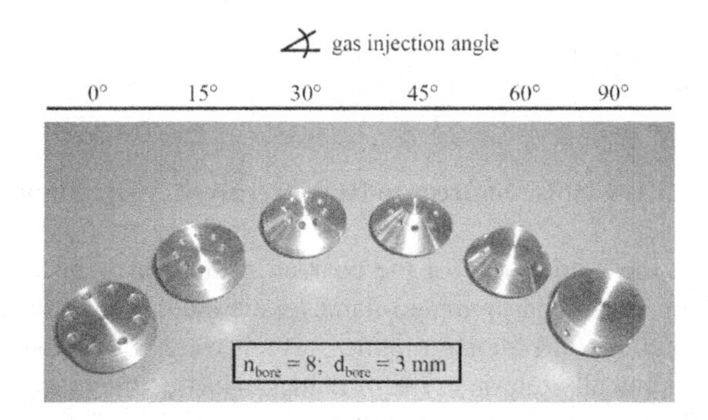

Fig. 2.12 Gas injection nozzles to realize different types of non-premixed flames

30° and 90° with variation of the air equivalence ratio. A decrease of air equivalence ratio should induce an increase of sound pressure level. But non-premixed flames show a remarkable behaviour: The sound pressure level is almost constant at SPL between 104 dB and 108 dB. It could be observed, that the non-premixed flames are partly louder than the equivalent premixed flame, which introduced the question whether there are other physical mechanisms influencing the combustion noise. A difference between the operating conditions of premixed and non-premixed flames is the existence of the pilot flame. The non-premixed flames have one central fuel

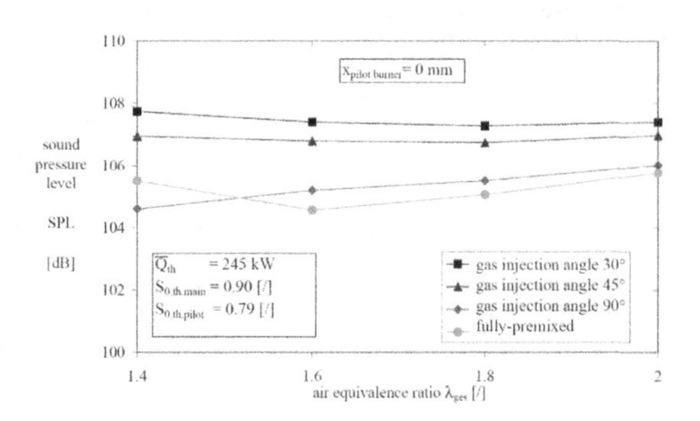

Fig. 2.13 Sound pressure level of non-premixed flames under variation of air equivalence ratio λ_{main} and the gas injection angle

supply and the different gas injection angles effect different stabilization points of the different flames. The next consequent step was the prevention of these fluctuations of the stabilization point by use of a premixed pilot flame.

Variation of Flame Stabilization: Influence of Pilot Flame

To investigate the influence of main flame stabilization by a premixed pilot flame, the same thermal load and the air equivalence ratio of the pilot flame as adjusted for investigations under premixed conditions of pilot and main flame were used. This enables to compare the premixed and non-premixed main flames with a stable, stationary ignition point of the main flame. The investigations showed the need of a locally stable ignition for investigations of the combustion noise, because the fluctuation of the flame ignition point causes an increase of noise up to 5 dB. But for such conditions a comparison with premixed flame cannot be done. Applying the premixed pilot flame in all cases, the non-premixed flames are always more quiet than the comparable premixed flame, which can be explained by lower reaction density of the non-premixed flames.

The experiment showed two regimes of non-premixed flames (figure 2.14): Louder

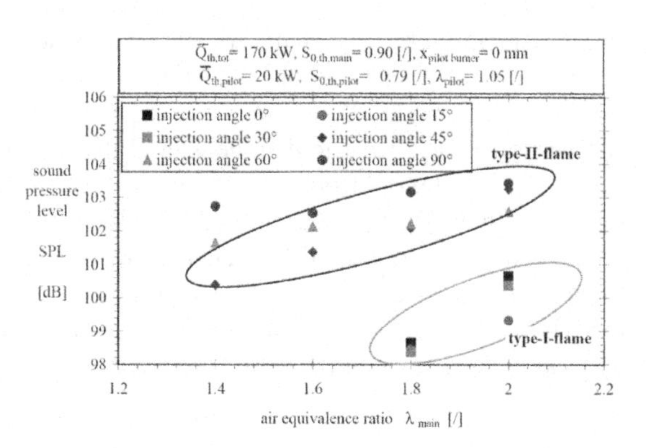

Fig. 2.14 Sound pressure level of non-premixed flames with premixed pilot flame under variation of air equivalence ratio λ_{main} and the gas injection angle

flames have an injection angle between 45° and 90° and look like premixed flames (blue, short flame zone) and the SPL is with values of 101-103 dB very close to premixed flames (SPL = 103-105 dB). The loudest non-premixed flames are the flames of a gas injection angle of 90°, leading to almost the combustion noise of fully-premixed flames (figure 2.15). The second regime with the smaller injection angles ($0-45^o$) is characterized by long, soot-forming flames with very low volumetric reaction densities. The SPL of this regime is very low at values of 96-100 dB (dependent from air equivalence ratio λ).

With the stable and time-independent ignition point of the non-premixed flame a possibility is shown to compare different types of non-premixed flames and to com-

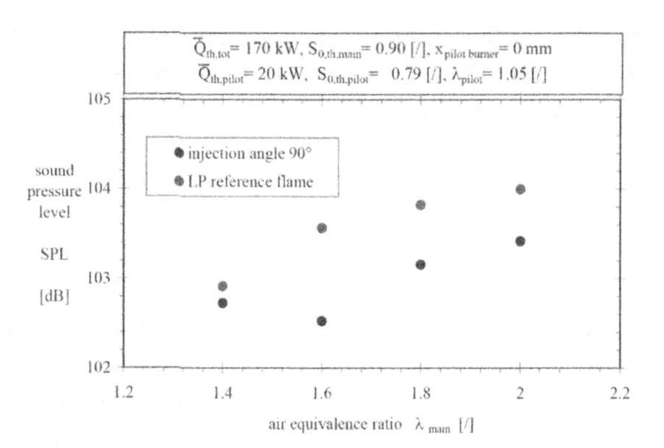

Fig. 2.15 Comparison of the sound pressure level of a premixed and non-premixed swirl flame

pare results with experiments for premixed operating conditions and another physical mechanism of noise generation especially under non-premixed conditions was detected and can be separated from the other noise sources.

2.3.1.5 Enclosed Premixed Swirl Flames

The test facility consists of the double-concentric swirl burner (figure 2.1) and a combustion chamber with an inner diameter $D_{cc} = 0.44$ m and a length l_{cc}= 0.5 m (figure 2.2). Figure 2.16 shows the dependence of the sound pressure level of combustion noise - measured at the wall of the combustion chamber (axial position $x_{microphone} = 0.25m$) - from the thermal load of the flame and the air equivalence ratio of the main flame (see chapter 2.2.1). The results showed that the operating parameters of the flames are limited, caused by the tendency of lean-premixed flames to get self-sustained combustion instabilities, being an order of magnitude louder than the stationary flames. In figure 2.17 the characteristics of a stationary and an oscillating flame are compared and the difference which is limited to the increase of

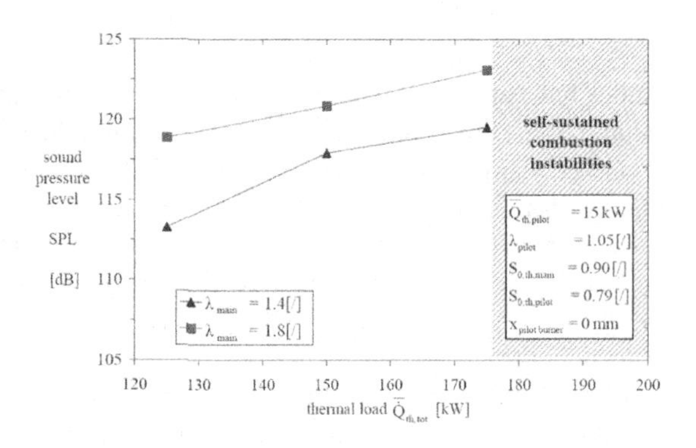

Fig. 2.16 Combustion noise of enclosed, premixed swirl flames

one preferential frequency - leading to an increase of the SPL of more than 6 dB - is obvious.

The investigations were made for a stationary operating point to compare results

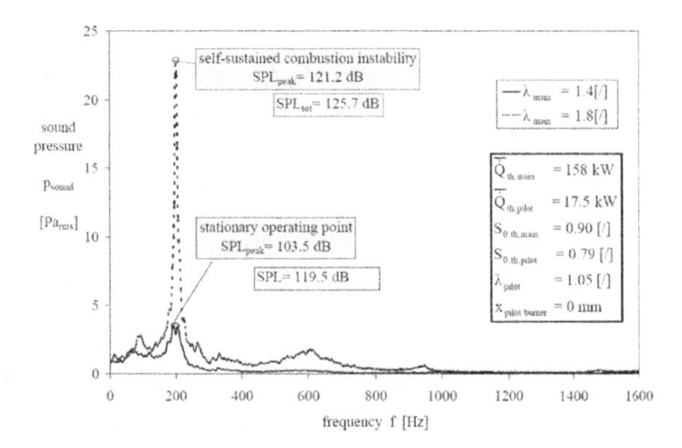

Fig. 2.17 Boundary between combustion noise and self-sustained combustion instabilities

with non-enclosed premixed flames (see chapter 2.3.1.3). The most important result is the level of the combustion noise in the chamber which is up to 20 dB louder caused by reflection of noise at the combustion chamber's wall and the higher level of temperature in the chamber with the formation of outer recirculation zones.

2.3.2 Numerical Simulation

2.3.2.1 Isothermal Flow

The simulations shown in the current section cover both the isothermal flow and the reacting flow with premixed flames. The x-axis is defined as the streamwise direction and $x = 0$ denotes the outlet plane of the annular main flow. In figure 2.18 the measured data and the results from the simulation are drawn. This gives an overview of the isothermal flow field displaying the time averaged radial profiles of the streamwise velocity component at four axial positions in the vicinity of the burner (x/D_0=0.05, 0.5, 1.0 and 1.5). The straight line shows the time averaged results of LES and the circles denote the measured data. On the axis a central, inner recirculation zone (IRZ) with the corresponding negative values for the axial velocity can be identified. The comparion shows very good agreement for both methods.

2.3.2.2 Dynamic Structures in the Isothermal Flow

To look more into details of the dynamic behavior of the isothermal flow, sound pressure data acquired with a microphone probe in the ambient of the swirling flow,

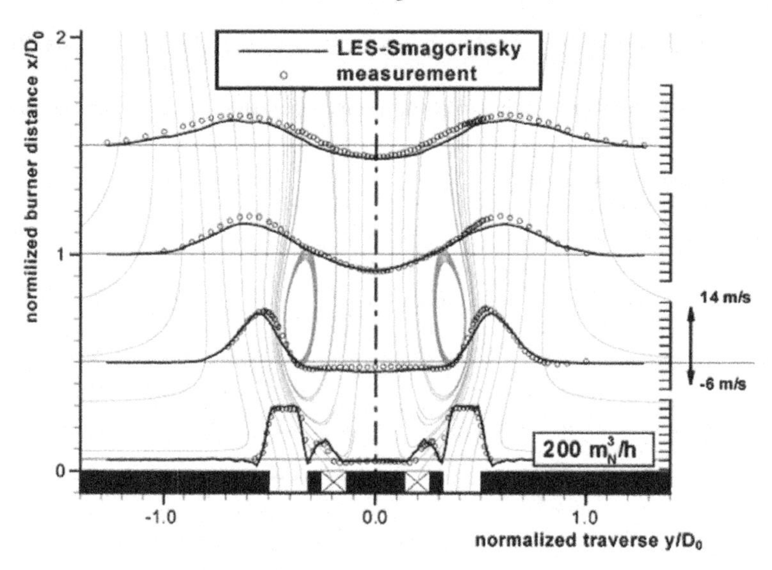

Fig. 2.18 Comparison of calculated mean streamwise velocity profiles with measurement at several axial positions in the isothermal flow case [18]

were analyzed with respect to their spectral representation. As clearly demonstrated in figure 2.19, the formation of periodic coherent structures (red line) in dependence of the chosen burner outlet geometry ($x_{lance} = -40mm$) does not effect the shape of the sound level spectra at all, but adds a considerable contribution at 56 Hz, resulting in an increase of the sound pressure level (SPL) from 75.7 dB to 80.7 dB. Figure 2.20 shows results of another LES calculation using the same computational

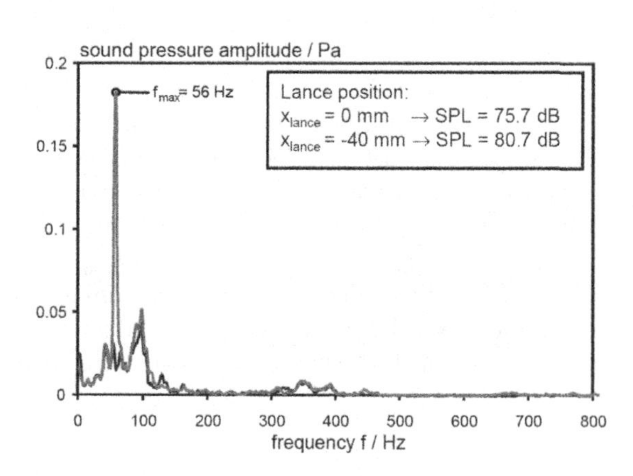

Fig. 2.19 Measured spectral distribution of sound pressure amplitudes and wavelength-integral sound pressure levels for two lance positions in the isothermal swirling flow. Blue line: $x_{lance} = 0mm$; red line: $x_{lance} = -40mm$ ($\dot{V}_{main} = 190m_N^3/h$, $\dot{V}_{pilot} = 10m_N^3/h$, probe position: $x/D_0 = 1$, $y/D_0 = 4.55$)

solution procedure, but utilizing a total volume flow of $\dot{V}_{main} = 650m_N^3/h$, where a spectral representation of the calculated static pressure at a monitoring point in the ambient flow is compared to an according measured normalized sound pressure level. The experiments are not directly comparable, as they were performed utilizing only $\dot{V}_{main} = 609 m_N^3/h$, which turned out to be the maximum inflow, that could be supplied by the compressor. Anyhow, the observed preferential frequencies compare very well. This indicates that the coherent structures result from an axial displacement of the central recirculation zone, which is characteristic for swirl flows. This supports the mechanism for the generation of coherent structures in swirl flames as already mentioned above and earlier found in literature ([10]).

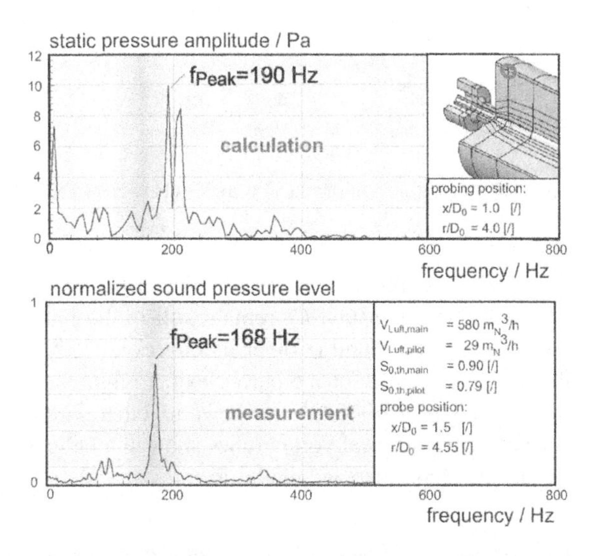

Fig. 2.20 Comparison of spectral resolved static pressure amplitude and normalized sound pressure level in the ambient flow of the isothermal swirling flow

2.3.2.3 Reating k with Premixed Flames

The LES-calculation of the swirled flame was run for about 20.000 time steps ($\approx 0.2s$). Figure 2.21 shows comparisons of the mean streamwise and tangential velocity profiles between measured data and LES results at two axial positions: $x/D_0 = 0.05$ and $x/D_0 = 0.5$. Since the flow field is charged with very strong swirl, the inner recirculation zone (IRZ) can also be detected in the premixed flame case. The results of both LES/Exp. are qualitative in good agreement. The plane $x/D_0 = 0.05$ lies 5mm above the burner exit, at this positions, sharp gradient of the velocity components forms, as the cold mixture penetrates into the steady ambience. Further downstream, due to the strong viscosity generated by the turbulent flow, the gradient becomes smoother. In the middle of the top right figure, the backflow can be detected again by the negative axial velocity. However, there is a difference directly above the pilot lance for the mean axial velocity, which was overpredicted in the simulation. This may be caused by the main inlet boundary condition, where the rotational velocity was not exactly known. The used relative coarse grid (1 mio. cells) may be an other reason. Also, uncertainties like definitions of the sub grid

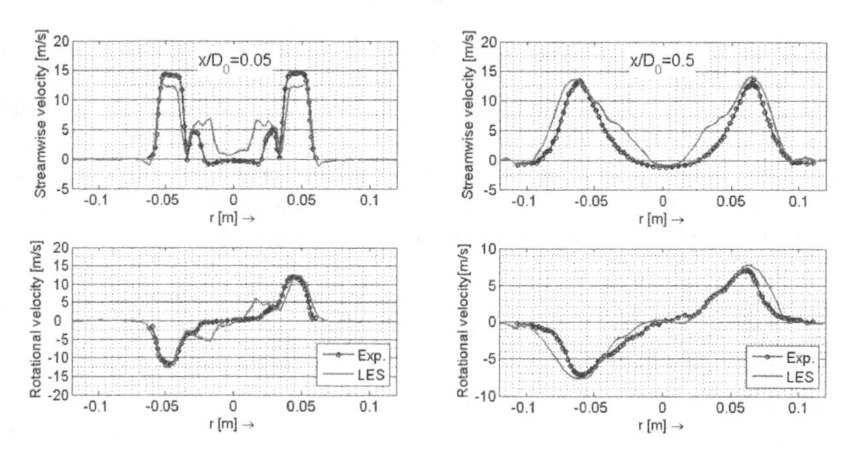

Fig. 2.21 Comparison of the mean velocities between LES and measurement

scale turbulent fluctuaions u'_{Δ} and the turbulent length scale Δ (see sec.2.2.2.2), or the use of a constant C_S remain still in the model. Nevertheless, similar problems can also be found in other literatures [30, 32, 14, 26], particularly for swirled flame in the vicinity of the burner exit. In figure 2.22, iso-surfaces of the vorticity $|\omega| = |rot \times \mathbf{u}| = 4000s^{-1}$ are shown together with contours of a slice at $x/D_0 = 3$. As expected, the local vortex tubes showed a helical structure and the iso-contours on the slice have also an annular form.

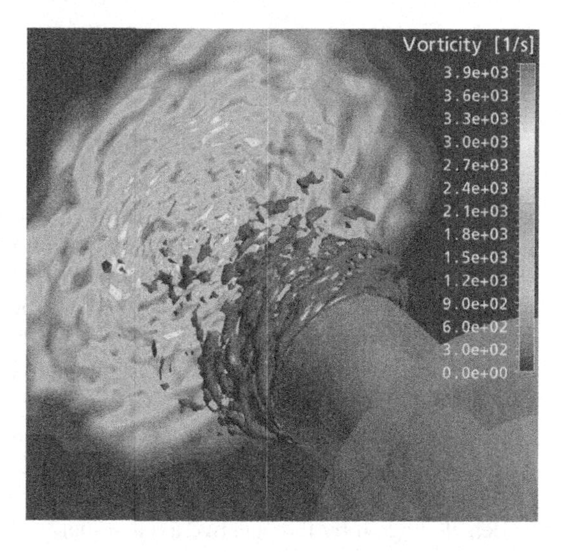

Fig. 2.22 Iso-surfaces of the vorticity $|\omega| = 4000s^{-1}$ and an axial slice of vorticity at $x/D_0 = 3$

2.3.2.4 Flame Dynamics

Figure 2.23 shows slices of the instantaneous progress variable and temperature for the open flame case with planar lance. The progress variable increases immediately from zero to one at the reaction zones where the cold mixture reacts to the hot products. At the same time, as the inner flow reaches the outlet of the burner, strong turbulence is generated due to the huge expansion of flow field, vortices shed from the burner mouth. Accordingly, the flame front becomes corrugated by the local turbulence. One notes, that the pilot flow has an air equivalence ratio of 1.05 (see sec.2.2.2.4, tab.2.1), which is very close to the stoichiometric mixture of methane and air. As a result, areas with the highest temperature are located near the burner axis directly above the pilot lance, the shape of this region is affected by the IRZ and the main jet. The IRZ causes a recirculation of the hot gases back to the reaction zones and supports the ignition stability of the flame. That is the main benefit of using swirled flows. Another reason is is that more intensive heat release can be achieved due to the high turbulence intensity. In figure 2.24, the instantaneous

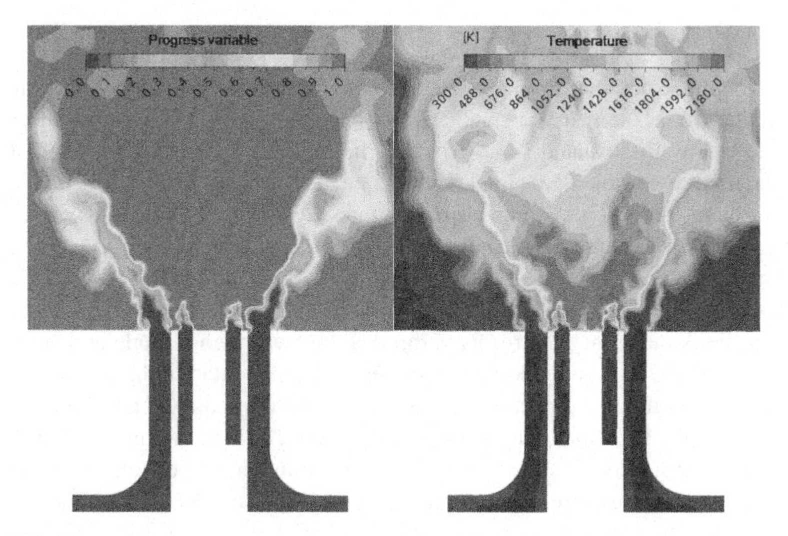

Fig. 2.23 Slices of the instantaneous progress variable (left) and the temperature (right)

contours of the sub grid scale turbulent fluctuations $u'_\Delta = C_S \Delta |\tilde{S}_{ij}|$ (sec.2.2.2.2) are shown. These are particularly large at the reaction zone and the shear layers, because the shear strain rate S_{ij} is very high in these regions. The flow is strongly accelerated at the reaction zone due to the expansion effect. The black curves in figure 2.24 represent the iso-contour lines of a selected reaction rate for θ by $\dot{\omega} = 80 kg/m^3/s$. As can be seen, the flame front is stabilized by the IRZ at the corners at the exit of the burner mouth. As only small eddies are generated and shed from the burner, these will be completely burned up by passing through the flame front and make the flame front corrugated. Further downstream, as the turbulence becomes large scaled,

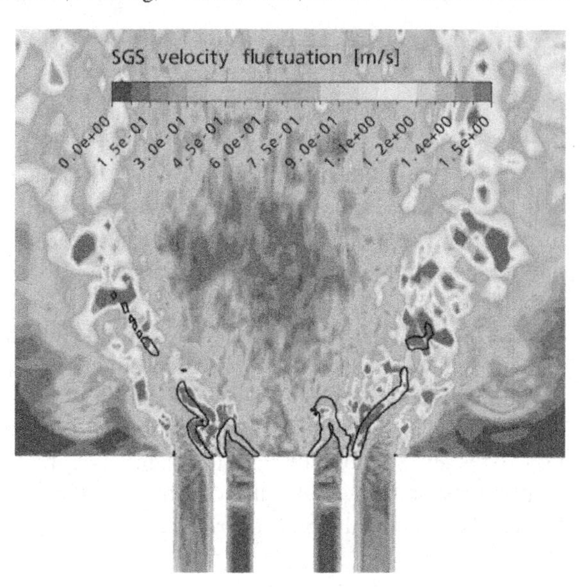

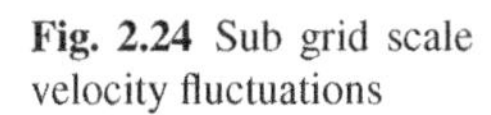
Fig. 2.24 Sub grid scale velocity fluctuations

these can not be completely burned by the flame front due to the higher fluctuations. The flame front is teared and small flame spots form in the burned gas. These can be detected by the isolated iso-contours in figure 2.24. In the context of the LES, the maximal resolvable flame thickness will be the local grid size Δ. The theoretical laminar flame front with about 0.1mm thickness can not be resolved on the coarse LES-grid. However, the TFC combustion model within the framework of LES filters implicitly the sharp gradients at the reaction zones through several LES-cells. In this case, the modeled eddies in the sub grid can penetrate into the flame and thicken the flame front. The resolved large eddies will then wrinkle and corrugate the flame. Thus, the simulated flames in our case are located in the thickend flame and corrugated flame regimes of the flame structure-diagram. On one hand, the assumption of the LES-modeling has caused this conclusion; and on the other hand, since the reacting flow has a very high Reynolds-number, a wide range of turbulent scales exists and therefore, there is turbulence with scales smaller than the laminar flame thickness.

2.3.2.5 Coherent Structures in Premixed Flame

In the case with a recessed pilot lance by -40 mm (figure 2.5), a coherent structure can be identified similar to the isothermal flow case (sec.2.3.2.2). This coherent structure has been identified and visualized using a phase locked averaging method in [19] for the isothermal flow case. As soon as the main flow reaches the exit plane of the pilot lance, the axial momentum of the main jet will decrease due to the expansion of the flow field. This causes the swirl instensity in this region to increase, so that the so formed IRZ can even overcome the combined inertial axial momentum of the inner flow. As a result, the flow points into the direction of the burner where it acts as flow perturbance. However, as the IRZ moves into the burner mouth, the outlet area of the burner will be reduced and this causes a higher axial momentum. Consequently, the swirl intensity will drop and the IRZ moves again to the outside of the burner. The IRZ becomes unstable in this case and moves backwards and forwards to the burner periodically. This is the mechanism of the coherent structure formation and it causes a very high tonal peak frequency in the spectrum of the pressure fluctuations [4, 10]. In figure 2.25, two calculated typical meridian cuts of the streamwise velocity fields are shown. Both fields represent a time delay of $\Delta t = 2.7ms$. The black lines denote isolines of the axial velocity by $u = -1m/s$. In the first snap shot (at the left of figure 2.25), the IRZ flows towards the burner due to very high swirl intensity and there is a large region with negative axial velocity at the burner mouth. In the second image (on the right of figure 2.25), the IRZ moves outwards to the outside of the burner and no backflow is observed near the burner mouth. This makes one turnaround approximately 2x2.7ms (about 180Hz).

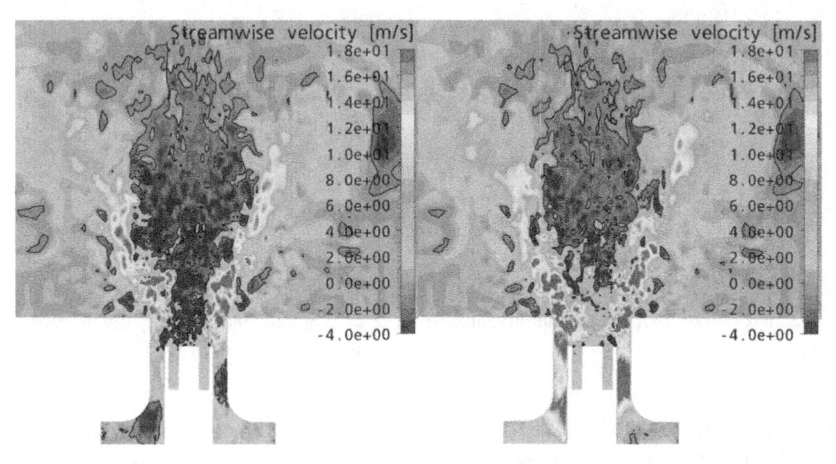

Fig. 2.25 Snap shots of the axial velocity near the burner mouth, black curves denote isolines by $u = -1m/s$

2.3.2.6 Analysis of the LES Result

In section 2.2.2.3 the acoustic analogy has been introduced and an order of magnitude analysis of the Lighthill tensor has been performed. In figure 2.26 and figure 2.27, the two major contributions to the Lighthill tensor are compared for the same point in time by a slice view. These are on the one hand the aeroacoustic source $S_{aero} = \frac{\partial^2}{\partial x_i \partial x_j}(\rho v_i v_j)$ and on the other hand the combustion noise source $S_{cn} = \frac{\partial^2}{\partial x_i \partial x_i}(p' - c_0^2 \rho')$, defined in equation 2.9 and equation 2.10. As can be seen, the main source regions are located close to the nozzle exit of the burner, where the flow domain expands and combustion occurs. Due to very strong variation of the density, the combustion noise source is located solely at the thin reaction zones. Moreover, the expansion of the inflows into the free domain and the chemical reaction causes strong turbulence and acoustic sources were generated, which are evident from figure 2.26, where one can detect both the reaction zones and vortex structures along the jet flame. The magnitude of S_{cn} is about 3 order higher than the magnitude of S_{aero}. This is reasonable since the overall observed Mach numbers are smaller than 0.07 and $S_{aero}/S_{cn} \propto Ma^2$. In fact, for subsonic flow with large reynolds number, the combustion process causes much stronger density variation than the turbulent motions. It is noteworthy, that large values of the acoustic sources

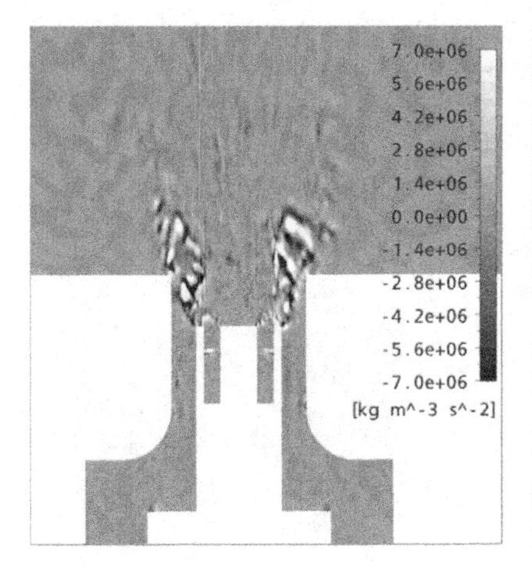

Fig. 2.26 Contour plot of the aerodynamic source S_{aero}.

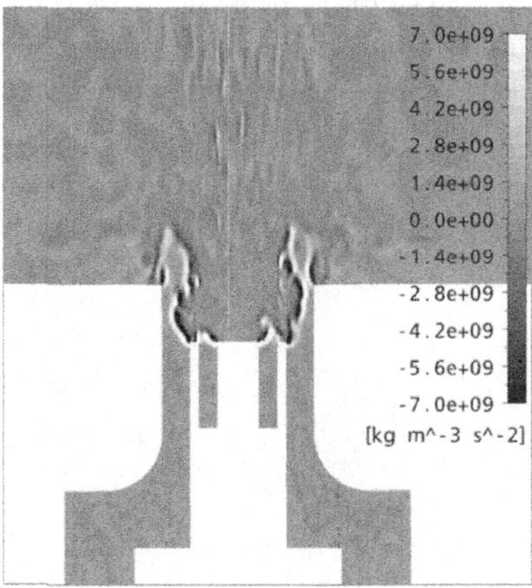

Fig. 2.27 Contour plot of the combustion noise source S_{cn}.

do not directly lead to higher noise level at farfield. However, these can be used as input data for other CAA methods (Computational Aero-Acoustics). Bui et al. [7, 8] have proposed the APE (Acoustic perturbation equations) to compute combustion generated noise using acoustic sources calculated by a LES simulation and found out, that the sound pressure level caused by the chemical reaction is significant larger than the one caused by the turbulent flow [7]. It is also proved by the acoustical measurements [4, 6], that the SPL of the reacting flow is much higher than the isothermal flow. Figure 2.28 gives an overview of the instantaneous vortex sound power according to Bamberger [2]

$$P_{vortex} = -\int \rho(\omega \times \mathbf{u})\mathbf{v_{ac}}dV, \tag{2.13}$$

with the density ρ, the vorticity ω, the acoustical velocity $\mathbf{v_{ac}}$. A positive net power delivers acoustic energy. As expected, the highest integrated vortex power are located along the shear layer of the jet. During the LES several monitor points (MP) in the flow domain have been used to follow the temporal progress of the flow vari-

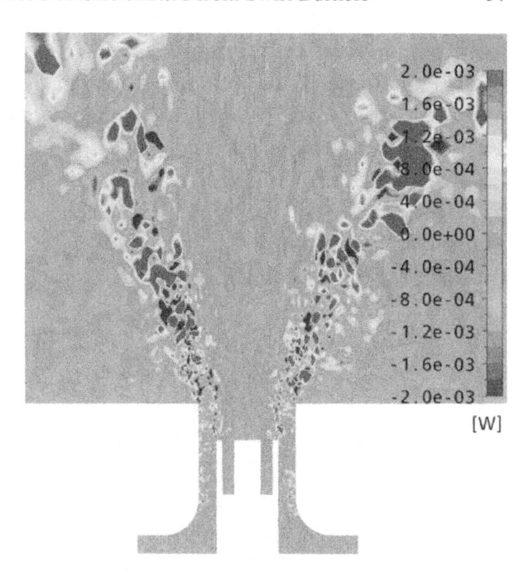

Fig. 2.28 Acoustic power of vortex given bei equation 2.13

ables. In figure 2.29, a frequency analysis of the monitor point (as indicated in the small slice) is made for the pressure und density fluctuations, this MP lies in the main annular flow. A distinct preferential frequency at 150Hz can be identified from the spectra of ρ' and p', which is known as the eigen frequency of the burner. In this case, the burner system acts as a Helmholtz-resonator due to the narrowed cross section in the annular main flow. The gas column in the burner throat works as inertial mass and oscillates via the gas volume, which acts as spring upstream in the burner. The corresponding flow structure will be transported downstream to the outside of the burner and can be detected in the main jet (figure 2.30). It also caused periodic fluctuation of the local mixture fraction in the reaction zone, and as a result, the reaction rate fluctuates with the same frequency.

In figure 2.30, a spectral analysis of the monitor point MP5 in the swirled jet (as indicated in the scmall slice) is shown, on the left, for ρ' and p'; and on the right, for

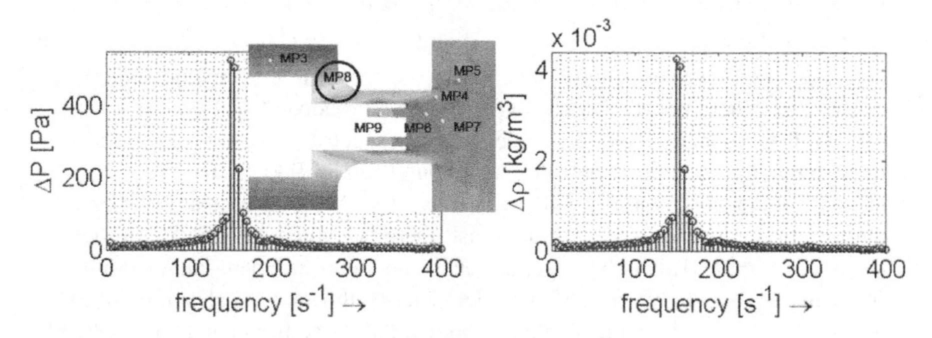

Fig. 2.29 Spectra of the monitor point MP8 in the burner inner flow.

the two main sources S_{aero} and S_{cn} in the Lighthill's equation for p' (2.11). Besides of the eigenfreqency of the burner, another dominant frequency at 180Hz arises in the spectra of ρ' and p' in the jet flame. This is caused by a coherent structure. The IRZ becomes unstable and moves in and out from the annular main flow periodically (sec.2.3.2.5). The mechanism of the coherent structure formation can be found in [10]. Other preferential freqencies with small peak pressures are refered to the vortex shedding frequencies from the main jet and reflection due to the reflecting boundaries. Unlike ρ' and p', there are no dominant frequencies identified

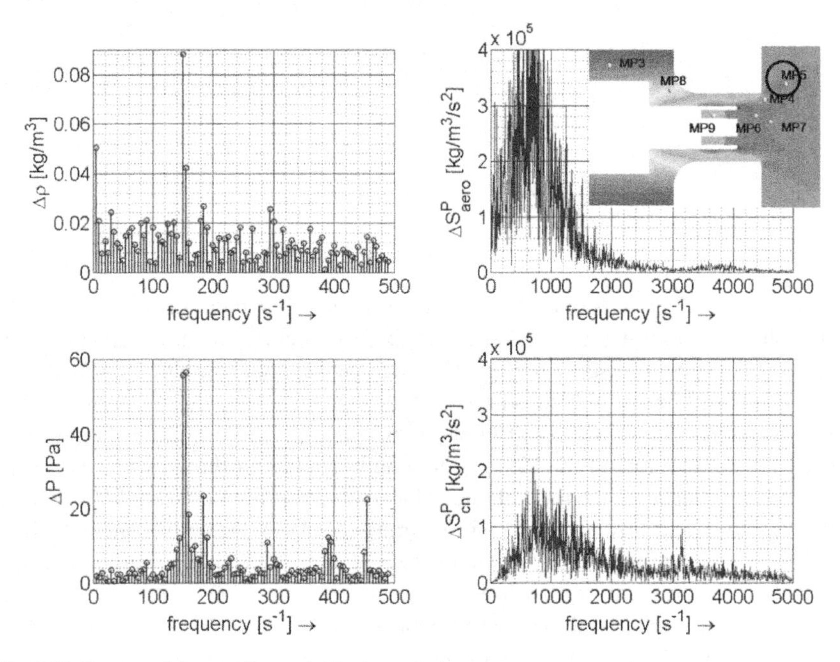

Fig. 2.30 Spectra of the monitor point in the main jet.

in the spectra of the two source terms. As strong turbulence is generated in the jet, the formulations of $S_{aero} = \frac{\partial^2}{\partial x_i \partial x_j}(\rho v_i v_j)$ and $S_{cn} = \frac{\partial^2}{\partial t^2}(p'/c_0^2 - \rho')$ implicitly represent the turbulence energy cascade, where vortices with a wide range in time and space exist. Therefore, the noise sources make a broadband spectrum from low to very high frequencies. Like a turbulent energy spectrum, the sources increases at first from $0Hz$, reach the peak values by about $700Hz - 900Hz$, and decrease again at high frequencies. Vortices with more than $f = 2500Hz$ turnover frequency do almost not contribute to S_{aero}. In contrast, there is a magnitude of S_{cn} in the high frequency region. This is because, although the small turbulent eddies (with high frequencies) carry only very little turbulent energy, the length scales of these eddies are still larger than the flame thickness, so that the flame front can be interacted by these small eddies and causes S_{cn}.

2.4 Conclusions

The investigations shown in the presented article cover detailed noise measurements of isothermal swirl flows and swirl flames (fully-premixed and non-premixed). The spectral analysis of pressure fluctuations shows explicit preferential frequencies for the isothermal and the reacting case (premixed conditions). As the source, coherent structures were detected. The underlying physical mechanism [10, 25] is the same for the isothermal and reacting flow.
The results of detailed noise measurements enable to correlate the total sound pressure level of swirl flames with its sources namely, turbulent flow (75 dB), combustion noise (+25 dB) and coherent flow structures (+2.5 dB), the latter being caused by an unfavorable burner exit geometry, that is, unfortunately, very common in industrial burner designs. The formation and reaction of these flow structures lead to an increase of SPL up to 5 dB for isothermal flow and up to 3 dB for swirl flames. Their prevention results in a significant decrease of combustion noise.
The part of the total sound emission caused by the turbulent flow (isothermal or with reaction) is given by the turbulent swirl flow with its benefits e.g. the fast, turbulent mixing or the flame ignition stability, and cannot be reduced significantly by design modifications. The additional noise, generated by reacting coherent structures (here: oscillation of the axial position of the inner recirculation zone), indeed, is avoidable without any loss of the above mentioned benefits of turbulent swirl flames.
The investigation of non-premixed flames showed the necessity of a stable and stationary ignition zone to prevent additional noise through fluctuations of the ignition point leading to an increase of the combustion noise of 3 dB. Numerical simulations (LES) have also been performed to study the flow field of this burner with and without reaction. The TFC (turbulent flame speed closure) model has been proposed in combination with the Schmid model to compute premixed flame. The comparisons of the LES results with the measured mean flow variables showed good agreement. Coherent structures have been identified in the experiment using the burner exit geometry with a recessed pilot lance and it has been numerically proved. The Lighthill's acoustic analogy was taken into account to evaluate the noise sources generated by the swirl burner. The contributions of the different components of the Lighthill tensor have been illustrated and compared by visualization of the LES result. The noise sources generated by the combustion process are significantly larger than the sources caused by the turbulence. The results of these LES-simulations will be used as input data to calculate the combustion noise using the CAA method (Computational Aero-Acoustics). A spectral analysis for several monitor points within the flow domain has been made to analyse the flow/combustion dynamics. Dominant frequencies in the spectra of the density and the pressure fluctuations were identified to be the eigenfrequency of the burner and the frequency of the oscillating inner recirculation zone (IRZ).

Acknowledgements The authors gratefully acknowledge the financial support by the German Research Council (DFG) through the Research Unit FOR 486 "Combustion Noise".

References

[1] Bai T, Cheng, XC, Daniel BR, Jagoda JI, Zinn BT (1993) Vortex shredding and periodic combustion processes in a Rijke type pulse combustor, Combustion Science Technology, 94, 245-258

[2] Bamberger A (2004) Vortex sound of flutes observed with Particle Image Velocimetry. ICA conference, Kyoto

[3] Bender C, Büchner H (2005) Mechanismen der Lärmentstehung in freibrennenden und eingeschlossenen Drallflammen, VDI-Berichte: 22. Deutscher Flammentag-Verbrennung und Feuerungen, 1888, 311-317

[4] Bender C, Büchner H (2005) Noise emissions from a premixed swirl combustor, Proceedings of Twelfth International Congress on Sound and Vibration (ICSV 12), CD-ROM.

[5] Beer JM, Chigier NA (1972) Combustion aerodynamics, Applied Science Publisher, London

[6] Brick H, Piscoya R, Ochmann M, Költzsch P (2005) Prediction of the Sound Radiated from Open Flames by Coupling a Large Eddy Simulation and a Kirchhoff-Method. Proc. Forum Acusticum, 85-89, Budapest

[7] Bui TP, Schröder W, Meinke M (2007) Acoustic perturbation equations for reacting flows to compute combustion noise. International Journal of Aeroacoustics, volume 6, nr.4

[8] Bui TP, Meinke M, Schröder W (2004) A Hybrid Approach to Analyze the Acoustic Field Based on Aerothermodynamic Effects. Proc. Joint Congress CFA/DAGA'04, Strasbourg, France, 121-122

[9] Büchner H (1992) Entstehung und theoretische Untersuchungen der Entstehungsmechanismen selbst-erregter Druckschwingungen in technischen Vormisch-Verbrennungssystemen, PhD Thesis, University of Karlsruhe

[10] Büchner H, Lohrmann M (2003) Coherent Flow Structures in Turbulent Swirl Flames as Drivers for Combustion Instabilities. Proc. Intern. Colloquium on Combustion and Noise Control

[11] Cabana M, Fortune V, Jordan P (2006) A look insight the Lighthill source term. 12th AIAA/CEAS Aeroacoustics Conference. Cambridge, MA, USA, AIAA-2006-2484

[12] Catlin J B, Day W H, Goom K (1999) The Pratt & Whitney Industrial Gas Turbine Product Line, Proc. of Power Gen Conference

[13] Colin O, Ducros F, Veynante D, Poinsot T (2000) A thickened flame model for large eddy simulations of turbulent premixed combustion. Phys. Fluids. 12, 7

[14] Duchamp de Lageneste L, Pitsch H (2001) Progress in large-eddy simulation of premixed and partially-premixed turbulent combustion. Center for Turbulence Research, Annual Research Briefs

[15] Fröhlich J (2006) Large Eddy Simulation turbulenter Strömungen, ISBN-10 3-8351-0104-8

[16] Gupta AK, Lilley DG, Syred N (1984) Swirl Flows, Abacus Press, Kent(U.K.)

[17] Habisreuther P, Bender C, Petsch O, Büchner H, Bockhorn H (2004) Calculated and Measured Turbulent Noise in a Strongly Swirling Isothermal Jet, Proceedings Joint Congress CFA/DAGA, 1179-1180
[18] Habisreuther P, Bender C, Petsch O, Buechner H, Bockhorn H (2006) Prediction of Pressure Oscillations in a Premixed Swirl Combustor Flow and Comparison to Measurements. Flow Turbulence and Combustion
[19] Habisreuther P, Lischer T, Cai W, Krebs W, Zarzalis N (2007) Visualisation of statistically periodic coherent structures in turbulent flow using a phase locked averaging method. Progress in computational fluid dynamics
[20] Hermesmeyer, Prade, Gruschka, Schmitz, Hoffmann and Krebs(2002) V64.3A Gas Burner Natural Gas Burner Development, Proceedings of ASME Turbo Expo
[21] Keck O, Meier W, Stricker W, Aigner M (2002) Establishment of a confined swirling natural gas/air flame as standard flame: Temperature and species distribution from laser Raman measure-ments, Combustion Science Technology, 174(8), 117-151
[22] Kühlsheimer C, Büchner H (2002) Combustion Dynamics of Turbulent Swirling Flows, Combus-tion and Flame, 131 (1-2), 70-84
[23] Leuckel W, Fricker N(1976) The characteristics of swirl-stabilized natural gas flames. Part I: Dif-ferent flame types and their relation to flow and mixing patterns, J. Inst. Fuel, 49, 103-112
[24] Lighthill M J (1952) On sound generated aerodynamically I. General Theory. Proc. R. Soc. A211, 564-587
[25] Lohrmann M, Büchner H (2000) Periodische Störungen im turbulenten Strömungsfeld eines Vor-misch-Drallbrenners, Chem.-Ing. Technik 72, 512-515
[26] Pitsch H, Duchamp de Lageneste L (2002) Large-eddy simulation of premixed turbulent combustion using a level-set approach. Proceedings of the Combustion Institute, Volume 29
[27] Roux S, Lartique G, Poinsot T, Meier U, Berat C (2005) Studies of Mean and Unsteady Flow in a Swirled Combustor using Experiments, Acoustic Analysis and Large Eddy Simulations. Combustion and Flame (141) S.40-54
[28] Schadow K, Gutmark E, Parr T, Parr K, Wilson K, Crump J (1989) Large-scale coherent structures as drivers of combustion instability, Combustion Science Technology, 64, 167-186
[29] Schmid H P, Habisreuther P, Leuckel W (1998) A Model for Calculating Heat Release in Premixed Turbulent Flames, Combustion and Flame 113, pp. 79-91
[30] Selle L, Lartigue G, Poinsot T, Kaufmann P, Krebs W, Veynante D (2002) Large-eddy simulation of turbulent combustion for gas turbines with reduced chemistry. Center for Turbulence Research, Proceedings of the Summer Program
[31] Smagorinsky J (1963) General circulation experiments with the primitive equations I: The basic experiment. Mon. Weather Rev. 91, 99-164

[32] Wang P, Bai XS (2005) Large eddy simulation of turbulent premixed flames using level-set G-equation. Proceedings of the Combustion Institute 30, 583-591

[33] Zhang F, Habisreuther P, Hettel M, Bockhorn H (2008) Modeling of a Premixed Swirl-stabilized Flame Using a Turbulent Flame Speed Closure Model in LES. Flow, Turbulence and Combustion, Accepted.

[34] Ziegler G(1991) Entflammung magerer Methan/Luft-Gemische durch kurzzeitige Bogen- und Glimmentladung. PhD Thesis, University of Stuttgart

Chapter 3
Modeling of noise sources in combustion processes via Large-Eddy Simulation

Anna Schwarz, Felix Flemming, Martin Freitag and Johannes Janicka

Abstract The focus of this chapter is the definition and development of the interface between CFD and CAA appproaches. Hybrid approaches for the investigation of noise resulting from flow phenomena are widely used and well accepted in aeroacoustics. Especially at low Mach number flows, the fluid dynamic and acoustic length scales are separated by more than an order of magnitude.
Further objective of this chapter is the development of LES models for non-premixed, partially premixed and premixed flames. In this context H3 and HD flames have been investigated for validation of non-premixed flames using the steady flamelet model. A model gas turbine was chosen for validation of partially premixed flames in respect to the flame surface density model and Tecflam burner was used for simulation of premixed flames using the G-equation model. Finally, a LES/CAA hybrid approach was developed for open flames.

3.1 Introduction

Hybrid approaches for the investigation of noise resulting from flow phenomena are widely used and well accepted in aeroacoustics [28], [18]. Especially at low Mach number flows, the fluid dynamic and acoustic length scales are separated by more than an order of magnitude. This allows the application of specialized techniques for each domain, namely the source region using computational fluid dynamics (CFD) and the acoustic propagation region all the way into the far field using computational aeroacoustics (CAA). Especially for turbulent combustion systems with moderate velocities the low Mach number approximation is applicable and yields satisfying results [20]. The utilization of CFD such as the LES method is preferable since the turbulent spectrum does not have to be modeled completely. Therefore, conclusions

Institute of Energy and Power Plant Technology, Department of Mechanical Engineering, Technical University of Darmstadt, Petersenstrasse 30, 64287 Darmstadt, Germany, e-mail: schwarz@ekt.tu-darmstadt.de · e-mail: janicka@ekt.tu-darmstadt.de

A. Schwarz, J. Janicka (eds.), *Combustion Noise*, DOI 10.1007/978-3-642-02038-4_3,

of the sound output of such configurations can directly be drawn. In the stationary RANS approach for the CFD part the turbulent spectrum and therefore the noise producing structures have to be modeled completely, thus limiting the general applicability of the hybrid approach, compared to LES. Finally, a direct approach such as DNS of the flow field is beyond any admissible computational costs for technically relevant systems.
Investigating the potential of Large Eddy Simulation for the prediction of different configurations with premixed, partially premixed and non-premixed flames is currently an important aspect in the scientific community. Technical applications are gas turbines in the aircraft and stationary applications in the automotive industry. As computational power grows, simulations become more and more important in different fields.

3.2 Theoretical Background

Combustion processes are always attached to the density changes, so that weighted density formulations are necessary. The structure of filtered conservation equations for mass, momentum and scalar read:

$$\frac{\partial \overline{\rho}}{\partial t} + \frac{\partial}{\partial x_j}(\overline{\rho}\widetilde{u_j}) = 0 \tag{3.1}$$

$$\frac{\partial}{\partial t}(\overline{\rho}\widetilde{u_i}) + \frac{\partial}{\partial x_j}(\overline{\rho}\widetilde{u_i}\widetilde{u_j}) = \frac{\partial}{\partial x_j}\left[\overline{\rho}\widetilde{\nu}(\frac{\partial \widetilde{u_j}}{\partial x_j} + \frac{\partial \widetilde{u_j}}{\partial x_j}) - \frac{2}{3}\overline{\rho}\widetilde{\nu}\frac{\partial \widetilde{u_k}}{\partial x_k}\delta_{ij} + \overline{\rho}\tau_{ij}^{sgs}\right] - \frac{\partial \overline{p}}{\partial x_j} + \overline{\rho}g_i \tag{3.2}$$

$$\frac{\partial}{\partial t}\overline{\rho f} + \frac{\partial}{\partial t}(\overline{\rho}\widetilde{f}\widetilde{u_j}) = \frac{\partial}{\partial x_i}(\frac{\overline{\rho}\widetilde{\nu}}{\sigma}\frac{\partial \widetilde{f}}{\partial x_j}) + \frac{\partial}{\partial x_j}(\overline{\rho}J_j^{sgs}) \tag{3.3}$$

In these equations the unknown subgrid scale stresses τ_{ij}^{sgs} and J_j^{sgs} appear which have to be modeled. The focus of the work in this sub-project is on the applications and the development of the models for combustion systems. The determination of unknown subgrid scales was not the research subject and so reverted to the standard models. LES results presented in the next chapter were obtained using the Smagorinsky model and dynamic determination of model coefficients using the Germano procedure [15].
Depending on the complexity of the geometry, two computer codes were used. The computer code FLOWSI is a code with low complexity and high computational efficiency. This program is appropriate for model development. The second computational code, FASTEST-3D, has been used for application-oriented simulations. High geometrical flexibility is the main advantage of this code, however the computational time is 8-10 times larger compared to the code FLOWSI. Numerical characteristics of both simulation programs have been summarized in Figure 3.1.

	FLOWSI	FASTEST-3D
Grid	1 block cartesian/cylindrical, staggered grid	block structured, boundary adapted grid
Pressure correction	projection method	SIMPLE
Timestepping method	explicit 3 step Runge-Kutta method	implicit, Crank-Nicholson with SIP-Solver or explixit
space discretization of momentum equation	partially 4th order for convective terms	2nd order CDS or flux blending for convection
space dicretization of scalar transport	TVD scheme with different flux limiters	
SGS model	Smagorinsky Ansatz, Eddy diffusivity Ansatz, dynamic procedures, scale similarity	
parallelization	Area segmentation with MPI Message Passing Library	

Fig. 3.1 Characteristics of the computational codes.

3.2.1 Non-Premixed Flames

Steady Flamelet Model

The idea of a turbulent flame consisting of an ensemble of individually stretched laminar flamelets was proposed by Williams [45]. Later, Peters [31] and Kuznetsov [25] independently derived flamelet equations based on the mixture fraction, which were extended to an unsteady formulation. A good overview on the concept and its application to turbulent flows is given by Peters [32].

Since chemical reactions in combustion processes are very fast, they usually occur within a thin layer of a flame: the inner layer. The steady flamelet approach is based on the assumption, that the smallest turbulent scales, the Kolmogoroff scales, are considerably larger than the inner layer of the flame. The Kolmogoroff eddy therefore resembles a quasi laminar flow field. The statistics of the flame location or surface can then be used as a representative of the reactive scalars. By solving the transport equation for the mixture fraction 3.3, the reaction surface can be seen as the stoichiometric contour of f. Normal to this surface the unsteady or steady flamelet equations 3.4 (shown for the unsteady case) are solved and the obtained profiles in mixture fraction space are attached to this surface. This yields a description of the chemical properties of the flow. Since a non reactive scalar is used the transport equation does not contain a source term and hence the classical turbulence modeling strategies, for example the gradient flux assumption, can be applied.

$$\rho \frac{\partial Y_a}{\partial t} = \frac{\rho \xi}{2} \frac{\partial^2 Y_a}{\partial f^2} + \omega_\alpha \tag{3.4}$$

In equation 3.4 the scalar dissipation rate ξ was introduced, which is defined in 3.5. It can be interpreted as a characteristic diffusion time in the mixture fraction space and is a localized function of the mixture fraction field $\xi = \xi(x_i, t) = \xi(f(x_i, t))$. For larger scalar dissipation rates, the chemical system is deviating from the equilibrium solution, until at some point the flame extinguishes, since the radicals required to keep the flame burning diffuse too quickly from the inner layer. This inner layer corresponds to the stoichiometric mixture, where the chemical reaction occurs.

$$\xi = 2D \left(\frac{\partial f}{\partial x_i} \frac{\partial f}{\partial x_i} \right) \tag{3.5}$$

Since the assumption of unity Lewis number was required to derive the mixture fraction equation, the dependency of the diffusivity on the composition is omitted and D is used instead. By neglecting the unsteady conditions in 3.4 the chemical state can now be parameterized as a function of f and ξ in this steady flamelet function f.

Figure 3.2 shows the dependency of the temperature, the speed of sound, the scalar dissipation rate and the density on the mixture fraction for a diluted, non-premixed hydrogen flame, computed via a detailed one-dimensional chemistry simulation applying the steady flamelet model. This model has been applied to all cases throughout this work and will be presented in the following, together with other chemistry models. The order in which the models are described is related to their complexity. Along with equations for the flow field (see Eq. 3.1, 3.2 and 3.3), a balance equation

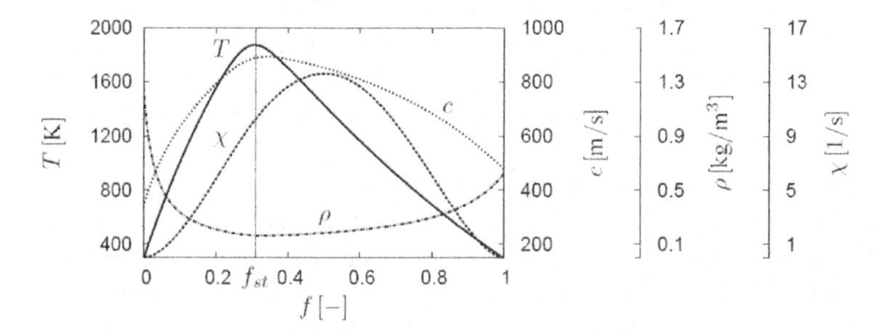

Fig. 3.2 Flamelet for a diluted, non-premixed hydrogen flame with a strainrate of $\alpha = 20s^{-1}$

for scalar quantities has to be solved. It reads:

$$\frac{\partial \overline{\rho}\widetilde{\xi}}{\partial t} + \frac{\partial}{\partial x_j}\left(\overline{\rho}\widetilde{u}_j\widetilde{\xi}\right) = \frac{\partial}{\partial x_j}\left(\overline{\rho}D\frac{\partial \widetilde{\xi}}{\partial x_j} - J_i^{\xi,sgs_i}\right) \tag{3.6}$$

As for the momentum equation, an unclosed term arises due to non-linearity of the convective term, named the subgrid scale scalar flux of mixture fraction and is closed in the same way as the Smagorinsky model, with an eddy-diffusivity type approximation.

$$J_i^{\xi,sgs} \approx -D_t\frac{\partial \widetilde{\xi}}{\partial x_i} \text{with} \quad D_t = \frac{\nu_t}{\sigma_t} \tag{3.7}$$

The turbulence-chemistry interaction is modeled with a presumed PDF method (Probability Density Function) [19]. The Favre-filtered thermochemical quantities are computed from the PDF integration assuming β-function form for the PDF of the mixture fraction.

$$\widetilde{\varphi} = \int_0^1 \varphi(\xi)\, P(\xi)d\xi, \tag{3.8}$$

3.2.2 *Partially Premixed Flames*

In technical combustion systems one deals in general with situations where premixed, partially premixed and non-premixed combustion occur simultaneously. Next-generation low-emission/low-fuel consumption combustion systems are characterized by multiple combustion regimes, i.e., "mixed modes" or "partially premixed" combustion [46, 17].
Janicka, Sadiki [20] and Pitsch [34] have reviewed the modeling and application of combustion LES. For non-premixed combustion the LES flamelet model shows a solid allround behavior and has established itself as quasi standard. A recent overview on techniques for premixed combustion was provided by Sadiki [38], Poinsot [36] and Vervisch [43].

Flame Surface Density Model

Originally the flame surface density model was developed for the prediction of purely premixed flames. In the frame of this model the flame front transport is described by means of a filtered reactive progress variable $\widetilde{c}$ [1].

$$\frac{\partial \bar{\rho}\tilde{c}}{\partial t} + \frac{\partial}{\partial x_j}(\bar{\rho}\tilde{u}_j\tilde{c}) = \tag{3.9}$$
$$-\frac{\partial}{\partial x_j}\left(\bar{\rho}_u\frac{S_L\Delta_c}{16\pi\sqrt{6\pi}} + \frac{\mu_t}{\sigma_t}\right)\frac{\partial \widetilde{c}}{\partial x_j} + 4\bar{\rho}_u S_L \Sigma,$$

In the source term in eq. 3.9 ρ_u is the density of unburnt gas, $\widetilde{S_L}$ is the laminar burning velocity and Σ is the flame surface density describing the flame front convolutions. An algebraic form for Σ proposed in [5] is used in this work:

$$\Sigma = \Xi \sqrt{\frac{6}{\pi}\frac{\widetilde{c}(1-\widetilde{c})}{\Delta_c}}. \tag{3.10}$$

Here, Ξ is the SGS flame front wrinkling factor calculated using the Charlett model [9], Δ_c is the filter size larger than the actual LES mesh.
In combination with the mixture fraction description usually used for prediction of diffusion flames, it is possible to describe partially premixed effects often observed in swirled flames. The coupling is made through a functional dependency of the burning velocity on the equivalence ratio, expressed in terms of the mixture fraction. The burning velocity appears as part of the source term in the progress variable equation. Due to the turbulence-chemistry interaction the mixture field fluctuations cause fluctuations in the laminar burning velocity and the combustion products concentrations. Filtered values of $\widetilde{S_L}$ and thermochemical variables corresponding to the burnt state are stored in a look-up table, used in computations. In order to extract the physical values of thermochemical filtered variables the conditioned (burnt/unburnt) variables have to be multiplied with the probability of being behind the flame front. In the context of the progress variable approach, the probability is directly related to $\widetilde{c}$. Hence,

$$\widetilde{\varphi} = \widetilde{c}\widetilde{\varphi^b}(\widetilde{\xi}, \widetilde{\xi''^2}) + \widetilde{\varphi^u}(1-\widetilde{c}) \tag{3.11}$$

where ^{b} and ^{u} denote conditioning on burnt and unburnt states, respectively. That means in the regions, where flammable mixture conditions are not available, the progress variable is zero and no reaction takes place. Another way to consider effects of partially premixing is using a reduced chemical mechanism such as the stabilization mechanism of lifted flames which is governed by transport and kinetics.

3.2.3 Premixed Flames

The main challenge in modeling premixed flames using LES comes from the fact that the reaction occurs at a layer which cannot be resolved by a typical filter width used in LES. Thus, the chemical reaction and its interaction with the flow field must be modeled completely. Since the early work of Damköhler, several different types of modeling concepts for premixed combustion have been derived.
One example of this is the artificially thickened flame model [36], [40] which uses a progress variable for the reaction. Its diffusivity is artificially increased until the flame front becomes resolvable within the LES filter width. Along with the modification of the diffusivity, the ratio between chemical and turbulent time scale changes. As a consequence, small eddies cannot interact with the flame front any

longer. Therefore, efficiency functions are introduced to model the interaction [9]. The second family of models relies on the flamelet concept, which assumes the reaction layer to be thinner than the smallest turbulent scales. A consequence of this assumption is that the flame front can be represented by an interface which moves with the laminar burning velocity. The most popular models based on this concept are the flame surface density model (FSD) [16] and the G-equation [45], [35]. For the FSD model a transport equation for the mean ratio of flame surface per volume is solved. It has been derived theoretically that the FSD concept is less adequate for simulations beyond the regime of the corrugated flamelets [32], [10]. The G-equation describes the interplay of flame speed and convective transport acting on the reaction layer. Therefore, the flame front must be tracked.
Several models for describing premixed combustion can be found in the literature and most of them have been successfully applied to different configurations. Although most of them, such as Eddy Break Up Model [41], PDF Transport Equation Model [37], Linear Eddy Model (LEM) [23], Conditional Moment Closure (CMC) [24], Bray Moss Libby Model (BML) [6], Flame Surface Density Model (FSD) [29], Artificially Thickened Flame Model (ATF) [8] and Marker Field Method [4], are not the subject of this contribution, so a short overview of the G-equation model which is applied in this work, should be given in this chapter.

G-equation

As a model for premixed combustion the filtered G-Equation approach was used in the research Initiative "Combustion Noise" [14]. The equation describes the interplay of burning velocity s_T and convective transport u_u^u ($*_u$ indicates unburnt). Because of the fact that the burning velocity s_l is only defined on the flame front, also the G-equation is valid only on the flame surface. The filtered flame front is represented by the G_0-iso-surface and separates the domain into perfectly mixed unburnt (fresh) and burnt gas.

$$\frac{\partial \hat{G}}{\partial t} = (u_u^u \cdot n + s_T) \left| \nabla \hat{G} \right| \tag{3.12}$$

The unresolved wrinkling of the flame front is equivalent to an increase in the flame surface. This remains to be modeled. Under the assumption that the ratio of laminar to turbulent surface area is proportional to the ratio of laminar to turbulent flame propagation speed, the turbulent burning velocity is modeled using:

$$s_T = s_L \left(1 + \kappa\alpha + C \left(\frac{u'_\delta}{s_l} \right)^{\frac{3}{4}} \right) \tag{3.13}$$

Here κ is the locally resolved curvature, α the resolved Markstein length and the model constant C is set to unity. Peters [32] and Pitsch [35] used the thermal diffusivity D instead of $s_l\alpha$, for a turbulent burning velocity valid in the regimes of the corrugated flamelets and the thin reaction zones. From theoretical arguments $D \sim s_L l_f$ and $\alpha \sim l_f M$ can be derived, where M denotes the Markstein number. Hence Eq. 3.13 and the definition given in [32] are altered by the Markstein number, which is in the order of unity. The laminar flame speed, used in 3.13, depends only on the thermochemical state of the premixed gas [44].

Regarding premixed combustion, commonly used LES models assume the system

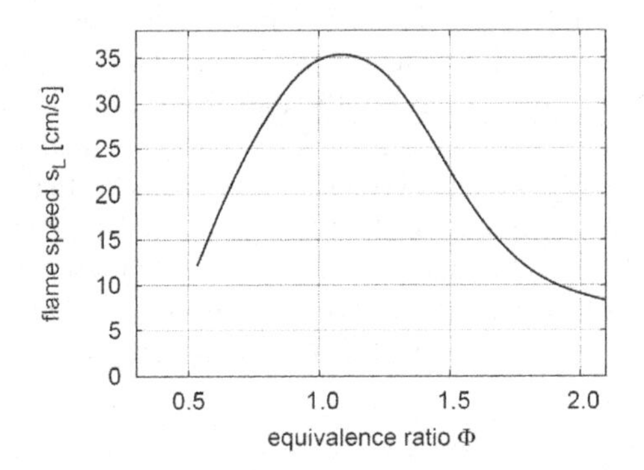

Fig. 3.3 Laminar burning velocity as a function of the equivalence ratio.

to consist of two ideal states: unburnt and burnt. Since the configuration is unconfined the classical G-equation approach is violated. In fact, the configuration consists of three thermochemical states, unburnt gas, burnt gas and ambient air. To give consideration to this problem an additional scalar, representing the mixture fraction, is transported. This scalar is used to calculate the local laminar burning velocity. Due to the fact that the mixing can only dilute the premixed gas, only the left wing of the parabolic curve in Fig. 3.3 is considered [44].

Below values of $\phi \approx 0.5$ reaction will not occur. This is modeled as follows. Firstly, the thermodynamical properties (temperature, density and viscosity) of the fluid, crossing the G_0-front, will not change. Secondly, the burning velocity is set to zero in regions where ϕ falls below a value of 0.5.

3.2.4 LES/CAA Hybrid Approach

Using LES, one can extract acoustic sources from the instationary flow field and investigate the propagation of sound through the flame, in terms of the temporarily and strongly changing fluid properties inside a turbulent flame. The combustion noise was investigated using an LES/CAA hybrid approach, combining the reactive

LES with a wave propagation method. Here the LES is a tool to analyze a secondary effect of the turbulent flame, namely its noise prediction.
The proposed LES/CAA hybrid approach is based on the observation, that for a turbulent flame the global and the local Mach number within the flame are usually very small, i.e. $Ma \leq 0.1 \ll 1$. This is due to two effects. First, the flow speed of the fuel at the nozzle exit is usually rather small, since the flame is supposed to burn stably at the nozzle. Second, the high temperature in the flame region increases the speed of sound by roughly a factor of three. This allows the flow field to be treated independently of the acoustic propagation, as long as there is no feedback of the acoustic to the flame, which is the case for open configurations.
Exploiting the difference in characteristic speeds for the turbulent flow field and the acoustic propagation allows the use of specialized techniques for both phenomena. Combining these techniques in a highly integrated LES/CAA hybrid approach enables good optimization of the complete simulation. This enhances the quality of prediction. The disadvantage of a zonal approach, namely introducing spurious frequencies of the discrete control surface between the two techniques combined, is not evident in this case. The proposed approach is computationally more efficient than a fully compressible simulation, because different optimized grids and numerical schemes can be applied to both the CAA and the LES regimes. Additionally, the Courant-Friedrich-Lewy (CFL) criterion is dominated by the speed of sound, while the dominant transport phenomena is related to the convective speeds in the flame. An estimate yields simulation approximately four times more efficient using the LES/CAA hybrid approach.
The sound propagation into the far field is usually evaluated using several Computational Aero-Acoustic techniques. The three approaches differ in the way the information on the flowfield is used. The Acoustic Pertubation Equations (APE) utilize the volumetric information of the CFD simulation to extract the sound sources (see chapter 7.2). The Boundary Element Method (BEM) computes the sound propagation by means of fluctuations on a defined control surface outside the source region (see chapter 4.2). Finally, the Equivalent Source Method (ESM) arranges artificial noise sources inside a control surface in such a way that their distribution of acoustic fluctuations on the control surface is equivalent to the results of the flow simulation (see chapter 4.2).
Using a reactive LES with a low Mach number approximation to compute a turbulent flame enables the temporally resolved extraction of all required properties for the propagation of acoustic waves - speed of sound, density, flow velocities and the acoustic sources. Using the classical wave equation in terms of pressure disturbance p, the acoustic field around the flame can be described by an appropriate simulation technique. The evaluation of the components of the Lighthill tensor showed that turbulent reacting flows (the source term related to the density) dominate the tensor by approximately two orders of magnitude. Hence, the wave equation reduces to

$$\frac{\partial^2 \rho}{\partial t^2} - c_0^2 \frac{\partial^2 \rho}{\partial x_i \partial x_i} = \frac{\partial^2}{\partial x_i \partial x_i} \left[\rho u_i u_j - d_{ij} + (p - c_0^2 \rho) \delta_{ij} \right] = \frac{\partial^2 T_{ij}}{\partial x_i \partial x_j} \tag{3.14}$$

By an integral method [20], [45], the equation is transformed into a system of first order hyperbolic differential equations using $q = [p_t : -p_{x_i}]^T$ as acoustic state vector [26], [27]:

$$q_t + Aq_{x_1} + Bq_{x_2} + Cq_{x_3} = s \tag{3.15}$$

In Eq. 3.15 A, B, and C are the coefficient matrices and s is the source vector. Solving the system of equations numerically has the advantage, that temporally varying fluid properties can be incorporated into the solution procedure and therefore the acoustic propagation through the turbulent flame can be captured correctly [3]. The solution provides temporally and spatially resolved information on the acoustic field in all three dimensions.

There are generally several issues related to the coupling of the different solution techniques. The type of the interface, the spatial resolution, and the time stepping between LES and CAA need to be adapted or matched. Both solvers rely on a volumetric description of the problem. Therefore the fluid properties needed by the acoustic solver are interpolated onto the acoustic grid which is embedded into the mesh of the LES. This interpolation is performed lineally after each time step of the LES and followed by the computation of the required acoustic source on the cartesian grid.

Combining LES (including chemical reactions) with CAA methods for propagating sound into the far field is a promising technique for the investigations of combustion noise for small Mach numbers. The strategies presented are intended to describe a "one way coupling". The influence of the acoustics on the reacting flow is not considered. The CAA techniques are used as a post-processing method for the LES results.

Two main issues arise on the coupling of CFD with CAA. First, the interface domain of surface needs to be defined. Since the CAA grid is usually much coarser than the CFD grid, the interpolation between the CFD and the CAA grids needs to be consistent. Second, the time-stepping needs to be matched for the two techniques.

The two approaches need to solve different time scales and have different numerical limitations of stability. As mentioned before, the CAA methods rely on instationary information within a given volume, encapsulating all sources, e.g. APE, or on a surface enclosing the source region, e.g. BEM and ESM.

For flow simulations with an explicit time integration scheme, a limitation for the size of the time-step is the CFL condition, $\vec{u}\Delta t_{LES}\Delta x < \alpha$, where the CFL number α depends on the time discretization scheme. A CFL number of unity corresponds to a fluid particle that is convected by the velocity **u** exactly from one cell center to the next Δx in one single LES time-step, Δt_{LES}. In order to remain as efficient as possible, the CFL criterion is usually evaluated after every time-step during the LES to advance with the maximum time-step possible.

An acoustic method works best if the time-step is kept constant, since every interpolation in time introduces artificially high frequencies. If the required acoustical time-step, Δt_{CAA}, is clearly smaller than the minimum of the varying time-step of the LES, Δt_{CAA} <min (Δt_{LES}), one can fix the time-step of the LES at a given point in time so that $\Delta t_{CAA} = \Delta t_{LES,cst} < \Delta t_{LES}$. From this point onwards an acoustic sample is generated every LES time-step. This allows a fast evolving LES to overcome the initialization effect of the simulation. Such a technique is depicted by the solid line in Figure 3.4. The second case, where the acoustical time-step is clearly larger than the LES time-steps, Δt_{CAA} >max(Δt_{LES}), is shown by the dashed line in Figure 3.4. Here, a number Δn of LES time-steps goes by for each acoustic sample (indicated by the dots in the figure). The accuracy of the acoustic time-stepping increases with the number of non-constant LES time-steps between two samples. The actual number of LES time-steps between two acoustic samples is not constant, whereas the acoustic time-step is almost constant. Problems arise if the required acoustic time-step is of the same magnitude as the LES time-steps $\Delta t_{CAA} \approx \Delta t_{LES}$.

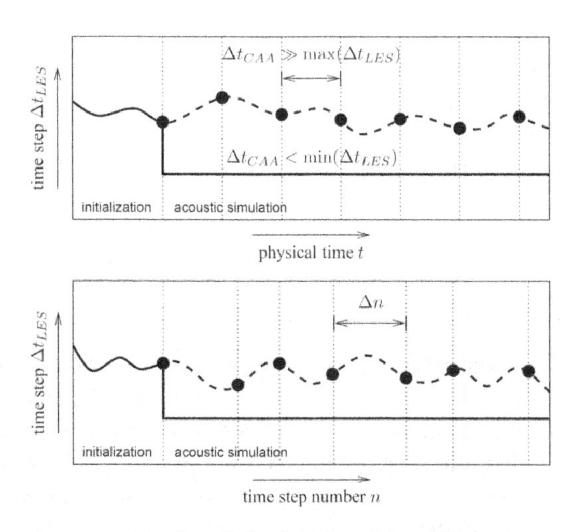

Fig. 3.4 Time-step Δt_{LES} as a function of physical time t(a) and of time-step number n(b) for two different coupling strategies.

3.3 Results and Analysis

3.3.1 Open, Non-Premixed Jet Flames

In the research Initiative "Combustion Noise" two turbulent non-premixed jet flames were treated. The properties of these flames are summarized in Table 3.1. These configurations have been investigated in detail by the TNF workshop [2]. As well as velocity data obtained by LDV measurements, the chemical composition was measured in detail using a combination of Raman/Rayleigh and LIF measurement

Table 3.1 Investigated jet flame configurations

Name	Fuel [vol.%]	f_{st}	U_{bulk}	U_{coflow}	D	Re
HD	23/77 H_2/N_2	0.583	$36.3m/s$	$0.2m/s$	8mm	16000
H3	50/50 H_2/N_2	0.310	$34.8m/s$	$0.2m/s$	8mm	10000

techniques [33], [42]. This complete set of experimental data allows a full validation of the current LES results. Furthermore, noise intensity measurements were performed by collaboration between TU Dresden, TFH Berlin and TU Darmstadt [12]. Further acoustical appliances can be found in [7].
Both configurations were simulated using Flamelet models and Presumed-PDF Ansatz. The Smagorinsky model with dynamic Germano procedure has been applied. Because of a high stoichiometric mixture fraction of the HD flame this configuration is relatively short and easy to handle with regard to LES. Due to the operation of this flame close to the flammability limits an increase of the Reynolds number was not possible.
The simulation was performed with the computational code FLOWSI.

3.3.1.1 HD Flame

After careful sensitivity analysis regarding the influence of parameters such as grid resolution, filtering, sampling window, inflow condition and mean inflow profile (see [21]) on the results, final comparisons have been accomplished to the experimental data [42]. These simulations were computed on a grid with 257 x 60 x 10 cells in axial, radial and tangential directions. The size of the domain was 24D in length and 7.5D radial (where D is the diameter of the nozzle). For the mean inflow velocity, the profile of a fully developed turbulent pipe-flow was set. To avoid the destabilizing influence of the classical inflow condition, the immersed boundary conditions were applied in this case and artificial turbulence was superimposed on the mean inflow velocity. Figure 3.5 shows the axial development of various quantities. It shows that the mean axial velocities and the jet break-up are predicted well (a), although the fluctuation is high (b). This deviation can eventually be diminished by reducing inflow turbulence or by refining the grid. Figure 3.5 (c) shows the decay of the mean mixture fraction, which agrees well with the experimental data. The mixture fraction fluctuation (d) is over-estimated, which can be explained by the high velocity fluctuation. Figures (e) and (f) show computational results for the density only as no experimental data are available. Figure (g) and (h) give the OH mass fraction. This radical only exists near the flame front and is therefore an efficient marker. Although the absolute levels deviate strongly, the profiles show that the location of the flame is well captured. Finally, the mean temperature and its fluctuation are shown in Figures (i) and (j). As expected from the mixture fraction data, mean and fluctuation are predicted well.
Figure 3.6 gives the radial profiles of the mixture-fraction mean and fluctuations at

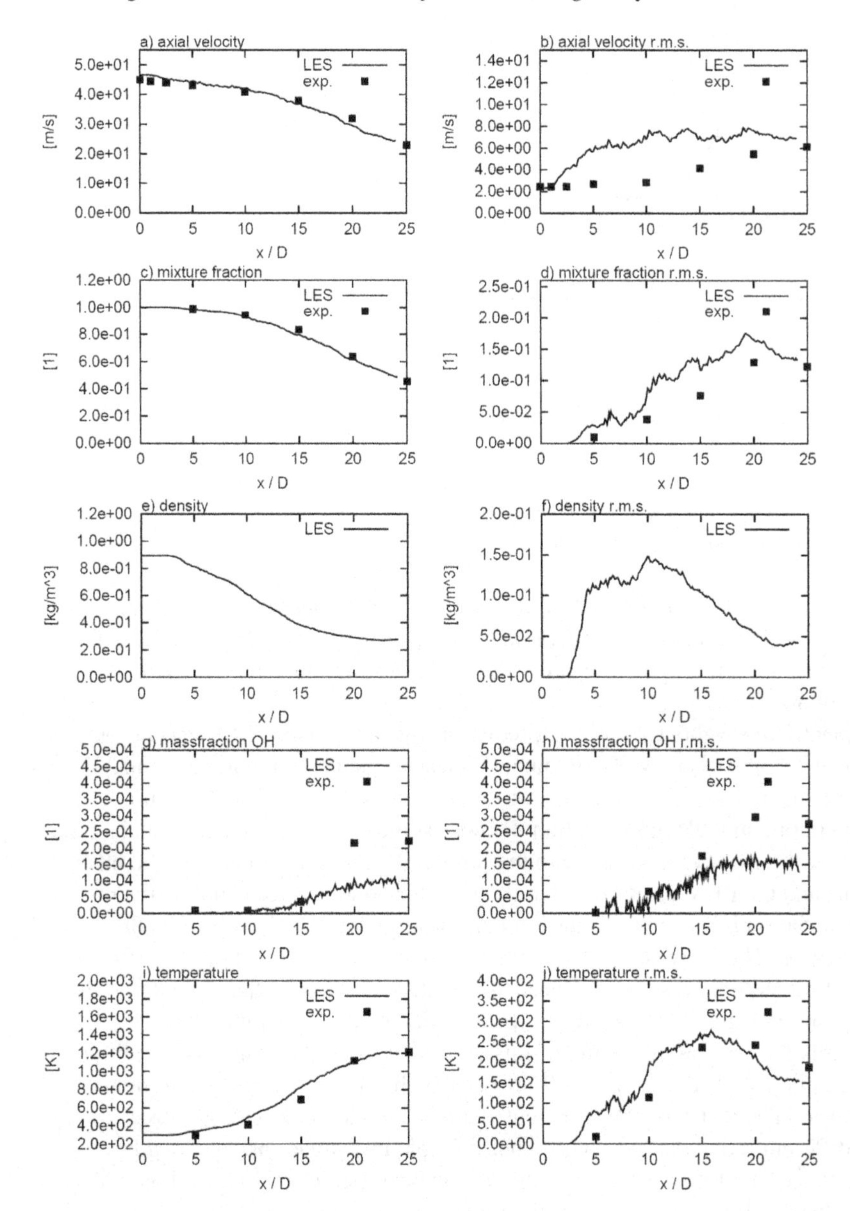

Fig. 3.5 Results for the flame HD along the centerline.

$x/D = 20$. The match at $R = 0$ was already shown to be excellent and the entire profiles are predicted well. For the mixture fraction fluctuations, Fig 3.5 (d) has shown that the values on the axis were computed too high. However, Fig. 3.6 (b) shows that

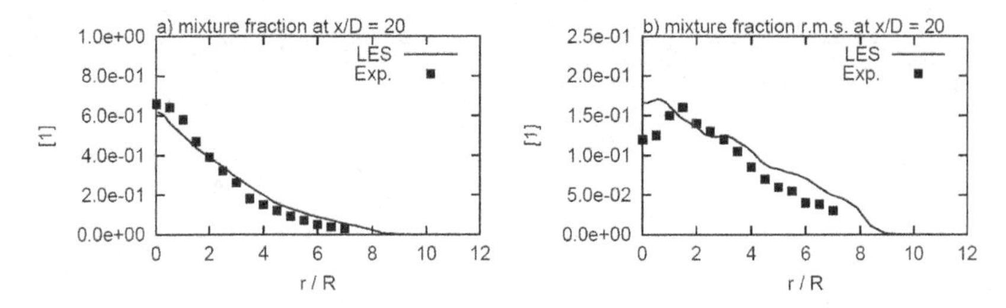

Fig. 3.6 Mixture fraction versus radius at $x/D = 20$.

this problem only occurs locally, i.e. on the axis.

3.3.1.2 H3 Flame

The LES has been performed on a cylindrical grid consisting of 1024 x 32 x 60 $\approx 2 \cdot 10^6$ cells in axial, circumferential and radial directions. The corresponding dimensions are 1m x 2π x 0.18m or 125 x 22.5 nozzle diameters in axial and radial directions.

A quantitative validation of the prediction by the reactive LES can be obtained through a comparison with experimental measurements. In Figures 3.7 and 3.8 axial and radial statistics are presented for the quantities axial velocity, mixture fraction, temperature and OH mass fraction. Besides some derivations for the standard deviation of the axial velocity close to the nozzle, the agreement between the measurements and the LES results is generally good for both the mean and the fluctuations. The deviation is related to an insufficient resolution of the shear layer at the edge of the nozzle. The increased level of velocity fluctuation results in a stronger mixing than the experiments suggest. Hence, the mixture fraction is slightly underpredicted, resulting in a small discrepancy in the axial temperature distribution for $x/D \leq 40$. Towards the outflow plane the results are affected by the boundary condition and hence, the profiles for $x/D = 80$ do not match with experiments any more. Nevertheless, the width of the turbulent flame is captured, as can be observed from the different radial distributions. Other possible investigations, like length scales or statistics of the local flame orientation, have been published in [22]. For further results, especially the variations of the Reynolds number, see [11].

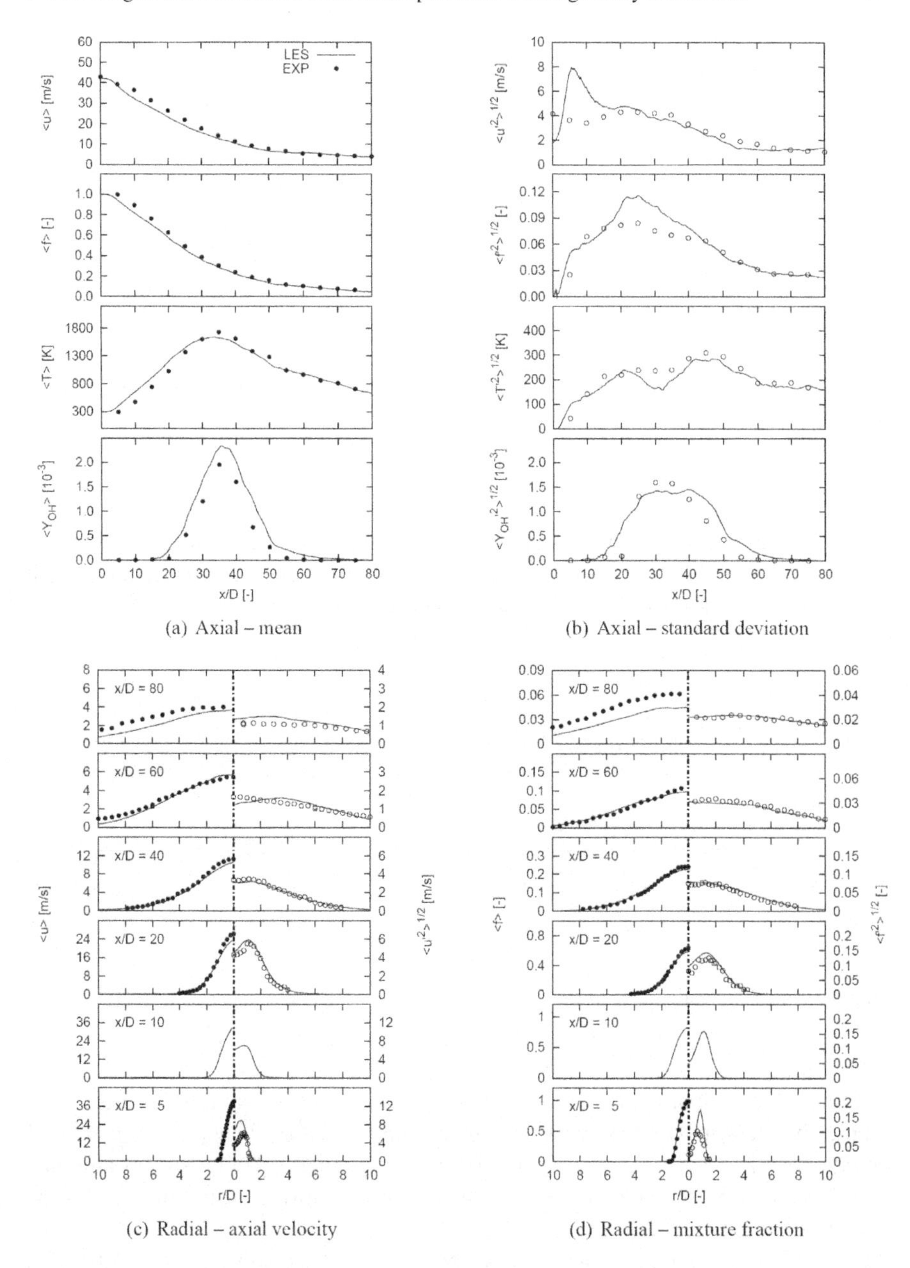

(a) Axial – mean

(b) Axial – standard deviation

(c) Radial – axial velocity

(d) Radial – mixture fraction

Fig. 3.7 Statistics for the H3 flame.

3.3.2 Model Combustor (Partially Premixed Flames)

The investigated configuration consists of a burner, an exit nozzle of Laval-shape and an exhaust duct to model turbine guide vanes. The swirl contains an inner and

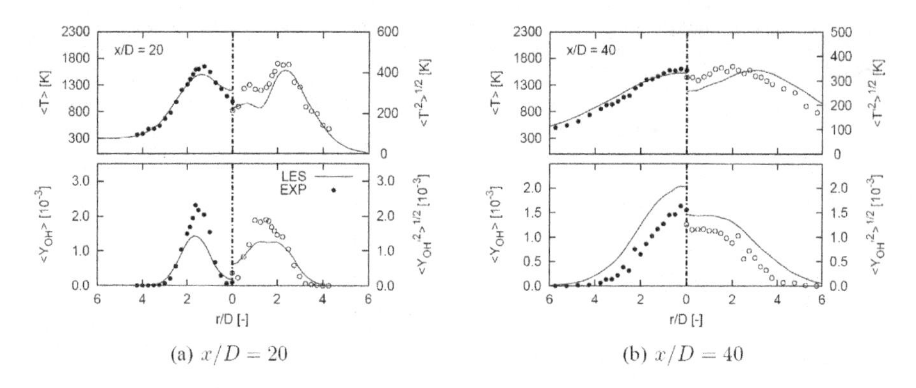

(a) $x/D = 20$ (b) $x/D = 40$

Fig. 3.8 Radial statistics for exemplary chemical properties of the H3 flame.

an outer co-rotating airflow. Gas is injected as fuel through an annular slot between the two airstreams. The combustion chamber has a diameter of 100 mm and length of 113 mm. At an operating point the mass flow rate of air amounts to 13 Nm^3/h,

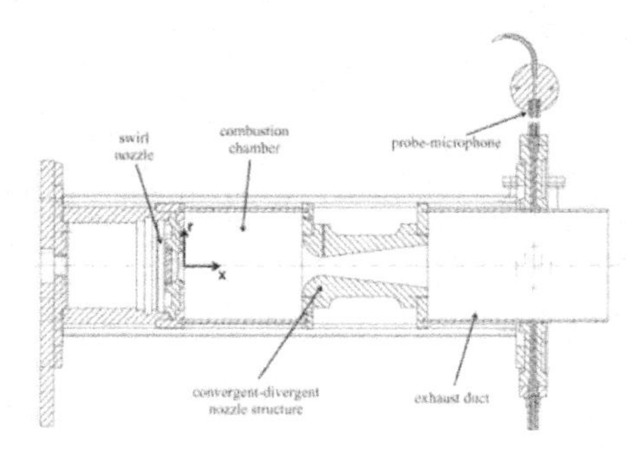

Fig. 3.9 Model combustion chamber.

the mass flow rate of methane is 1 Nm^3/h which corresponds to a thermal capacity of the chamber of 10.05 kW with an equivalence ratio of 0.726 and a nominal swirl number of $S_n = 0.55$. The air-flow split between the inner and outer swirler amounts to $m_i/m_o = 0.75$. The velocity measurements were performed using the LDV technique and an Ar-Ion laser [30].

The reacting LES computations were performed on a grid resolved with $2.3 \cdot 10^6$ CV. Pure methane is used as fuel in numerical simulations. The experimental data mass flows were set as the inlet boundary conditions. Turbulent fluctuations at the inlet are not prescribed relying on the features of LES to generate the turbulent fluctuations inside the swirler. No-slip boundary conditions were performed on any solid walls in the computational domain. At the exit plane of the outlet nozzle Neumann boundary condition was used.

A collocated grid with a cell-centered variable arrangement and implicit time stepping scheme was used. For space discretization the central-differencing scheme is performed. To ensure smooth and bounded solutions of scalars it was necessary to introduce some numerical dissipation, which was achieved with a total variation diminishing (TVD) scheme. From the large class of TVD schemes a flux limited formulation has been used in FASTEST-3D applying the Flame Surface Density Model described in chapter 3.2.2. In the case of combustion simulations, second order Crank-Nicholson method for the equations was used. A typical value for the simulations performed in this work is $\Delta t = 3.5 \cdot 10^{-6}$. The coupling of the pressure and velocity in FASTEST-3D is achieved by the SIMPLE algorithm. The resulting set of linear equations is solved iteratively using the strongly implicit procedure of Stone (SIP).

Although the configuration under investigation was designed as a non-premixed

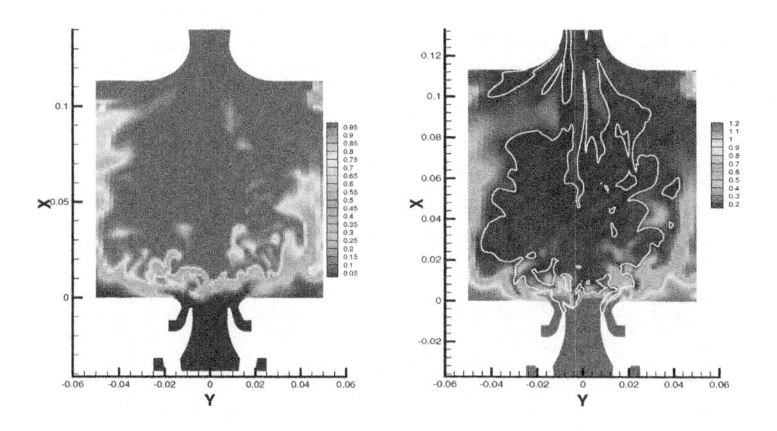

Fig. 3.10 Instantaneous field of the filtered reactive progress variable $\widetilde{c}$ (left) and instantaneous density field and isoline of the stoichiometric mixture fraction (right).

combustor, the LES simulations revealed that the flame is not burning attached to the nozzle, but is lifted and stabilized in the recirculation zone. The reaction zone, introduced by means of the filtered progress variable $\widetilde{c}$, can be observed in Fig. 3.10 (left). The structure of the flame is illustrated by an instantaneous distribution of the density together with the isolines of stoichiometric mixture fraction, Fig. 3.10 (right). It is observed here that the density change takes place not along the stoichiometrical mixture fraction but approximately 1 cm above the burner.

The samples taken from the simulations were time averaged and profiles of statistical quantities were extracted along the radial and axial lines according to the positions measured experimentally [30]. Comparison of simulated and measured flow field quantities show in general good agreement. Quantitative results of the axial averaged velocity and its fluctuations are shown in Fig. 3.11. At the position

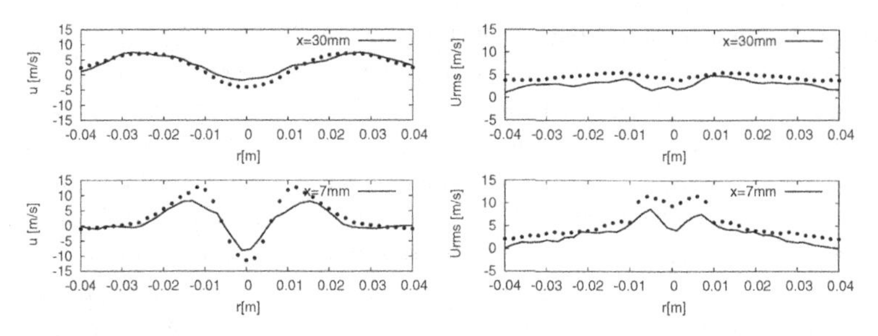

Fig. 3.11 Radial profiles of averaged axial velocity (left) and radial profiles of fluctuations of axial velocity (right).

x=7mm the propagation of axial velocity in radial direction is in general in good agreement with measured values. Nevertheless, in spite of the fact that the geometry of the burner was properly reproduced, the length of the recirculation zone as well as the level of fluctuations is still slightly underpredicted. The peak values of the axial velocity could not be achieved by the simulations. At x= 30mm, as well the qualitative as the quantitative distribution of the simulated velocity fields is in a very good agreement with the experiment. The width and length of the recirculation zone could be very well reproduced by the simulations. The comparison of the axial profiles of axial and tangential velocities at radial position r = 0, -5, -12 mm are shown in Fig. 3.12. It is obvious that the main features of the reacting flow could

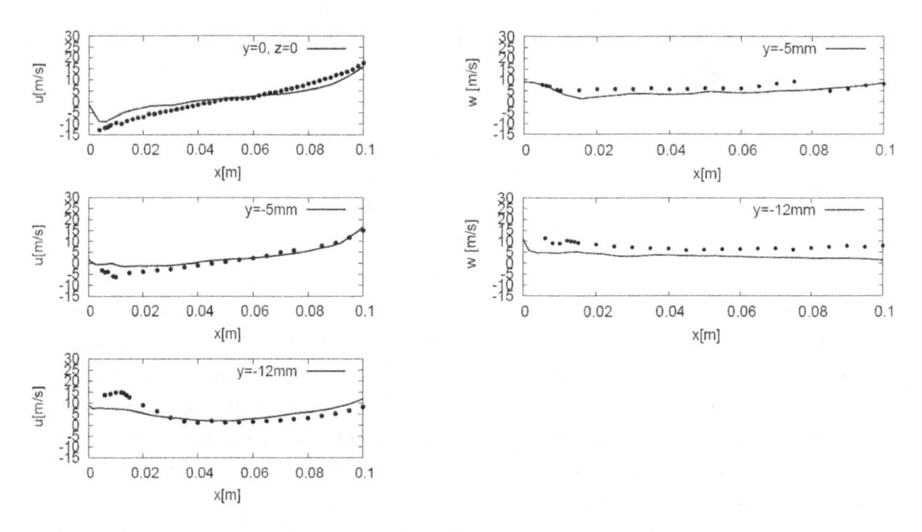

Fig. 3.12 Axial profiles of averaged axial component of the velocity (left) and axial profiles of averaged tangential component of the velocity (right).

be captured by this model. The complete system shows robust behavior and yields very reasonable results for reacting flows.

3.3.3 Tecflam Burner (Premixed Flames)

The investigated configuration is based on an unconfined premixed swirl burner. The experimental database includes airflow measurements under isothermal and reacting conditions. The setup consists of a movable block type swirler, which feeds an annulus from where a mixture of air and natural gas enters the measurement section at ambient pressure and temperature. A sketch is given in Fig. 3.13. The Reynolds number (10000) is calculated from the bulk velocity and the bluff body diameter. The burner gas runs with an equivalence ratio $\Phi = 0.833$. The geometrical swirl number [3] is set to $S = 0.75$ in the current case. A co-axial air flow of $0.5 m/s$ surrounds the swirler device. More details of the setup and the swirler device can be found in [39]. The extension of the computational domain in axial (x) and radial

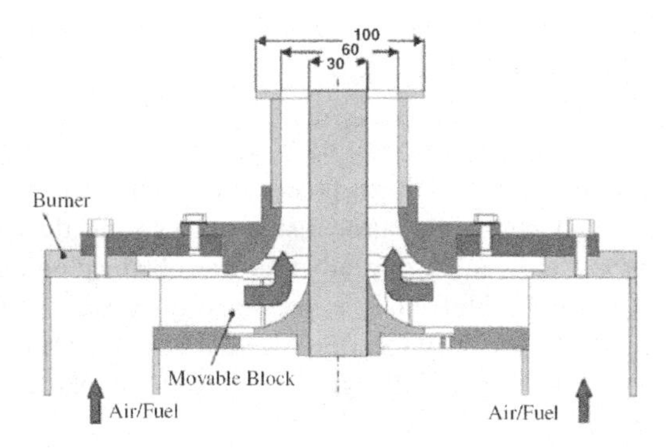

Fig. 3.13 Sketch of the swirler device.

(r) direction is 12D x 8D. Here D represents the bluff body diameter (30mm). The coarse grid resolves the domain with 18 x 64 x 120 (x, ϕ, r) points, whereas the number doubles for the fine grid in each direction (360 x 128 x 240). The filtered conservation equations are solved using a finite volume technique on a cylindrical mesh with variables located on a staggered grid. The G-Equation approach was used to model premixed flames in combination with the FLOWSI-Code. The G-variable as well as the mixture fraction and the density are stored cell centered in the pressure cells. The dynamic version of the Smagorinsky model has been used to model the residual stresses of the momentum equation [15]. The scalar equation is closed using a gradient flux assumption. In the next section general flow features under reacting conditions will be identified and discussed. Further information about modeling and flow results under isothermal conditions as well as the comparison of the isothermal and reacting case can be found in [14]. Figures 3.14 and 3.15 present

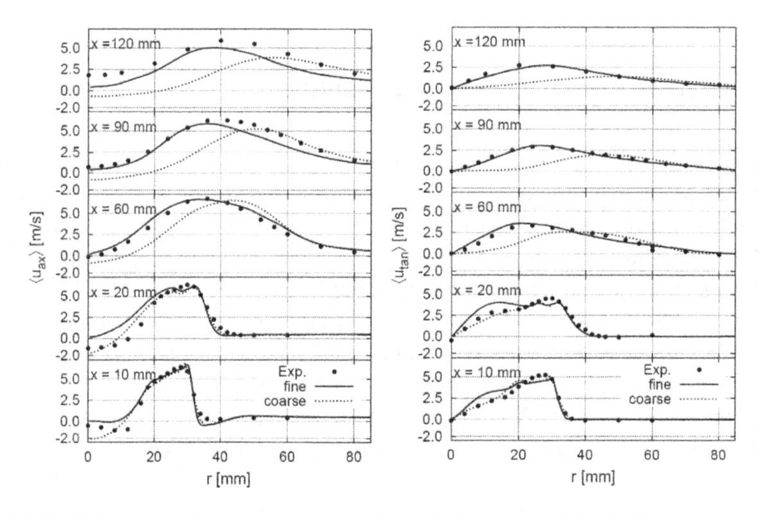

Fig. 3.14 Radial profiles of the mean stream-wise and circumferential velocity for several axial heights for the premixed swirling flame.

the mean velocity components and the turbulent kinetic energy of the flow at several axial heights. An overall evaluation of the results shows some disagreement for the coarse grid but a promising quality of prediction was obtained on the fine grid. The recirculation zone above the bluff body is predicted quite well on the coarse grid. However, this seems to be the only location where the coarse grid is superior compared to the results obtained on the fine grid. The resolution plays a significant role at positions far downstream from the bluff body.

The prediction of the mean radial velocity is encouraging except for a moderate over-prediction of the peak value at x = 10 and 20 mm. The progress of the predicted turbulent kinetic energy correlates almost everywhere with the experimental results. The exception is a small increase which is located in the vicinity of the average flame front position (x= 10, 20mm, r = 20, 30mm). Such an effect was not detected experimentally. Additionally, the grid is obviously too coarse to resolve the outer shear layer of the nozzle at r= 30 mm precisely. Another explanation might be the imperfect inlet conditions.

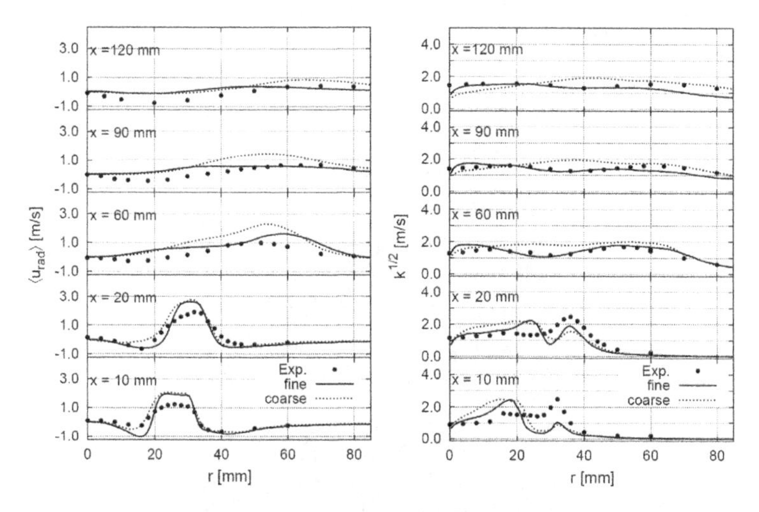

Fig. 3.15 Radial profiles of the mean radial velocity and the turbulent kinetic energy for the premixed swirling flames.

3.3.4 LES/CAA Coupling

The LES/CAA hybrid approach is tested for the H3 flame. The LES is performed on a staggered cylindrical grid with 512 x 32 x 60 $\approx 10^6$ cells in axial, circumferential and radial directions. The complete domain has a physical size of 0.5 x 2π x $0.18m$ which corresponds to roughly 60D x 20D in axial and radial directions.
The acoustic solver is working on a collocated equidistant grid consisting of 200 x 100 x 100 cells in x, y and z direction. A conservative estimate of 20 points per wavelength required to capture the acoustic accurately results in a maximum frequency of 8.5 kHz resolvable on the acoustical grid, assuming the speed of sound in air. The Nyquist frequency of the temporal resolution is 500 kHz and therefore well above the spatial resolution. To omit numerical oscillations, the extracted acoustic time series have been low pass filtered using a cutoff frequency of 50 kHz prior to further analysis. For further information see [13].
The acoustic source, given by the right-hand side of Eq. 3.14 is highly instationary and strongly localized within the flame and its reaction zone, as can be seen in the snapshot in Fig. 3.16. Nevertheless, it is not purely connected to the stoichiometric value of the mixture, which has been highlighted in the figure. Here most of the heat release would be expected. Also the line of a constant speed of sound of 700 m/s is somewhat connected to the combustion noise sources, since it coincides with the range of the steep gradients in the density/mixture fraction relation on the lean and rich side of the flame, respectively. This can be emphasized by considering the relation $\rho = \rho(f)$ and $\frac{d\rho}{dt} = \frac{d\rho}{df}\frac{df}{dt}$. The speed of sound of 700 m/s corresponds to ranges, where $\frac{d\rho}{df}$ is large. This consideration does not hold for the turbulent case

directly, but it gives some hint on the distribution of the acoustic sources. Using the

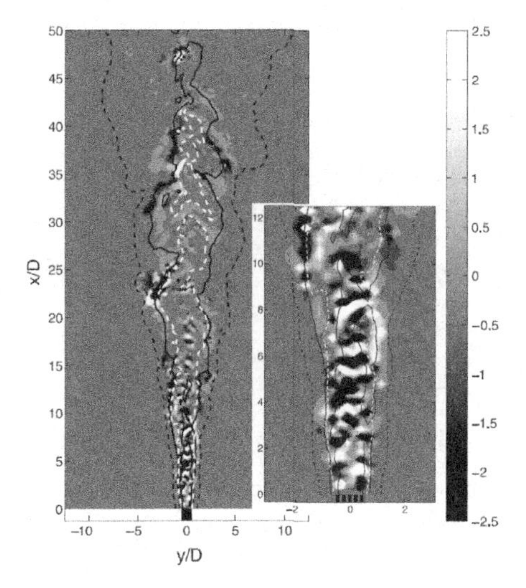

Fig. 3.16 A snapshot of the acoustic source used as input for the acoustic solver. The isolines correspond to a constant speed of sound of 350 m/s(—) and 700m/s (-), and the stoichiometric mixture $f_{st} = 0.31$ (white line).

acoustic source in addition to the temporally resolved properties of the fluid enables the description of the propagation of acoustic waves in the vicinity of the flame. In Fig. 3.17 the snapshot of the first component of the solution vector provides the best qualitative insight into the acoustic field. In a global sense, the flame seems to act as a cylindrical source radiating sound, as expected. It can be seen, that the radiation is not in phase along a constant radius. Scaling the sound intensity by

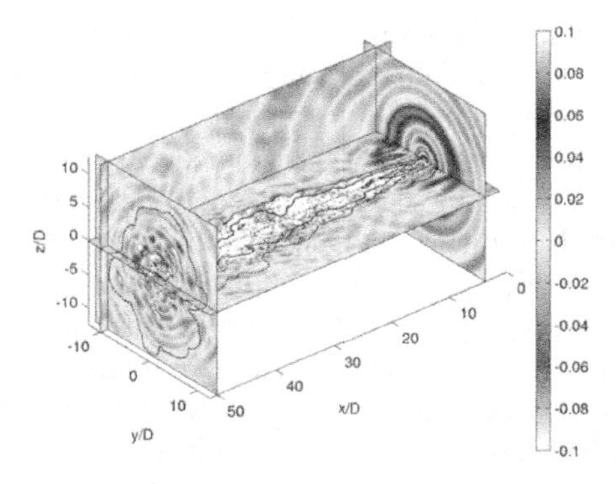

Fig. 3.17 The q_1 component of the CAA solution is shown for a single time step. Again, the isolines correspond to a constant speed of sound of 350 m/s (—) and 700 m/s (-).

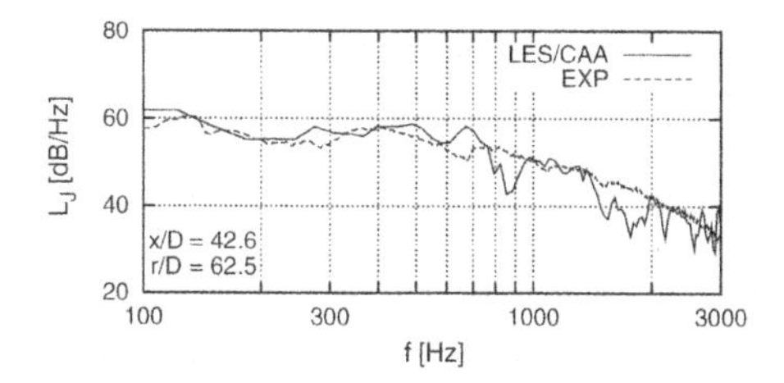

Fig. 3.18 A comparison of the scaled, frequency resolved, sound intensity level, computed with the LES/CAA approach and the experimental measurements at the same axial and radial position (x/D; r/D) = (42.6; 62.5) .

means of the proportionality $J(r) \approx 1/r^2$ and $J(r_1)/J(r_2) = r_2^2/r_1^2$, a comparison to intensity measurements for the same flame at the position (x/D; r/D) = (42.6; 62.5) can be performed. In Fig. 3.18 the sound intensity spectrum computed by the current LES/CAA hybrid approach is plotted against the experimental findings. The agreement in the low frequency range is evident, while the high frequency range is reproduced well. Despite the remaining uncertainties, the combination of an LES containing a low Mach number approximation with a wave equation approach for the acoustic propagation gives encouraging results.

3.4 Summary and Conclusions

Large Eddy Simulation has revealed itself as an appropriate tool for the description of configurations such as diffusion, partially premixed and premixed flames. The following modeling approaches have been presented in this work:

- steady flamelet model for diffusion flames
- flame surface density model for partially premixed flames
- G-equation for premixed flames

The comparison between LES and experimental results show an excellent agreement. These results provide a strong motivation for further numerical investigations of acoustics using the APE (see chapter 7.4), BEM and ESM (see chapter 4.3) approaches.
An efficient hybrid technique has been proposed to describe the noise emissions of turbulent non-premixed flames. While the turbulent reacting flow field was solved by means of an incompressible LES, employing a low Mach number approximation and providing all required fluid properties and sources, the acoustic pressure field was described by a wave propagation algorithm, capable of handling inhomogeneous fluid properties. Instead of implementing the hybrid technique from scratch, two well validated and often applied tools have been coupled. Coupling different

tools allows the exchange between them rather easily, as long as the interface provides all required data.
This technique can potentially be extended to capture the onset of thermo-acoustic instabilities as they arise in a lean premixed gas turbine combustor required to meet the pollution limitations required by the legislation. Since both simulation techniques employed are based on the time domain and capture the underlying physical phenomena, the hybrid technique can be predictive and might therefore be used to evaluate prototypes prior to their experimental testing. This can reduce costs of extensive experimental test runs by reducing the number of prototypes that need to be investigated.

Acknowledgements Financial support by the German Research Foundation (DFG) through the Research Unit FOR 486 "Combustion Noise" is gratefully acknowledged. The authors also would like to express their gratitude to E. Schneider and A. Kempf for the work they have accomplished for this project.

References

[1] Akula R (2006) Study of the performance of different subgrid-scale models and large eddy simulation of premixed combustion. PhD Thesis, Technical University Darmstadt
[2] Barlow R (ed) (1996-2004) Proceedings of the TNF Workshops, Sandia National Laboratories, Livermore, CA, `www.ca.sandia.gov/TNF`
[3] Beer J, Chigier N (1972) Combustion aerodynamics. Applied Science Publisher London
[4] Bilger R (2004) Marker field method for turbulent premixed combustion. CF 138:188–194
[5] Boger M, Veynante D, Boughanem H, Trouve A (1998) Direct numerical simulation analysis of flame surface density concept for large eddy simulation of turbulent premixed combustion. PCI pp 917–925
[6] Bray K, Libby P, Moss J (1994) Laminar flamelets in turbulent flames. Academic Press, London pp 63–113
[7] Bui TP, Meinke M, Schröder W, Flemming F, Sadiki A, Janicka J (2005) A hybrid method for combustion noise based on LES and APE. In: AIAA Paper 2005-3014
[8] Butler T, O'Rouke P (1977) A numerical method for two dimensional unsteady reacting flows. PCI 16:1503–1515
[9] Charlette F, Menevau C, Veynante D (2002) A power low wrinkling model for LES of premixed turbulent combustion. Part I: Non-dynamic formulation and initial tests. CF 131:155–180

[10] Düsing M, Sadiki A, Janicka J (2006) Towards a classification of models for the numerical simulation of premixed combustion based on a generalized regime diagram. CTM 10:105–132

[11] Flemming F (2006) On the simulation of noise Emissions by Trubulent Non-Premixed Flames. PhD Thesis, Technical University Darmstadt

[12] Flemming F, Nauert A, Sadiki A, Janicka J, Brick H, Piscoya R, Ochmann M, Költzsch P (2005) A hybrid approach for the evaluation of the radiated noise from a turbulent non-premixed jet flame based on large eddy simulation and equivalent source & boundary element methods. In: 12th International Congress on Sound and Vibration, Lisbon, Portugal, ICSV12

[13] Flemming F, Sadiki A, Janicka J (2007) Investigation of combustion noise using a LES/CAA hybrid approach. PCI 31:3189–3196

[14] Freitag M, Janicka J (2006) Investigation of a strongly swirled unconfined premixed flame using LES. PCI 31

[15] Germano M, Piomelli U, Moin P, Cabot W (1991) A dynamic subgrid-scale eddy viscosity model. Phys Fluids A 3(7):1760–1765

[16] Hawkes E, Cant R (2001) Implications of a flame surface density approach to large eddy simulation of premixed turbulent combustion. CF 126:1617–1629

[17] Haworth D (2000) A pdf/flamelet method for partially premixed turbulent combustion. Proc of the summer Program, CTR pp 145–157

[18] Hüttle T, Wagner C, Delfs J (2002) Proceedings of the international workshop on "LES for acoustics". Göttingen, Germany

[19] Janicka J, Kollmann W (1978) A two-variable formalism for the treatment of chemical reactions in turbulent H_2-air diffusion flames. Proc Combust Inst 17:421–430

[20] Janicka J, Sadiki A (2005) Large eddy simulation of turbulent combustion systems. PCI 30:537–547

[21] Kempf A (2003) Large Eddy Simulation of Non-Premixed Turbulent Flames. PhD Thesis, Technical University Darmstadt

[22] Kempf A, Flemming F, Janicka J (2005) Investigation of lengthscales, scalar dissipation, and flame orientation in a piloted diffusion flame by LES. Proc Combustion Inst 30:557–565

[23] Kerstein A (1988) Linear-eddy modeling of turbulent scalar transport and mixing. CST 60:391–421

[24] Klimenko A, Bilger R (1999) Conditional moment closure for turbulent combustion. PEC 25:595–687

[25] Kuznetsov VR (1982) Effect of turbulence on the formation of large superequilibrium concentration of atoms and free radicals in diffusion flames. Mehan Zkidkosti Gasa 6:3 – 9

[26] LeVeque RJ (1997) J Comput Phys 131:327–353

[27] LeVeque RJ (2002) Finite volume methods for hyperbolic problems. CUP Cambridge, UK

[28] Lyrintzis A (2003) Surface integral methods in computational aeroacoustics-from the (cfd) near-field to the (acoustics) far field. IJAA 2(2):95–128

[29] Marble F, Broadwell J (1977) The coherent flame model for turbulent chemical reactions. Project SQUID Technical Report TRW-9-PU

[30] Olbricht C, Flemming F, Sadiki A, Janicka J, Bake F, Michel U, Röhle I (2005) A study of noise generation by turbulent flow instabilities in a gas turbine model combustor. In: ASME Turbo Expo 2005, 6-9 June, GT2005-69029, Reno-Tahoe, ASME

[31] Peters N (1980) Local quenching of diffusion flamelets and non-premixed turbulent combustion. Western States Section of the Combustion Institute, paper WSS 80-4, Spring Meeting, Irvine, CA

[32] Peters N (2000) Turbulent Combustion. Cambridge University Press

[33] Pfuderer D, Neuber AA, Früchtel G, Hassel EP, Janicka J (1996) Turbulence modualtion in jet diffusion flames: Modelling and experiments. CF 106:301–317

[34] Pitsch H Large eddy simulation of turbulent combustion systems. PCI 30:537–547

[35] Pitsch H, de Lageneste L (2002) Large-eddy simulation of premixed turbulent combustion using a level-set approach. PCI 29:2001–2008

[36] Poinsot T, Veynante D (2001) Theoretical and Numerical Combustion. Edwards

[37] Pope S (2000) Turbulent Flows. Cambridge University Press

[38] Sadiki A Aand Maltsev, Wegner B, Flemming F, Kempf A, Janicka J (2006) Unsteady methods (URANS and LES) for simulation of combustion systems. Int J Thermal Sciences 45:760–773

[39] Schneider C, Dreizler A, Janicka J (2005) Fluid dynamical analysis of atmospheric reacting and isothermal swirling flows. FTC 74(1):103–127

[40] Selle L, Lartigue G, Poinsot T, Koch R, Schildmacher KU, Krebs W, Prade B, Kaufmann P, Veynante D (2004) Compressible large eddy simulation of turbulent combustion in complex geometry on unstructured meshes. Combust Flame 137:489–505

[41] Spalding D (1971) Mixing and chemical reaction in steady confined turbulent flames. PCI 13:649–657

[42] Tacke M (1998) Zur Stabilität angehobener turbulenter Diffusionsflammen. PhD Thesis, Technical University Darmstadt

[43] Vervisch L, Veynante D (2002) Turbulent combustion modelling. PECS 28:193–266

[44] Warnatz J (1992) Resolution of gas phase and surface combustion chemistry into elementary reactions. PCI 24:553–579

[45] Williams FA (1964) Combustion Theory, The Fundamental Theory of Chemically Reacting Flow Systems. Addison-Wesley

[46] Xue H, Aggarwal S (2002) Asessment of reaction mechnismus for counterflow methan-air partially premixed flames. AIAA 40(6):1236–1240

Chapter 4
Modelling of the Sound Radiation from Flames by means of Acoustic Equivalent Sources

Rafael Piscoya, Haike Brick, Martin Ochmann and Peter Költzsch

Abstract This chapter addresses the calculation of the far field sound radiation of turbulent flames based on a coupling of a Large Eddy Simulation (LES) with two acoustic methods, the Equivalent Source Method (ESM) and the Boundary Element Method (BEM). The numerical aspects of the coupling, including the choice of a coupling interface, data sampling at coarsened grids and Fourier Transform of the coupling variables from time to frequency domain, are surveyed. The simulation results for the radiated sound power of an open turbulent non-premixed jet flame are compared to measurement data. The influence of ground effects on the sound radiation is briefly discussed. Considering a second configuration, where the exhaust of a combustion chamber opens to an adjacent temperature field, the Dual Reciprocity Boundary Element Method (DRBEM) is applied for the simulation of the sound propagation through a non-homogeneous medium. The sensitivity of the radiated sound field to the temperature profile of the surrounding field is investigated. As the results show, the presented hybrid approach is able to predict effectively the sound radiation of flames and the DRBEM still expands the possibilities of the hybrid methodology for the numerical simulation of combustion noise.

4.1 Introduction

Hybrid methods are widely used in the area of aeroacoustics to determine the sound radiation from non-reacting turbulent flows [10, 14]. They are characterized by the application of specialized techniques to the different domains, i.e. the flow-field solution in the non-linear source region is evaluated by a Computational Fluid Dynam-

TFH Berlin - University of Applied Sciences, Department of Mathematics, Physics and Chemistry, Luxemburger Str. 10, 13353 Berlin, Germany, e-mail: piscoya@tfh-berlin.de, brick@tfh-berlin.de, ochmann@tfh-berlin.de · Peter Költzsch
Technische Universität Dresden, Institut für Akustik und Sprachkommunikation, Mommsenstr. 13, 01062 Dresden, Germany

A. Schwarz, J. Janicka (eds.), *Combustion Noise*, DOI 10.1007/978-3-642-02038-4_4,

ics (CFD) code whereas the sound field in the propagation region is computed by means of a pure computational (aero)acoustical method. In this work, the application of a hybrid technique to reactive turbulent flows, particularly to turbulent flames is presented. The novelty of this approach is the use of two well known methods of computational acoustics, the ESM and the BEM, as acoustic coupling methods. Both methods are based upon the homogeneous Helmholtz equation and require only one acoustic quantity on a control surface, which is provided by the CFD calculation. This control surface (Kirchhoff surface) must enclose all sound sources and lie in a homogeneous medium. The BEM has been used in aeroacoustics, for example for aeroacoustic analyses of lifting bodies (wings and rotors) [15, 18], of a circular cylinder in a cross flow [17] and of noise from engine inlets [32].

The ESM is a technique that have been successfully used to describe the sound radiation of vibrating bodies and the sound scattered from solid structures in a homogeneous and quiescent medium. An extension of the application of equivalent sources to aeroacoustic problems was made by Holste [9] who used the method to predict the radiated sound field of a prop fan model. In the present study, the velocity distribution at the Kirchhoff surface serves as input data for the computation of the radiated sound power of the flame in frequency domain and the radiation patterns applying the ESM and BEM.

Section 4.2 gives a brief description of each of the acoustic methods and the CFD code that simulated the flame. The analysis of the different numerical aspects of the coupling and the results are presented in section 4.3.

4.2 Theoretical Background

4.2.1 Hybrid Approach

The acoustic far field produced by turbulent flames cannot currently be determined only by a CFD simulation because the computational cost would be enormous. Instead, hybrid methods, in which the turbulent reactive flow in the source region and the acoustic far field are computed separately, are more effective and require less computational time and resources. In these approaches, the computational domain is divided into two regions, the combustion zone which is solved with the CFD method and the radiation zone which is solved with the acoustic method (Fig. 4.1). The sound propagation in the acoustic domain is described by the homogeneous Helmholtz equation

$$\nabla^2 p + k_0^2 p = 0. \tag{4.1}$$

Some approaches consider even an intermediate domain between source and radiation regions [39]. Details about CFD methods and calculations have been given in chapter 3.2. We mention only briefly which code provided the data for the acoustic calculations that are shown here. The focus in this chapter is on the methods that predict the radiated far field from the CFD data. For the computation of the far field,

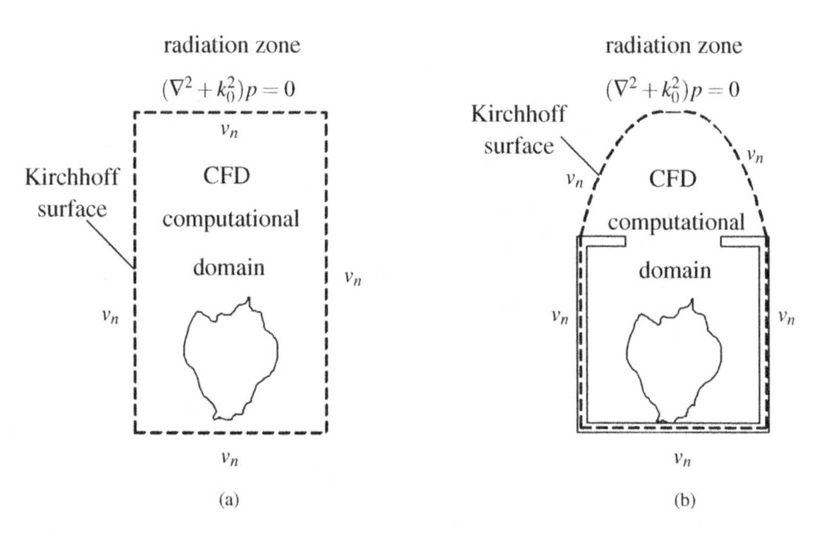

Fig. 4.1 Hybrid approach for a) open flame; b) enclosed flame.

different approaches are known. In Fig. 4.2 some of the CFD and acoustic methods are listed. The hybrid technique, that is presented here, is highlighted to distinguish it from other possibilities. For the acoustic calculation, we present two methods that have been successfully used to predict the sound radiation from vibrating structures, namely the ESM and the BEM. Both methods express the sound field as the sum of the sound fields coming from individual "equivalent sources". While in the ESM these equivalent sources are located inside the vibrating structure and can be of multipole nature (monopoles, dipoles, quadrupoles, etc.), in the BEM they are located at the surface of the structure and consist only of monopoles and dipoles. For vibrating structures, the surface of the body is well defined and the sound field is computed outside the surface. For turbulent flames, a stationary surface can not be defined but an arbitrary control surface (also known as Kirchhoff surface) can be introduced. Assuming that the Helmholtz equation holds in the region outside the control surface, the sound field can be obtained from a prescribed velocity or pressure distribution on the control surface. These acoustic boundary conditions should be provided by the CFD code. The advantage of the ESM and BEM over other acoustic methods is that they require information only at the control surface, so fewer data has to be processed and the far field can be directly calculated. The CFD data have to be preprocessed before the acoustic computation as shown in Fig. 4.2. A description of the data processing is given in the next section.

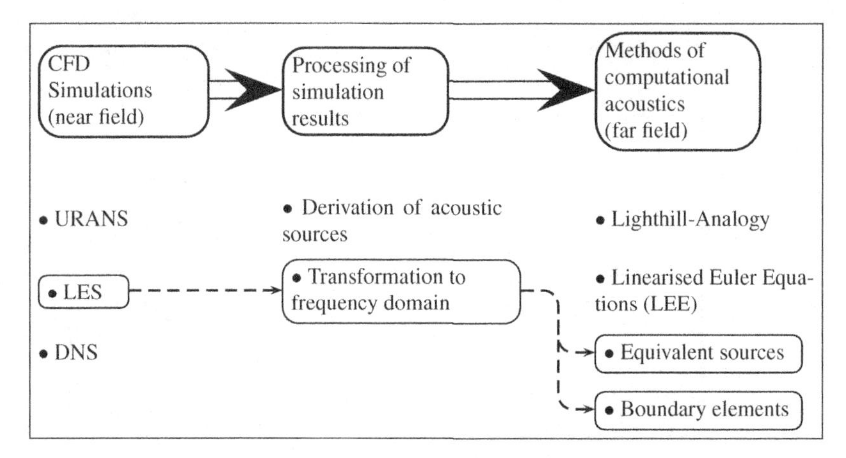

Fig. 4.2 Scheme of the main tasks in the hybrid approach.

4.2.2 Equivalent Source Method (ESM)

The principle of the ESM is to replace the original sound source with a system of equivalent multipoles whose amplitudes are determined in such a way that a boundary condition is satisfied [19]. When the boundary condition is exactly reproduced by the equivalent sources, their radiated sound is identical to the sound radiated by the original source. Normally the boundary condition is not completely fulfilled and an error arises. The strengths of the multipoles are then obtained by minimization of this error. The simplest equivalent sources are described by spherical wave functions, which are solutions of the Helmholtz equation and satisfy the Sommerfeld radiation condition at infinity. The most general expression for the sound pressure generated by Q multipole sources with order up to N, which are located at positions given by the vectors $\mathbf{r_q}$, is

$$p(\mathbf{r}) = \sum_{q=1}^{Q} \sum_{n=0}^{N} \sum_{m=0}^{n} \frac{h_n^{(2)}(k\,|\mathbf{r}-\mathbf{r}_q|)}{h_n^{(2)}(kR)} P_n^m(\cos\gamma_q)(A_{nm}^{(q)}\cos m\delta_q + B_{nm}^{(q)}\sin\delta_q), \qquad (4.2)$$

where P_n^m are the Legendre functions, $h_n^{(2)}$ are spherical Hankel functions of the second kind and γ_q is the angle between the vector $(\mathbf{r}-\mathbf{r}_q)$ and the z-axis and δ_q is the angles of the projection of $(\mathbf{r}-\mathbf{r}_q)$ onto the x-y-plane and the positive x-axis. The spherical wave functions in equation (4.2) have been normalized with respect to $h_n^{(2)}(kR)$, with R as a typical dimension of the source. This normalization brings a favourable effect in the numerical operations involving the system matrix. The total number of spherical functions and thus the number of equivalent sources in expression (4.2) is $N_{tot} = Q(N+1)^2$.
For simplicity, equation (4.2) is written as

$$p(\mathbf{r}) = \sum_{i=1}^{N_{tot}} c_i \psi_i(\mathbf{r}), \tag{4.3}$$

where the subindex i represents a combination of q, n and m, and c_i and ψ_i are the amplitude and wave function for that q, n, m combination.
The amplitudes c_i are determined by minimizing the velocity error ε at the control surface using the method of weighted residuals [19, 20]. The velocity error is defined as

$$\varepsilon = \frac{\mathrm{j}}{\omega\rho} \sum_{i=1}^{N_{tot}} c_i \frac{\partial \psi_i}{\partial n} - v_n, \tag{4.4}$$

where v_n is the known velocity distribution, $\partial/\partial n$ is the derivative in the direction of the outward normal $\mathbf{n}$ on the control surface. The boundary error ε is multiplied by weighting functions w_l and the integral of the product over the surface S is set to zero. This yields the following equations

$$\frac{\mathrm{j}}{\omega\rho} \sum_{i=1}^{N_{tot}} c_i \int_S \frac{\partial \psi_i}{\partial n} w_l dS = \int_S v_n w_l dS \quad l = 1, 2, \ldots N_{tot}. \tag{4.5}$$

The weighting functions used for the calculations are the complex conjugate normal derivatives of the wave functions $w_l = \partial \psi_i^* / \partial n$. The selection of these functions implies a minimization of the velocity error in the least square sense. With the discretization of the surface and assuming constant values over each element, equation (4.5) is transformed into a system of linear equations $Ax = b$ that has to be solved. A is a $N_{tot} \times N_{tot}$ matrix and x is the vector of amplitudes c_i.
Once the strengths c_i are known, the sound field can be computed at any point outside S using Eq. (4.3). All other related acoustic quantities like particle velocity, intensity and power are directly derived from Eq. (4.3). A characterization of the nature, location and strength of the sound sources originated in the combustion zone is not the priority of this paper. Although the ESM replaces the flame with sources in the combustion region, these sources are in principle fictitious. A correspondence between the real and the equivalent sources can be tested by looking for optimal position of the sources. This aspect of finding optimal positions of the equivalent sources has been studied in [29, 30].

4.2.3 Boundary Element Method (BEM)

The fundamental derivations of the boundary element method in acoustics can be found for example in [5, 23, 38]. Here merely a brief outline of the used BEM is given. In this work, the direct BEM approach is applied. Starting point of the direct BEM is the Helmholtz integral equation (HIE) for exterior field problems,

$$C(\mathbf{x})p(\mathbf{x}) = \int\limits_S \left(p(\mathbf{y}) \frac{\partial g(\mathbf{x},\mathbf{y})}{\partial n(\mathbf{y})} - \frac{\partial p(\mathbf{y})}{\partial n(\mathbf{y})} g(\mathbf{x},\mathbf{y}) \right) dS, \tag{4.6}$$

where

$$g(\mathbf{x},\mathbf{y}) = \frac{e^{-\mathrm{j}k|\mathbf{x}-\mathbf{y}|}}{4\pi|\mathbf{x}-\mathbf{y}|} \tag{4.7}$$

is the Green's function in the three dimensional free space. $C(\mathbf{x})$ is equal 1 for $\mathbf{x}$ in the acoustic domain, i.e. in the exterior domain, 1/2 on the surface of the radiating structure and 0 for $\mathbf{x}$ in the interior domain. In a medium at rest, pressure p and particle velocity $\mathbf{v}$ are related by the equation

$$\mathbf{v} = \frac{\mathrm{j}}{\omega\rho} \nabla p,$$

hence in Eq. (4.6), $\partial p/\partial n$ can be replaced by $-\mathrm{j}\,\omega\rho v_n$:

$$C(\mathbf{x})p(\mathbf{x}) = \int\limits_S \left(p(\mathbf{y}) \frac{\partial g(\mathbf{x},\mathbf{y})}{\partial n(\mathbf{y})} + \mathrm{j}\rho\omega v_n g(\mathbf{x},\mathbf{y}) \right) dS. \tag{4.8}$$

Similar as in case of the ESM, the surface is discretized by a set of N constant elements and the the integral is replaced by a finite sum. The accuracy of the approximation is assured by the six-element-per-wavelength rule [16]. Since g represents the sound field from a monopole source and its derivative $\partial g/\partial n$ the sound field from a dipole source, the sound pressure at points outside S is given by the sum of monopole and dipole sources located at the surface with strengths proportional to v_n and p respectively. The discretized integral equation (4.8) can be transformed into a system of equations

$$(C\bar{I} - H)\hat{p} - \mathrm{j}\,\omega\rho G\hat{v}_n = 0, \tag{4.9}$$

where $\bar{I}$ is the identity matrix, H and G are the system matrices and $\hat{p}$ and $\hat{v}_n$ the acoustic variable vectors. After rearranging (4.9) in a way, that the unknown variables (either $\hat{p}$ or $\hat{v}_n$) stay on the left-hand side, the resulting system of equations can be solved by standard complex matrix solvers. Once both acoustic variables are known, Eq. (4.8) can be numerically evaluated for every $\mathbf{x}$ outside S. The diagonal elements of the system matrices, which become singular, are treated as it is described in [23]. The Helmholtz integral equation (4.6) for an exterior Neumann problem, where $\partial p/\partial n$ is prescribed on the surface of the radiating structure, does not have a unique solution at the characteristic eigenfrequencies of the associated interior Dirichlet problem. To avoid the non-uniqueness of the system of equations at these eigenfrequencies, the CHIEF-method (Combined Helmholtz Integral Equation Formulation) suggested by Schenck [34] can be applied. In this formulation, additional collocation points (the CHIEF-points) are located in the interior domain of the object, where the interior Helmholtz integral formulation must be satisfied. With this overdetermination, the surface HIE and interior HIE are solved simultaneously to enforce the finding of the unique surface solution.

4.2.4 Numerical Simulation of the Flames

Simulated data from a turbulent jet flame were provided by our research partner at the Technical University Darmstadt. The data came from a Large Eddy Simulation (LES) of an open jet flame combusting a mixture of H_2 and N_2. The parameters of the simulated flame are shown in Table 4.1. The LES code solves the transport

Table 4.1 Parameters of the flame; fuel: hydrogen/nitrogen ratio, D: nozzle diameter, U_{jet}: flow velocity, $U_{co\text{-}flow}$: co-flow velocity, Re: Reynolds number, f_{stoic}: stoichiometric mixture fraction

fuel [vol%]	D [mm]	U_{jet} [m/s]	$U_{co\text{-}flow}$ [m/s]	Re	f_{stoic}
23/77 H_2/N_2	8	36.3	0.2	16,000	0.583

equations for mass, momentum and conserved scalar mixture fraction. A low Mach number approximation, where the density is not constant but at the same time is not a function of the pressure, is used. The governing equations are solved on a staggered cylindrical grid. The mean inflow profiles are superposed by artificially generated turbulence, while at the outflow a Neumann condition is applied. The circumferential outer boundary is described with a simplified momentum equation to allow for entrainment of fluid. A detailed description of this LES approach can be found in [7, 11].
The output of a LES are usually mean and standard deviations of different quantities of the flow field, like axial or radial distributions of velocity, mixture fraction or temperature. The fluid dynamic properties of the numerical results agree very well with experimental data. Corresponding evaluations are published in [6, 8]. For the acoustic calculations, an extraction of instantaneous values is needed. Hence, data samples at time steps of every 10^{-4} s were generated out of the LES run, which includes the velocity data in x-, y-, z-direction, density, temperature and mixture fraction at selected spatial points of the LES grid.

4.3 Results and Analysis

4.3.1 Numerical Aspects of the Hybrid Method

The coupling of the CFD and the acoustic methods is made at the closed control surface S that encloses the flame. Since the velocity at the control surface is provided by the LES, the radiated sound field can be determined by ESM and BEM. For the ESM, the amplitude of the equivalent sources must be obtained by minimizing the velocity error (4.4), for the BEM, the pressure at the surface has to be calculated via the set of equations (4.9). A direct transfer of the data from the LES to the ESM/BEM is not possible since the LES works in the time domain and the ESM or

BEM in the frequency domain. Also, the required time and spatial resolutions differ in both methods. A processing of the CFD data must be performed before they can be used by the acoustic method. In the following subsections, the operations and transforms carried out on the data are presented.

4.3.1.1 Generation of the Acoustic Mesh

The LES mesh is much finer than it is necessary for the acoustic requirements. Additionally, for acoustical purposes, a non-uniform surface mesh leads to an increase of computational costs without an increase of accuracy. Therefore the LES grids were coarsened in axial and radial directions until the resolution met the tangential resolution and the assembling of nearly uniform quadratic surface elements was possible. Fig. 4.3 illustrates this mesh generation.
Ten concentric cylindrical surface meshes with increasing radii were built, so that acoustical simulations on the basis of different enveloping surfaces were possible. The frequency limit for the validity of the model can be obtained considering the six-elements-per-wavelength rule. The velocity data at the grid points of the coars-

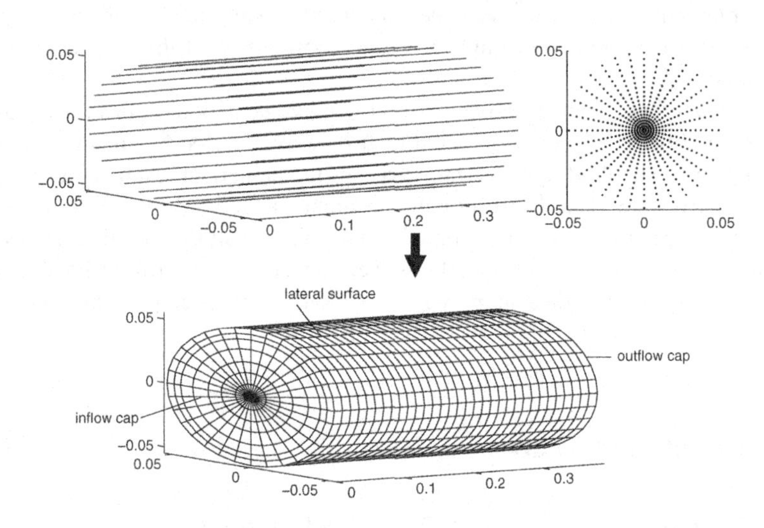

Fig. 4.3 Generation of an equidistant acoustical surface mesh (below) from the grid points of the LES (above).

ened mesh are extracted from the given LES data with different sampling techniques as described in the following section. The velocity on the inflow cap was set to zero because at this side, the adjacent burner represents a rigid boundary condition. With is zero setting it is also possible to eliminate the highly disturbing inflow boundary condition of the LES as described in [2].

4.3.1.2 The Downsampling

The mesh coarsening comes along with a downsampling of the given LES-Data. Since axial and radial resolutions were different, the grid in axial and radial directions had to be treated separately. The velocity data at the coarsened grid nodes can be sampled pointwise out of the finer mesh (in case the grid nodes of both meshes coincide) or can be sampled in conjunction with a convolution with a spatial filter. The simplest filter is a rectangle filter, which results in the so called simple moving average. We applied both, the pointwise and filtered sampling. A major or minor aliasing effect for the pointwise sampling is expected depending on the smoothness of the original normal velocity curve [24]. The filtering with a the rectangle window function along the coarsened grid coordinates represents a low-pass filtering of the original signal and should provide a small aliasing effect in the wavenumber spectrum. The wavenumber spectrum $V_n(k_x)$ results from the Fourier transform from the spatial to the wavenumber domain, which is defined as

$$V_n(k_x) = \frac{1}{L}\int_0^L v_n(x)e^{-\mathrm{j}k_x x}dx, \tag{4.10}$$

where L is the length of the cylinder. Fig. 4.4 shows the normal velocity in axial direction at the first node line of lateral surface of the tenth cylinder of the original and coarsened mesh. It can been seen, that the velocity distribution is quite smooth, the pointwise extracted data as well as the filtered data (moving average) follow well the curve of the given data. Nevertheless, in the wavenumber spectrum an aliasing effect can be observed in the higher wavenumber range for the pointwise sampling. The wavenumber spectrum of the coarsened mesh is limited to $10\,k/k_0$, $(k_0 = 2\pi/L)$ due to downsampling.
Fig. 4.5 shows the velocity and its spectrum for the same line at a later time step. The spatial velocity distribution appears again to be smooth enough to be well represented by the pointwise extraction as well as by the averaged data. Here, we find a wavenumber spectrum, which does not show an aliasing effect for both of the sampling methods. From the wavenumber spectra shown for one line for different time steps, we see that a resulting aliasing effect from a pointwise sampling depends strongly on the velocity curve of the originally given data. The pointwise extraction does not necessarily lead to an aliasing effect.
Fig. 4.6 shows the normal velocity distribution along a radial line at the outflow cap. Here, the averaging smooths the original curve, which shows some short wave fluctuations. A direct transform in the wavenumber domain is not possible for the velocity distribution in radial direction, because the mesh is not equidistant in this direction.

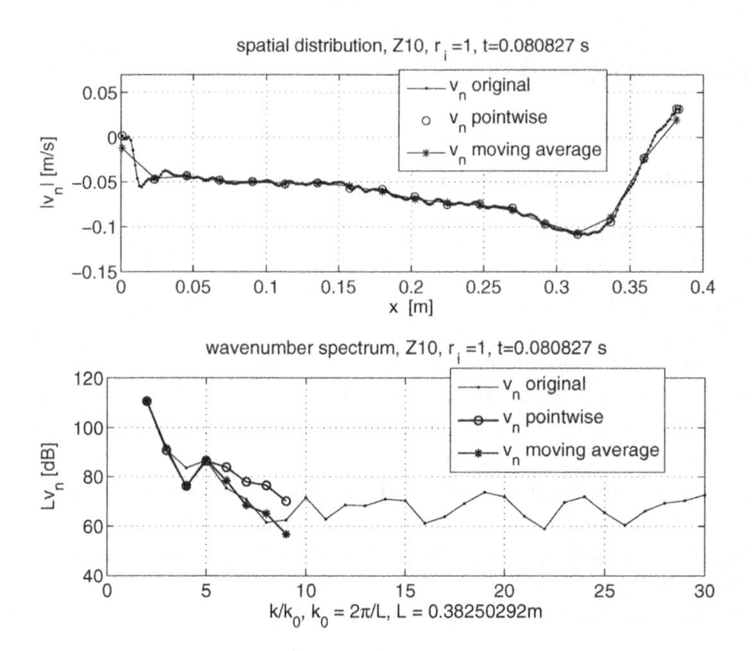

Fig. 4.4 Instantaneous velocity distribution in axial direction x at $t = 0.081$s for the axial line of the tenth cylinder at $\varphi = 0°$, *upper plot:* spatial distribution, *lower plot:* wavenumber spectrum.

An increase of the wavenumber spectra as seen in Fig. 4.4 can lead to a higher sound radiation in the level in the higher and middle frequency range. But it is difficult to estimate the general effect on the acoustic radiation because the spatial spectra vary for every time step and every axial node line. A comparison of the results with both approaches is shown in Fig. 4.7. The radiated sound power of the tenth control surface is presented and only a slight difference at low frequencies for both sampling methods can be observed. As discussed in [30] the outflow cap was found to determine mainly the amount of radiated sound power. To study the effect of the downsampling at the outflow cap, an acoustic calculation with a model keeping the original LES mesh at the outflow cap but with a coarsened lateral surface was performed. Fig. 4.8 shows the resulting sound power level for the filtered sampling and for the case of no downsampling on the outflow cap. In this case, the radiated sound power is overestimated by using the coarsened mesh and applying a simple moving average does not effectively suppress the occurring aliasing effect.
As the above results show, the computed radiated sound depends strongly on the used mesh. The sensitivity of the CAA-solutions to the use of fine or coarse grids within a hybrid approach has been also reported by one of our research partners in [3]. The length of coherent structures in fluid dynamics is small compared to the acoustic wavelengths and most of them may not radiate, but in order to obtain an accurate sound field they need to be well described or filtered out. Otherwise the

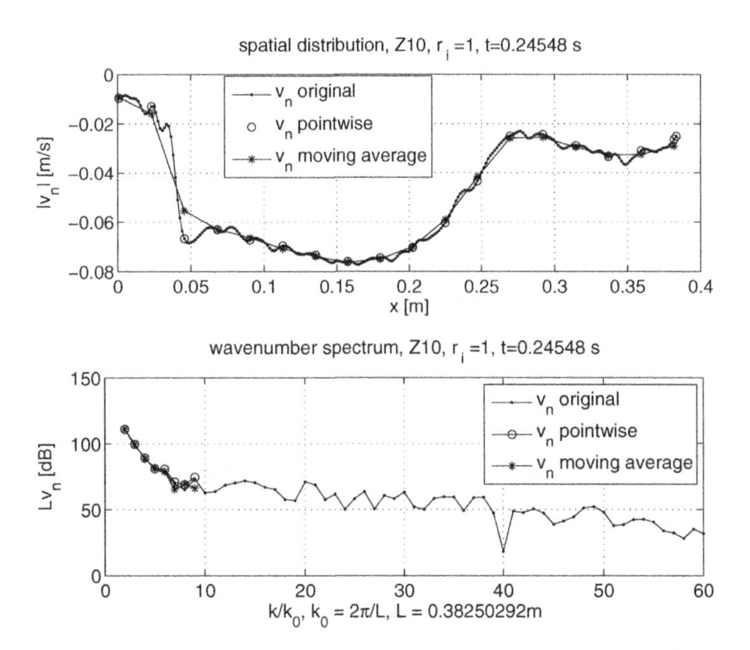

Fig. 4.5 Instantaneous velocity distribution in axial direction x at $t = 0.245$s for axial line the axial line of the tenth cylinder at $\varphi = 0°$, *upper plot:* spatial distribution, *lower plot:* wavenumber spectrum.

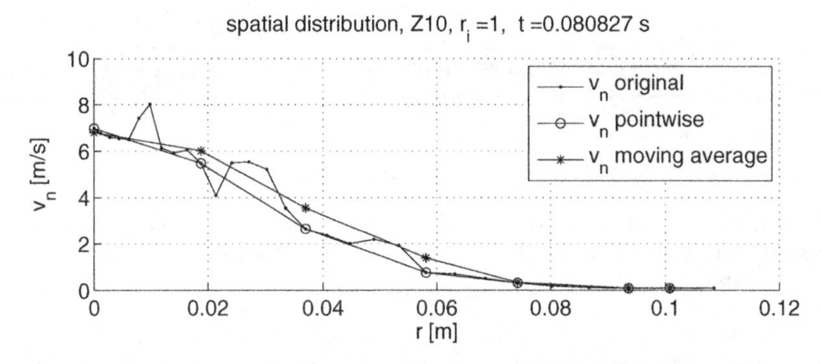

Fig. 4.6 Instantaneous velocity distribution in radial direction r at $t = 0.081$s for the radial line at $\varphi = 0°$.

sound radiation is incorrectly predicted by the hybrid method.

Further research work should be devoted to more elaborated filtering techniques.

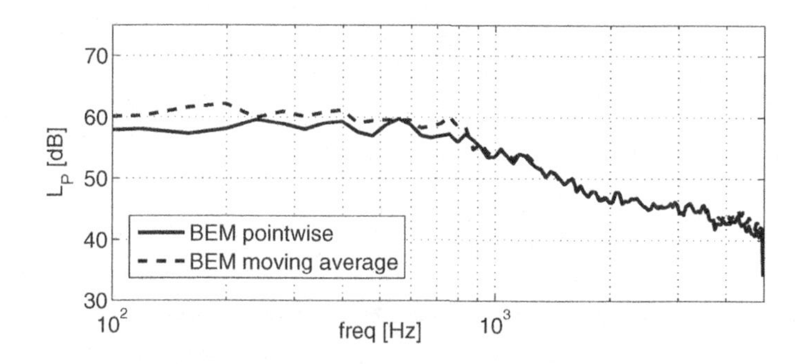

Fig. 4.7 Radiated sound power level L_p of the tenth control surface for a pointwise and filtered data sampling.

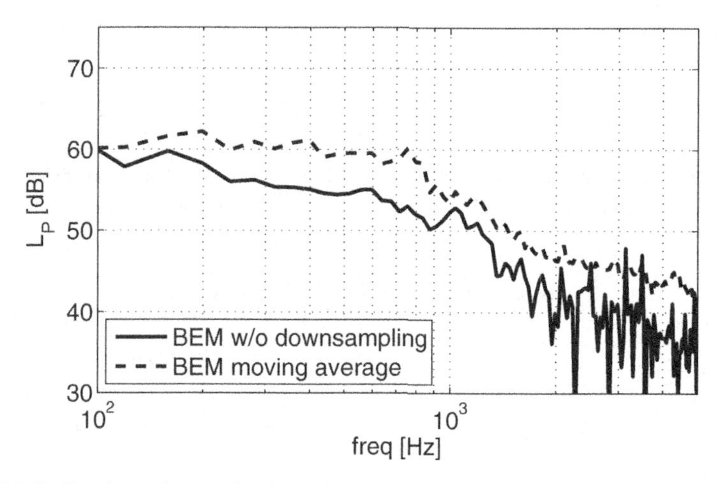

Fig. 4.8 Radiated sound power level L_p of the tenth control surface for a complete data sampling (without downsampling) at the outflow cap and filtered data sampling (moving average) on the coarsened mesh.

4.3.1.3 Determination of the Velocity Spectra with Reduction of the Variance

Since the LES works in the time domain, the LES data have to be first transformed from the time to the frequency domain by applying a Fourier Transform. The original time steps of the LES are in the range of 10^{-7}s. For the acoustic calculation, a data sampling of 10^{-4}s assures information in the frequency range of [0-5000] Hz. Since a downsampling in the time domain is performed, aliasing effects are also possible. But knowing that the types of flames investigated radiate in the low and middle frequencies, the instantaneous data were directly written out without any low pass-filtering.
Since only one time series was provided and the power spectrum of the whole signal

involves a high variance, the time series were split into sets of 250 samples, so that each set was used as input velocity for one sound field calculation. The predicted sound power and sound pressure level in the far field are the average over all these individual calculations. In order to increase the number of individual sets and there-

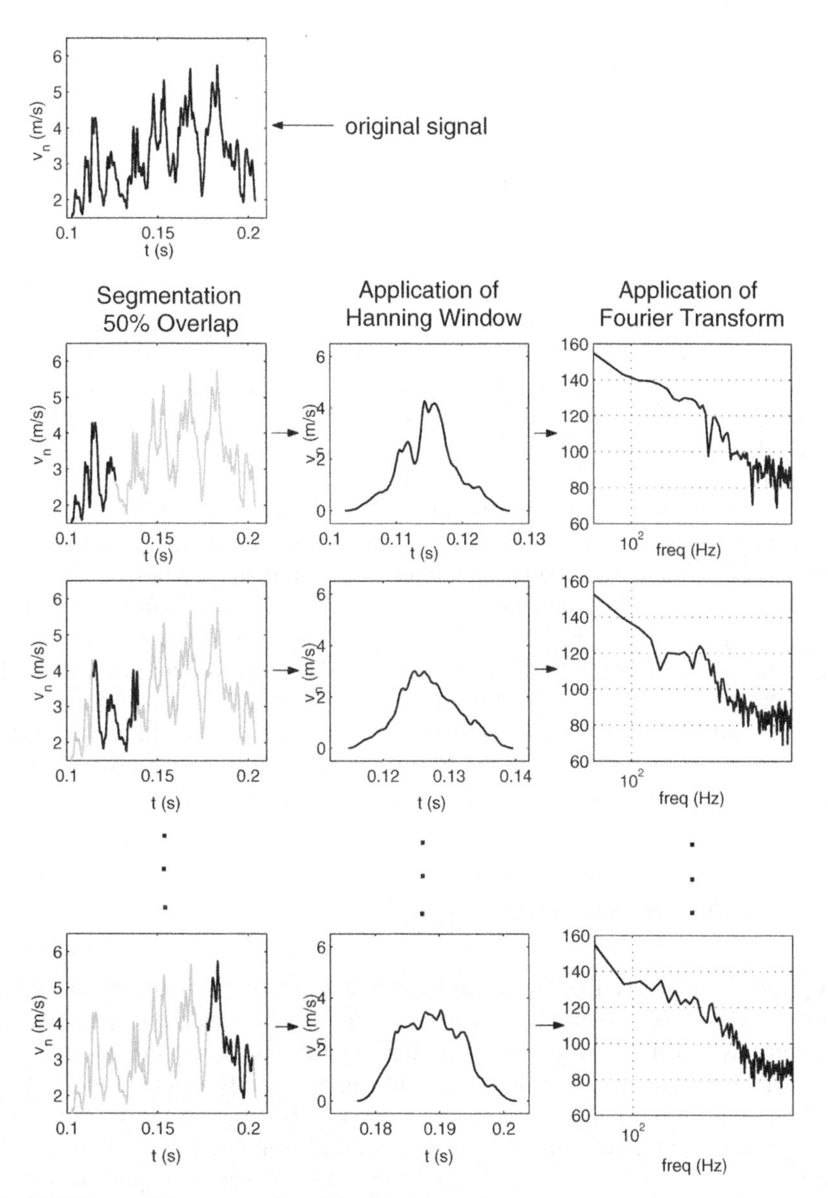

Fig. 4.9 Scheme of the computation of the velocity spectra.

fore decrease the variance of the averaged spectrum, an overlap between adjacent segments was considered. According to Welch [36], a 50% overlap along with the use of a Hanning window reduce the aliasing effects as well as keep the dependence between segments in a low level. The procedure to compute the velocity spectra is illustrated in Fig. 4.9. As a result of the segmentation, the frequency resolution is decreased. Since each segment had 250 time samples and the time step is 10^{-4}s, the frequency resolution is 40 Hz.

4.3.2 Location of the Control Surface

The location of the control surface is a very important point of this coupling strategy. The equations for the radiated sound pressure, Eq. (4.2) for ESM and Eq. (4.8) for BEM, are only valid if the control surface is placed in a region where the medium is uniform and at rest. If the source region is large and the control surface cannot be placed in the homogeneous region, both expressions yield erroneous results. A way for testing if the control surface is placed in a homogeneous medium is to study the radiated sound power of different control surfaces. The sound power is obtained by integrating the sound intensity over the control surface. In an homogeneous medium, the sound power is independent of the extent and location of the surface if it encloses all sound sources. Thus, by calculating the sound power using different surfaces, we expect to determine which surfaces are placed in the homogeneous medium. This idea was applied to the studied flame. Ten cylindrical control surfaces, denoted as Z1, Z2,..., Z10 and having the same length but different radii were used to compute the sound power. In Fig. 4.10, the curves of the sound power are compared. It can be seen that the curves are practically identical above 1000 Hz. Below that frequency, there are differences but a certain convergence could be found for the last two cylinders (Z9 and Z10). According to the above results, the last cylinder, which has the largest radius, is chosen for the computation of the sound radiation.

4.3.3 Inclusion of Ground Effects

In practical situations, the flames are located in a certain environment (room, laboratory, etc). Assuming that side walls and ceilings have some type of acoustical treatment, the scenario will be more similar to a half space problem than to a free field problem. The presence of the ground can be directly included in the ESM and BEM by considering image sources located at the other side of the plane. For the BEM, an appropriate Green's function $g_H(\mathbf{x},\mathbf{y})$ is introduced in Eq. (4.8). The Green's function in the presence of an infinite plane has a simple expression only in two ideal cases, where the acoustic impedance Z is infinite ($Z=\infty$) or zero ($Z=0$). The first case corresponds two an acoustically rigid plane and the second to a soft plane. The expression for the Green's function is given by

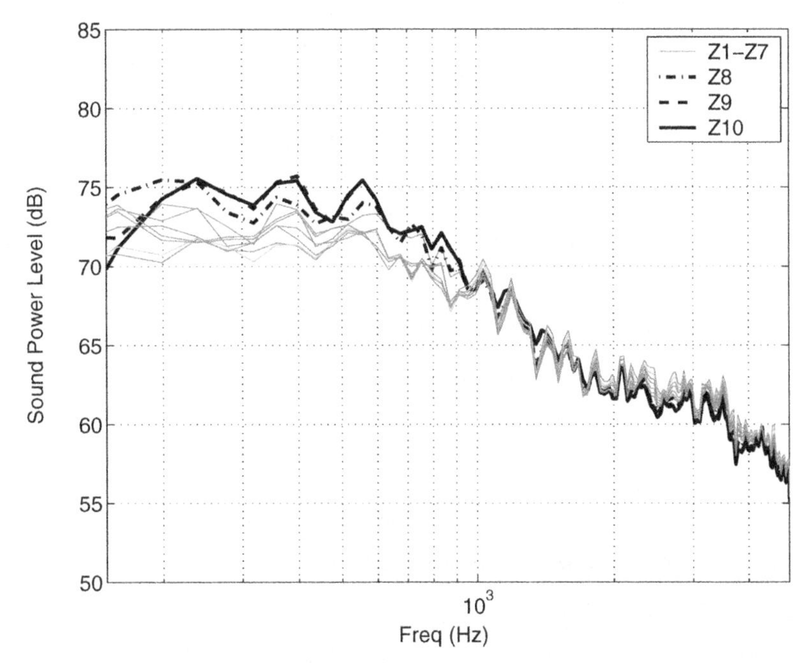

Fig. 4.10 Simulated and measured sound power level of the open flame.

$$g_H(\mathbf{x},\mathbf{y}) = \frac{e^{-\mathrm{j}k_0 R_1}}{4\pi R_1} + R\frac{e^{-\mathrm{j}k_0 R_2}}{4\pi R_2} \tag{4.11}$$

with $R_1 = |\mathbf{x} - \mathbf{y}|$, $R_2 = |\mathbf{x} - \mathbf{y}_i|$. The reflection coefficient R is +1 if the plane is rigid and -1 if the plane is soft. The vector $\mathbf{y}$ defines a point above the impedance plane and the vector $\mathbf{y}_i$ defines the position of its image point (see Fig. 4.11). A calculation of the sound field of the open flame with and without the two types of planes was performed. In Figs. 4.12 and 4.13, the radiation patterns and the curves of sound power of the flame for the three cases: free space, rigid plane and soft plane are shown. In the plots of the radiation patterns (Fig. 4.12), the superposition of direct and reflected waves cause a local increase and decrease of the sound pressure compared to the free space case. The radiated sound power is sensitive to the presence of the planes only in the lower frequency range (Fig. 4.13). The presence of a soft or rigid plane leads to a slight decrease or increase of the sound power level in this frequency range, respectively. Generally, the influence of these types of planes on the radiated sound power decrease with the distance h of the flame from the plane. In case $k_0 h \gg 1$, where k_0 is the wavenumber, the influence of an ideal rigid or soft plane on the radiated sound power can be almost neglected. Nevertheless, the directivity of the sound field is strongly affected by the presence of a plane, independently of h, and has to be considered when doing sound power measurements. In the higher frequency range, only a careful scanning of the intensity or

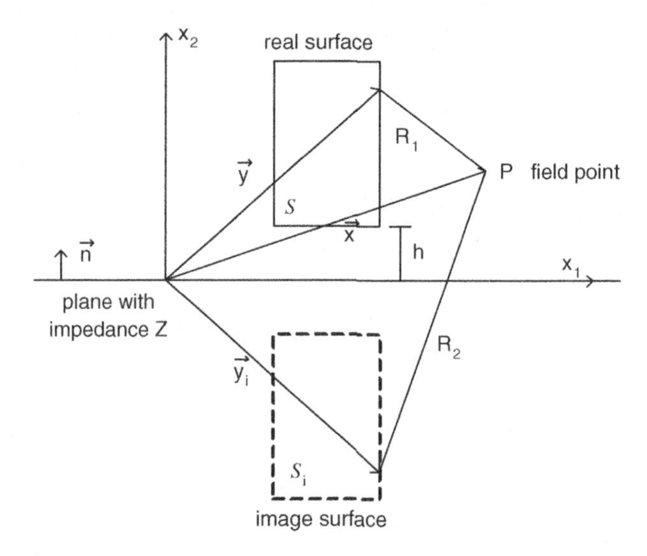

Fig. 4.11 Sketch of real and image surface in the presence of a reflecting plane.

of the sound pressure on an enveloping surface around the flame can give reliable results due to the mentioned effects. The presented simple image source ansatz is not suited to describe the sound propagation above an impedance plane with finite and complex impedance, which show damping and spring/mass characteristics. An appropriate Greens function, which is able to describe the sound propagation above an impedance plane and is in addition suitable for an implementation into a BEM code, is thoroughly discussed in [21] and [22].

4.3.4 Measurement of the Flame

The sound power of the investigated flame was experimentally determined. The jet burner was located in a facility without acoustic treatment, for this reason intensity measurements were preferred to pressure measurements to determine the sound power. The radiation patterns were not measured since free field conditions were not fulfilled. The jet flames were thin and long so that the temperature did not represent a problem at all. Only at the top of the flame, a few centimetres away from the flame axis, the sound intensity could not be measured. The measurements were made when all activities in the laboratory had finished and all other equipments were switched off in order to avoid the presence of disturbing sound sources. A measurement of the background noise was made and it was well below the flame noise above 100 Hz. A first test measurement confirmed the axial symmetry of the sound. Thus, the number of measured points was reduced to 28. Figure 4.14 shows a sketch of

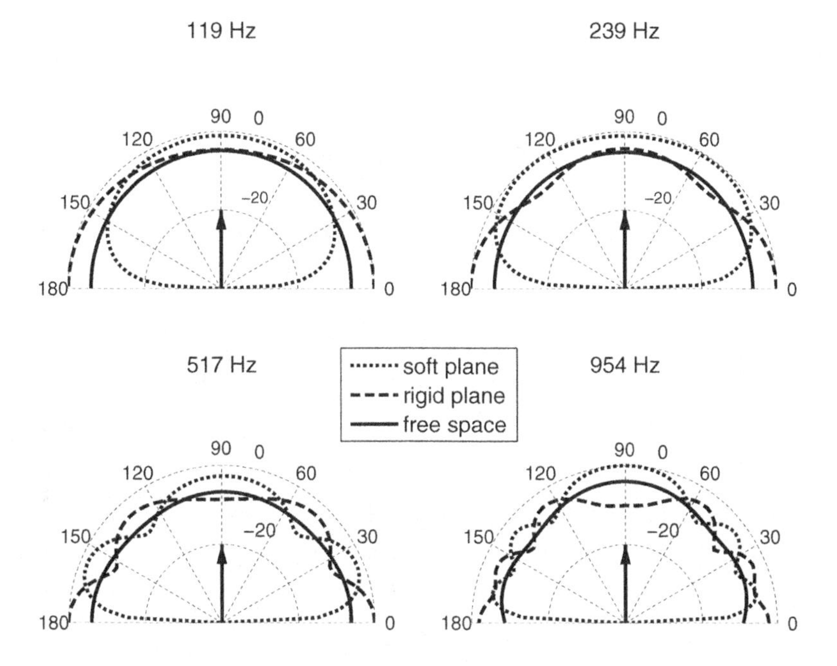

Fig. 4.12 Comparison of radiation patterns of the normalized sound pressure level. The arrow indicates the direction of the flow. The flames control surface is located on a soft plane, on a rigid plane or in free space, respectively.

the measurement grid and the burner and a comparison of the sound power of the isothermal jet without combustion (non-reactive flow) and the flame (reactive flow). A more detailed description of the measurement setup can be found in [30]. The differences in sound power level between reactive and non-reactive cases can be clearly seen. Between 100 Hz and 1000 Hz, the sound power level produced by the combustion is about 20 dB higher. This indicates that the principal mechanisms of sound generation are different for reactive and non-reactive flows.

4.3.5 Results of the Simulation and Comparison with the Measurement

The sound radiation of the diffusion flames was calculated using the velocity at the surfaces of the cylinder enclosing the flame. For each velocity spectrum (see section 4.3.1.3) a radiated sound field is computed and the resulting sound field is obtained by averaging all calculations

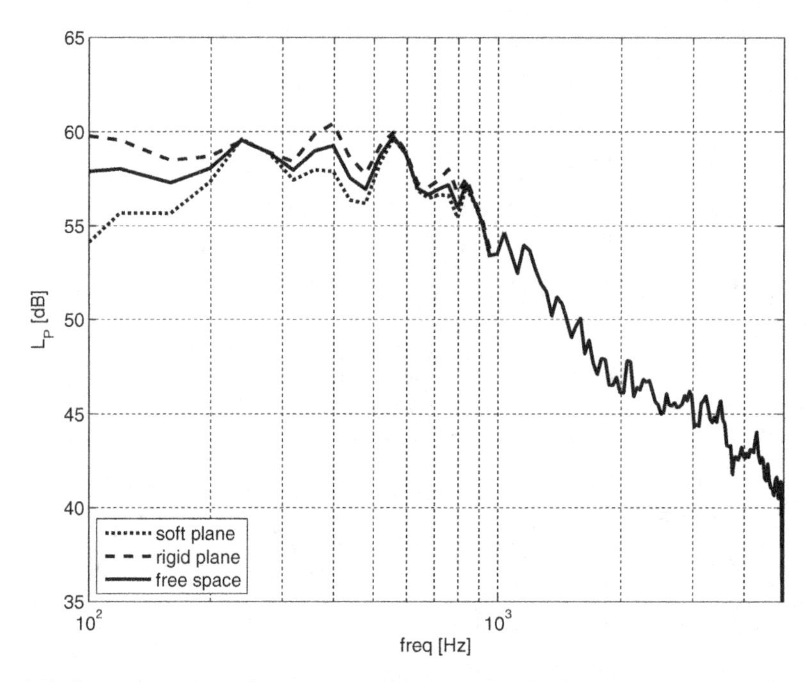

Fig. 4.13 Comparison of sound power spectra. The control surface is located above on a soft plane, on a rigid plane and in free space, respectively.

$$P_{avg}(f) = \frac{1}{N} \sum_{i=1}^{N} P_i(f). \tag{4.12}$$

The results of the ESM and BEM are very similar (see Fig. 4.15). Above 200 Hz, the difference in the sound power is not bigger than 2 dB. The radiation patterns show some differences which vary from one frequency to the other but remain always small. When compared to the experiment, the shape of the simulated curves agrees well with the measured curve at frequencies below 1 kHz but the amplitude is about 3 dB overestimated. Above 1 kHz, the slope of the decay differs a little from the measured decay. It is not easy to find the source of error in a method that has several steps and combines two very different techniques. Since the ESM and BEM results are very similar, it seems likely that the errors are caused by the input data. A more detailed analysis of the measurement and simulation data as it is presented in [30] reveals the strong influence of the velocity distribution at the outflow cap on the total amount of radiated sound power. The velocity data at the outflow cap are expected to contain non-radiating components caused by convecting vortical structures of the flow, which have not subsided sufficiently. Also, the discussed aliasing effects due to the spatial downsampling may contribute to the overestimation of the sound power level.

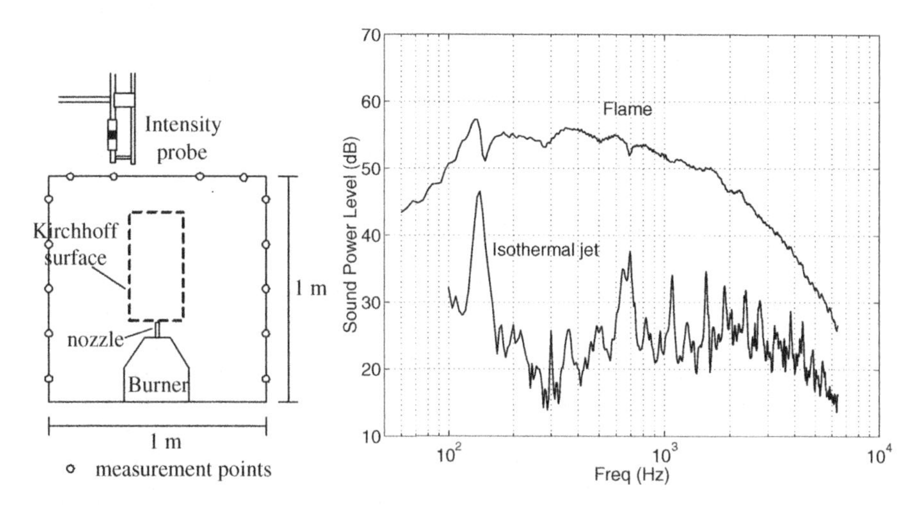

Fig. 4.14 Measurement of the flame; *left plot*: sketch of the measurement points; *right plot:* spectrum of the flame and the isothermal jet.

4.3.6 Sound Propagation in a Non-Homogeneous Medium

The actual hybrid approach suggests extending the Kirchhoff surface (CFD domain) as much as possible to assure the validity of the homogeneous Helmholtz equation outside the Kirchhoff surface, where a linear sound propagation can be assumed. This may not be always possible and a non-homogeneous region outside the control surface may remain. In other cases, a study of the influence of some inhomogeneity on the sound field may be desired. In both situations, sound propagation in a non-homogeneous medium has to be considered. Following the idea of the acoustic analogy, an inhomogeneous Helmholtz equation, which is valid inside the inhomogeneous region (Ω), can be derived

$$\nabla^2 p + k_0^2 p = q, \tag{4.13}$$

where q describes the inhomogeneity and can be a complex term. Outside Ω, the Helmholtz equation (4.1) holds. The flame models of Fig. 4.1 changes now to Fig. 4.16 To deal with this problem, we use the same approach as for the BEM, and the integral form of Eq. (4.13) inside Ω is given by [37]

$$\begin{aligned} C(\mathbf{x})p(\mathbf{x}) = &- \int_S \left(p(\mathbf{y}) \frac{\partial g(\mathbf{x},\mathbf{y})}{\partial n(\mathbf{y})} - \frac{\partial p(\mathbf{y})}{\partial n(\mathbf{y})} g(\mathbf{x},\mathbf{y}) \right) dS(\mathbf{y}) \\ &- \int_\Omega q(\mathbf{y}) g(\mathbf{x},\mathbf{y}) dV(\mathbf{y}). \end{aligned} \tag{4.14}$$

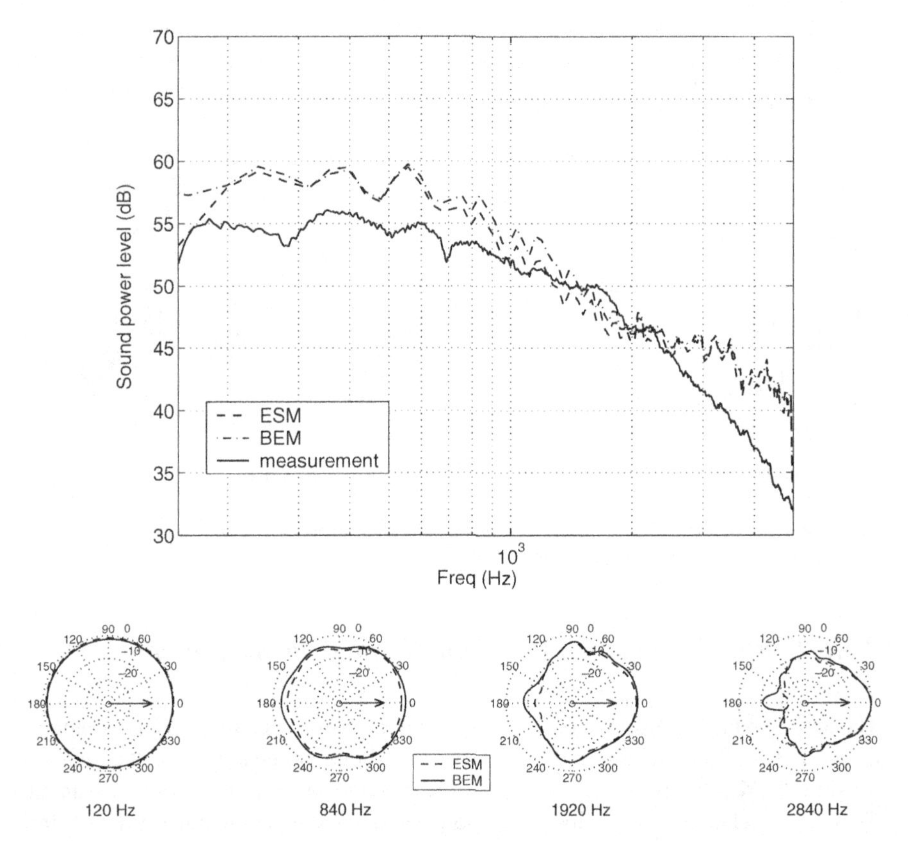

Fig. 4.15 *Upper plot:* Simulated and measured sound power level of the HD-flame. *Lower plot:* Directivity patterns of the flames at four different frequencies. The arrows indicate the direction of the gas outflow.

The normal vector $\mathbf{n}(\mathbf{y})$ points outside Ω. Comparing this expression with Eq. (4.6), we see that in Eq. (4.14) there is an additional volume integral over the source term q. The additional integral eliminates the advantage of the BEM approach of working only with surface integrals. For an acoustic calculation in frequency domain, the volume integral has to be evaluated for each field point and each frequency, increasing the computation time considerably. But if the Dual Reciprocity BEM (DRBEM) is applied, the evaluation of the volume integral is avoided by replacing it by a sum of surface integrals, as it is shown in the next subsection. For each frequency, a larger system of equations has to be solved only once and for the far field calculation, Eq. (4.6) can be used.

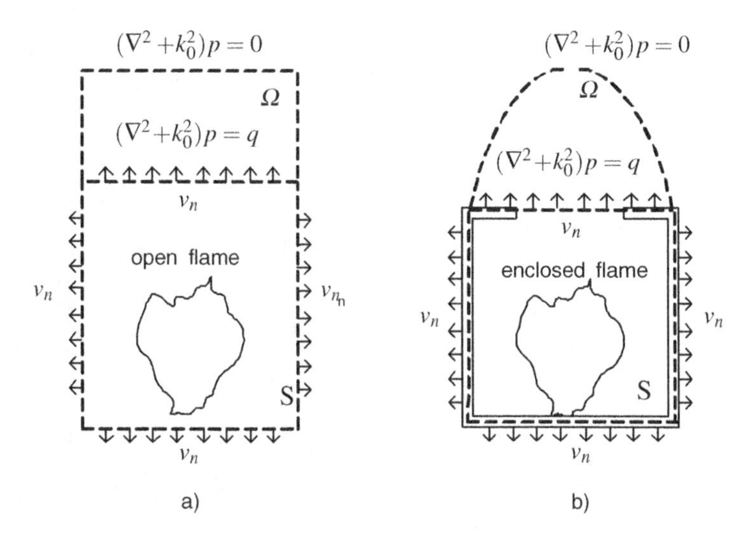

Fig. 4.16 Flame model including a non-homogeneous region Ω for an a) open flame; b) enclosed flame.

4.3.6.1 DRBEM

To find the sound field at all positions outside S, the propagation domain is divided into two regions. Region I corresponds to the volume Ω and region II to the homogeneous region. We subdivide S into two surfaces: S_0 and S_1. S_1 is the joint surface between Ω and the volume enclosed by S. S_0 is then $S \backslash S_1$. Volume Ω is enclosed by $S1 \cup S2$. The model is illustrated in Fig. 4.17 for the enclosed flame.

We have to solve a set of two differential equations, one in each region

$$\begin{aligned}(\nabla^2 + k_0^2)p_I &= q, \quad \text{in Region I,}\\ (\nabla^2 + k_0^2)p_{II} &= 0, \quad \text{in Region II.}\end{aligned} \tag{4.15}$$

with a set of boundary conditions that has to be satisfied,

$$\begin{aligned}v_{nI} &= v_S \quad \text{on } S_1,\\ v_{nII} &= v_S \quad \text{on } S_0,\\ p_I &= p_{II} \quad \text{on } S_2,\\ v_{nI} &= v_{nII} \quad \text{on } S_2.\end{aligned} \tag{4.16}$$

The integral forms of Eqs. (4.15) give us two sets of equations,

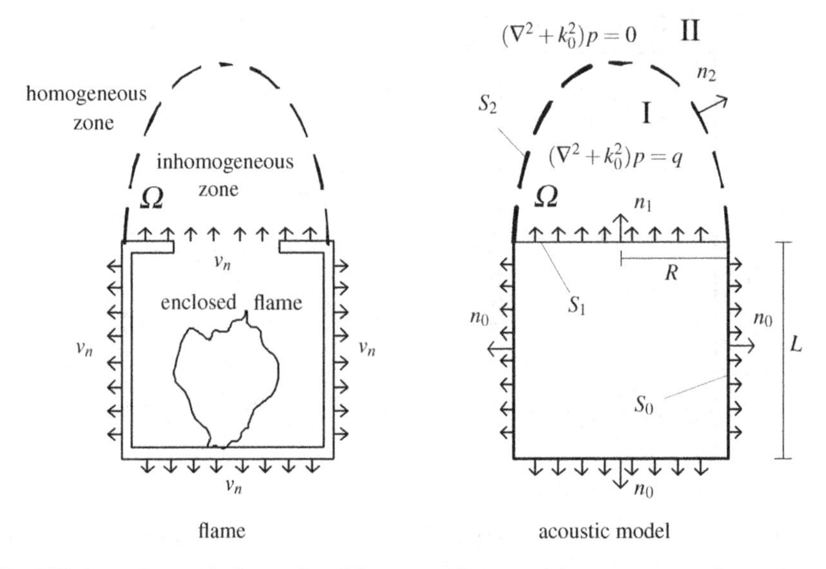

Fig. 4.17 Acoustic model of an enclosed flame considering an inhomogeneous region.

$$C_I(\mathbf{x})p_I(\mathbf{x}) = -\int_S \left(p_I^S(\mathbf{y}) \frac{\partial g(\mathbf{x},\mathbf{y})}{\partial n(\mathbf{y})} - \frac{\partial p_I^S(\mathbf{y})}{\partial n(\mathbf{y})} g(\mathbf{x},\mathbf{y}) \right) dS(\mathbf{y}) \tag{4.17}$$

$$-\int_\Omega b(\mathbf{y})g(\mathbf{x},\mathbf{y})dV(\mathbf{y}), \quad \text{Region I,}$$

$$C_{II}(\mathbf{x})p_{II}(\mathbf{x}) = \int_S \left(p_{II}^S(\mathbf{y}) \frac{\partial g(\mathbf{x},\mathbf{y})}{\partial n(\mathbf{y})} - \frac{\partial p_{II}^S(\mathbf{y})}{\partial n(\mathbf{y})} g(\mathbf{x},\mathbf{y}) \right) dS(\mathbf{y}), \tag{4.18}$$

$$\text{Region II,}$$

with

$$g(\mathbf{x},\mathbf{y}) = \frac{e^{-\mathrm{j}k_0 R}}{4\pi R}, \quad R = |\mathbf{x}-\mathbf{y}|. \tag{4.19}$$

The values of the two constants C_I and C_{II} are ”complementary”,

$$C_I = \begin{cases} 1, & \mathbf{x} \text{ in } \Omega, \\ 0.5, & \mathbf{x} \text{ on } S_1 \cup S_2, \\ 0, & \mathbf{x} \text{ outside } \Omega, \end{cases} \quad C_{II} = \begin{cases} 0, & \mathbf{x} \text{ in } \Omega, \\ 0.5, & \mathbf{x} \text{ on } S_0 \cup S_2, \\ 1, & \mathbf{x} \text{ outside } \Omega. \end{cases} \tag{4.20}$$

In the following the explicit notation of the points $\mathbf{x}$ and $\mathbf{y}$ is omitted in the formulas as far as it is possible without loss of clarity. The basic idea of the DRBEM is to expand the source distribution q in a series of functions f_j

$$q(\mathbf{x}) = \sum_j \alpha_j f_j(\mathbf{x}), \tag{4.21}$$

and to find some functions ψ_j that are solutions of an inhomogeneous Helmholtz equation with functions f_j as source term

$$(\nabla^2 + k_0^2)\psi_j = f_j. \tag{4.22}$$

The integral form of Eq. (4.22) gives an expression for the volume integral over $f_j g$ in terms of ψ_j

$$-\int_{\Omega} f_j g dV = C_I \psi_j + \int_{S_1 \cup S_2} \left(\psi_j^S \frac{\partial g}{\partial n} - \frac{\partial \psi_j^S}{\partial n} g \right) dS. \tag{4.23}$$

with $\mathbf{n}$ pointing outside Ω. By substituting Eqs. (4.23) and (4.21) into the Eq. (4.17), the final equation for p_I depending only on surface integrals is obtained

$$\begin{aligned} C_I p_I = & - \int_{S_1 \cup S_2} \left(p_I^S \frac{\partial g}{\partial n} - \frac{\partial p_I^S}{\partial n} g \right) dS \\ & + \sum_j \alpha_j \left(C_I \psi_j + \int_{S_1 \cup S_2} \left(\psi_j \frac{\partial g}{\partial n} - \frac{\partial \psi_j}{\partial n} g \right) dS \right). \end{aligned} \tag{4.24}$$

The systems of equations (4.24) and (4.18) together with the boundary conditions (4.16) have to be simultaneously solved. A matrix equation for Eqs. (4.18) and (4.24) can be obtained by discretizing surfaces S_0, S_1 and S_2. The accuracy of the results is increased if Eq. (4.24) is discretized also at L points inside Ω with $L \leq N$, where N is the number of collocation points at $S_1 \cup S_2$. Until now, the coefficients α_j have not been determined. Assuming the source distribution q is known at the N points at the surface $S_1 \cup S_2$ and at the L points in the volume Ω, and considering $M = L + N$ terms of the expansion (4.21), M coefficients α_j can be computed. They are expressed in matrix form as

$$\alpha = F^{-1} b, \tag{4.25}$$

where F is the $(M \times M)$ matrix of elements f_j and b the $(M \times 1)$ vector of the source values q. The matrix form of Eq. (4.24) is given by

$$C_I \bar{I} p_I + H_I p_I^S - G_I \frac{\partial p_I^S}{\partial n} = \left(C_I \bar{I} \psi + H_I \psi^S - G_I \frac{\partial \psi^S}{\partial n} \right) \alpha, \tag{4.26}$$

where H_I and G_I are again the system matrices and ψ and $\partial \psi / \partial n$ are matrices of the associated functions. A more detailed description of the matrix equations can be found in [31]. The matrix form of Eq. (4.18) is

$$C_{II}\bar{I}p_{II} - H_{II}p_{II}^{S} = -G_{II}\frac{\partial p_{II}^{S}}{\partial n}. \tag{4.27}$$

This technique was tested using a "spherical flame" with different source distributions. The source terms were chosen dependent only on the distance to the centre of the sphere because in that case, analytical solutions can be found for certain source functions. The numerical solutions showed very good agreement with the analytical ones [28, 31].

4.3.6.2 Treatment of Irregular Frequencies

As already mentioned in section 4.2.3, the Helmholtz integral equation for an exterior problem does not have a unique solution at characteristic frequencies related to resonances inside the boundary S [4]. To avoid problems at these irregular frequencies, the CHIEF method was implemented in the BEM routine. The number of CHIEF points required to avoid the irregular frequencies increases with the frequency, hence, the size of the matrix to be inverted and the computation time also increases. Another approach to tackle this problem is the Burton and Miller method (B&M). The method consists of a linear combination of the standard Helmholtz integral equation (4.6) and its normal derivative. Both equations are combined through a coupling constant whose optimal value have been found for high frequencies [1]. Since constant elements are being considered, the approach of Osetrov and Ochmann [25] was adopted and implemented. Following this procedure, the factor C_{II} is written as

$$C_{II} = 1 + \int_{S_0 \cup S_2} \frac{\partial G}{\partial n} dS \tag{4.28}$$

with $G = 1/(4\pi|\mathbf{x}-\mathbf{y}|)$. After inserting (4.28) in (4.18), the term $p_{II} \int_{S_0 \cup S_2} \frac{\partial g}{\partial n} dS$ is subtracted from both sides of the equation. Then, the resulting equation is differentiated in the direction of the normal $\mathbf{n}'$ to obtain the following equation

$$\begin{aligned} C_{II}\frac{\partial p_{II}}{\partial n'} + \int_{S_0 \cup S_2} \frac{\partial p_{II}^{S}}{\partial n}\frac{\partial g}{\partial n'} dS = & \int_{S_0 \cup S_2} (p_{II}^{S} - p_{II})\frac{\partial^2 g}{\partial n' \partial n} dS \\ & + p_{II} \int_{S_0 \cup S_2} \left(\frac{\partial^2 g}{\partial n' \partial n} - \frac{\partial^2 G}{\partial n' \partial n} \right) dS. \end{aligned} \tag{4.29}$$

Vector $\mathbf{n}'$ represents the normal at point $\mathbf{x}$, while $\mathbf{n}$ denotes the normal direction at point $\mathbf{y}$. $\partial/\partial n'$ and $\partial/\partial n$ are the normal derivatives. Discretizing (4.29) via collocation at the surface $S_0 \cup S_2$ ($C_{II} = 1/2$) leads to a matrix equation complementary to (4.27)

$$\left(\frac{1}{2}\bar{I} + H_{II}' \right) \frac{\partial p_{II}^{S}}{\partial n} = K p_{II}^{S}. \tag{4.30}$$

H'_{II} consists of the terms $\partial g/\partial n'$, while K comprises $\partial^2 g/\partial n'\partial n$ and $\partial^2 G/\partial n'\partial n$. Eqs. (4.27) and (4.30) are combined through a coupling constant χ that is chosen to be $\chi = j/k_0$, and a new matrix equation relating p^S_{II} and $\partial p^S_{II}/\partial n$ is obtained

$$\left(\frac{1}{2}\bar{I} - H_{II} - \chi K\right) p^S_{II} = -\left(G_{II} + \chi\left(\frac{1}{2}\bar{I} + H'_{II}\right)\right)\frac{\partial p^S_{II}}{\partial n}. \tag{4.31}$$

Finally, the variables p^S_I, $\partial p^S_I/\partial n$, p^S_{II} and $\partial p^S_{II}/\partial n$ are calculated solving simultaneously the matrix equations (4.26) and (4.31), together with relations (4.16) and (4.25). The effectiveness of the B&M method is shown in the next numerical example.

4.3.6.3 Example: The Enclosed Flame

Open flames are important because they allow studying and understanding the mechanisms of sound generation due to combustion processes. However, in practical applications flames are often located inside a combustion chamber. In this situation, the interaction between combustion and sound waves is stronger than in case of open flames and it constitutes a central subject of study for many researchers. From the perspective of noise control, the prediction of the sound field radiated to the outside of the combustion chamber can be considered as a main goal. The total sound field radiated from the enclosed flame is caused a) by the sound waves leaving the combustion chamber through the outlet (exhaust noise) and b) by the oscillations of the chamber walls induced by the interior acoustic and flow field.
The corresponding acoustic model (Fig. 4.17) consists of

- a closed surface $S_0 \cup S_1$ representing the system of the flame and the chamber. It is assumed that the normal velocity of the walls and the velocity of the fluid at the chamber exhaust are known,
- a region with an inhomogeneous medium of volume Ω outside the chamber exit, limited by the surface $S_1 \cup S_2$,
- an unbounded homogeneous medium surrounding the surface $S_0 \cup S_2$.

Flame and Combustion Chamber

In this example, a cylindrical combustion chamber of length L and radius R is assumed. The chamber walls are assumed to be rigid, thus, no vibrations arise and the normal velocity at the walls is zero. This assumption allows us to observe more clearly the effects of the temperature distribution on the radiated sound, but does not imply a limitation on the model, since an arbitrary velocity distribution at the walls can be taken into account. At the exit, a radial velocity distribution of the form

$$v_n(r) = e^{-r^2/R^2}, \quad 0 \leq r \leq R,$$

is imposed, presuming that the velocity of the fluid has a maximum at the axis. A similar velocity distribution can be found at the outflow area resulting from a Large-Eddy-Simulation of open turbulent flames, see Fig. 5 in [30].

Inhomogeneous Zone

An inhomogeneous region due to a locally varying temperature distribution with axial symmetry is considered. The hot region is assumed to have a length L_T, and its temperature is described by the function

$$T(x,r) = T_m e^{-\mu A r^2/(x_0 - x)}. \tag{4.32}$$

A and x_0 are constants and $\mu = \ln(T_m/T_a)$, where T_m and T_a correspond to the maximum and ambient temperature. In Fig. 4.18, the temperature distributions for two values of T_m are shown. The variation of the sound speed and the density with the

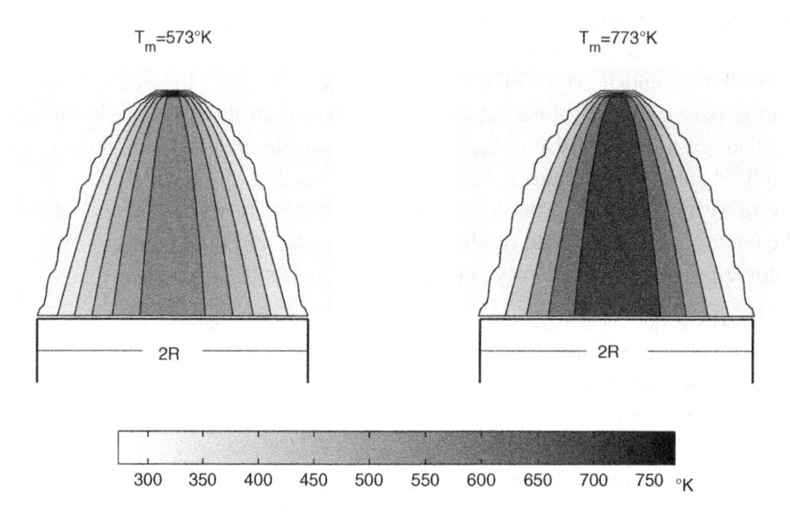

Fig. 4.18 Temperature distribution at the chamber exit.

temperature was obtained by using the relations

$$c = 20.05\sqrt{T(^\circ K)}, \quad \rho = 360.77819\, T(^\circ K)^{-1.00336}.$$

Wave Equation and Source Term

Considering only a temperature gradient, the differential equation describing the wave propagation in the hot region is given by [33]

$$\frac{1}{c^2}\nabla\cdot(c^2\nabla p_I)+k^2 p_I=0. \tag{4.33}$$

Inserting the relation for perfect gases $c^2=\gamma R_g T$ in (4.33), with constant γ and R_g, leads to

$$\nabla p_I+k^2 p_I+\frac{\nabla T}{T}\cdot\nabla p_I=0, \tag{4.34}$$

where the dot denotes the scalar product. The wave number $k=\omega/c$ also depends on the temperature. By adding and subtracting the term $k_0^2 p_I$ and rearranging the terms we obtain

$$\nabla p_I+k_0^2 p_I=q, \tag{4.35}$$

$$q=(k_0^2-k^2)p_I-\frac{\nabla T}{T}\cdot\nabla p_I. \tag{4.36}$$

For the homogeneous region, the differential equation is again the homogeneous Helmholtz equation (4.15). The integral equations are Eqs. (4.17) and (4.18). The source term (4.36) contains the unknown variable p_I and also its derivatives. In order to deal with the partial derivatives, p_I will also be expanded in a series of functions. The expansion functions can be the same functions f_j – as used for the source in Eq. (4.21) – but can be also other functions. For the expansion functions of p_I there is more freedom of choice, since they are not related to other functions like f_j which are associated to the functions ψ_j by (4.22).
If the expansion functions are chosen as d_j, p_I is given by

$$p_I(\mathbf{x})=\sum_j \beta_j d_j(\mathbf{x}), \tag{4.37}$$

the gradient of p_I is expressed as $\nabla p_I=\sum_j \beta_j \nabla d_j$ and

$$\frac{\nabla T}{T}\cdot\nabla p_I=\sum_j \beta_j u_j, \quad u_j=\frac{\nabla T}{T}\cdot\nabla d_j. \tag{4.38}$$

By discretising (4.37) and (4.38) at the same $M=N+L$ points as (4.24), a matrix expression for the source term b can be obtained

$$b=(k_0^2-k^2)p_I-UD^{-1}p_I, \tag{4.39}$$

where U is a matrix with elements $U(l,m)=u_m(\mathbf{x}_l)$ and D is a matrix with elements $D(l,m)=d_m(\mathbf{x}_l)$.

Calculation and Results

For the numerical computation of the sound field of the enclosed flame, the combustion chamber is modelled by a cylinder of length $L=0.5$ m and a radius of

$R = 0.22$ m. The cylinder has 768 elements. The inhomogeneous region is limited by a paraboloid of revolution. The effect of the temperature distribution is studied by varying T_m and L_T. Three different values of L_T are considered: $0.7R$, $1.4R$ and $2R$. The maximum temperatures investigated are 373°K, 573°K and 773°K. For $L_T = 0.7R$, the surface S_2 has 224 elements, for $L_T = 2R$, the surface S_2 has 448 elements. The number of interior points for the approximation of the source term is 200 for the shorter hot region and 400 for the larger one. The surface models and interior points are shown in Fig. 4.19. The interior points are regularly distributed at concentric circles in such a way, that the point density does not vary much in the inhomogeneous domain. The chosen expansion functions f_j and d_j are radial

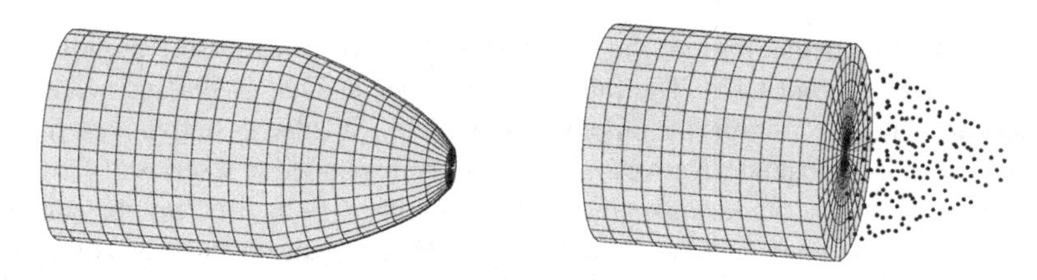

Fig. 4.19 Surface models and interior points ($L_T = 1.4R$).

functions as used in [26]

$$\begin{aligned} f_j(\mathbf{x}) &= 1 + r_j, \\ d_j(\mathbf{x}) &= 1 + r_j^3, \qquad r_j = |\mathbf{x} - \mathbf{y}_j|, \end{aligned} \tag{4.40}$$

where $\mathbf{x}$ refers to the field point and $\mathbf{y}_j$ to one of the points at the surface or in the interior volume. The associated functions ψ_j are solutions of the inhomogeneous Helmholtz equation (4.22)

$$\psi_j(\mathbf{x}) = \frac{1 + r_j}{k_0^2} - \frac{2}{k_0^4}\left(\frac{1 - \cos(kr_j)}{r_j}\right). \tag{4.41}$$

For this calculation, the B&M method has been implemented. Fig. 4.20 shows the spectrum of the sound power without B&M and with B&M. Without B&M (left plot), the curve of the sound power is not smooth at middle and high frequencies, and at two distinct frequencies $f_1 = 1320$ Hz and f_2=1920 Hz the sound power shows a jump. With the application of the B&M method (right plot), a smooth sound power curve with no irregular frequencies is obtained. The radiation patterns are also wrongly estimated at these frequencies if B&M is not applied. Fig. 4.21 presents the polar distribution of the sound pressure around the chamber at a distance of 5 m from the centre of the cylinder for the frequency f_1. The effect of the temperature distribution on the sound field of the enclosed flame is presented in Figs. 4.22-4.25,

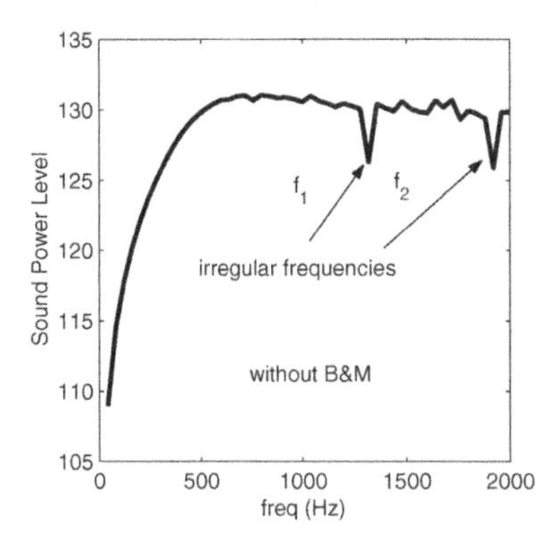

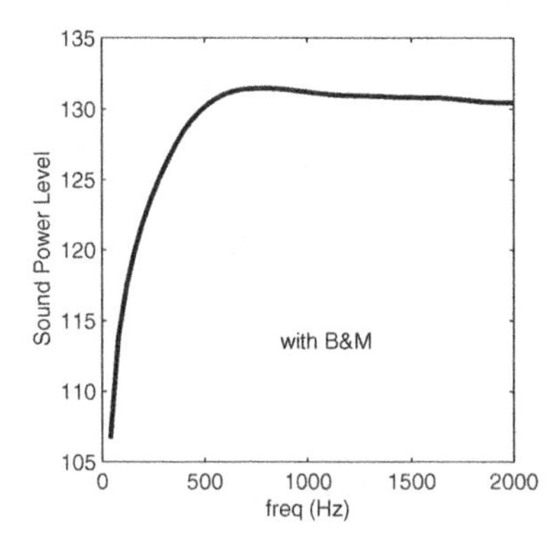

Fig. 4.20 Sound power without and with B&M.

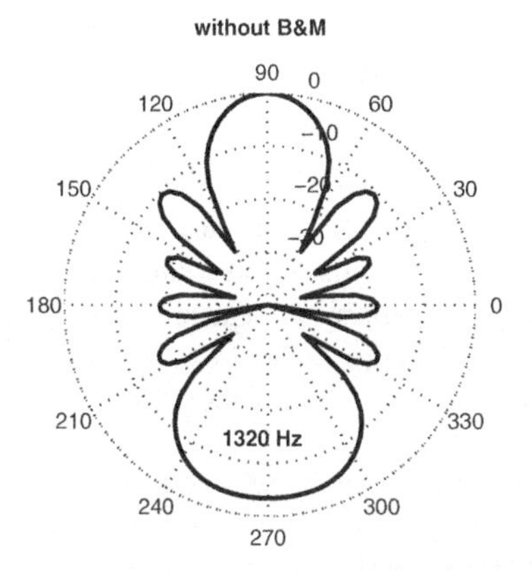

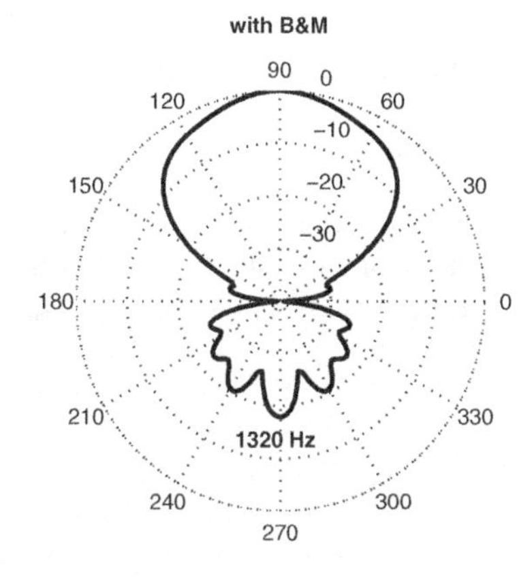

Fig. 4.21 Radiation pattern without and with B&M.

where all curves are compared to the case without any hot region. In Fig. 4.22, the curves of the sound power for three different temperature distributions are plotted. The upper curve corresponds to the case $\nabla T = 0$. The sound power is not affected by the ambient temperature field at very low frequencies. At middle and high frequencies the sound power decreases with increasing temperature of the hot region. This effect can be explained by considering that more energy is reflected back into the hot region if the temperature is increased. The length of the hot region appears to have less influence on the sound power than the temperature itself. In Fig. 4.23, the sound power for $T_m = 773°$K for three different values of L_T is shown. Only small differences can be seen especially at high frequencies. The influence of the temperature on the radiation patterns is illustrated in Fig. 4.24. The polar plots show the sound pressure level in the far field normalised by the maximum value at the axis. It can be observed that the higher the temperature, the broader the radiation patterns become. This effect is caused by the refraction of the sound waves in the hot region

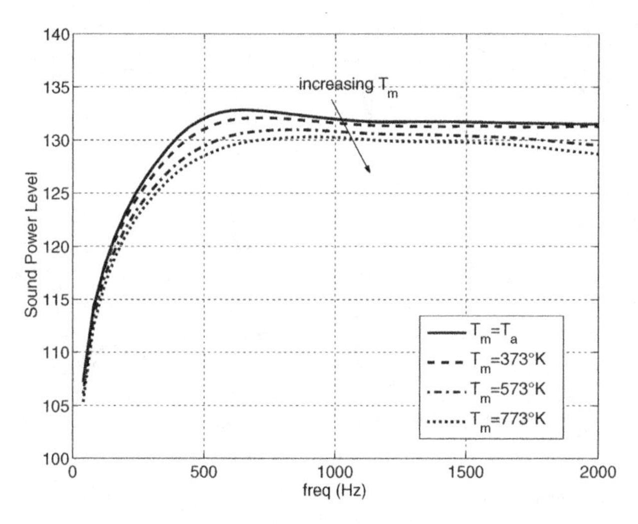

Fig. 4.22 Dependence of the sound power on the temperature T_m.

due to the changing sound speed. Such refraction effects were also reported from other numerical simulations [27] and from measurements of the sound radiation of open, turbulent flames [12, 13, 35]. The radiation patterns appear to be more sensi-

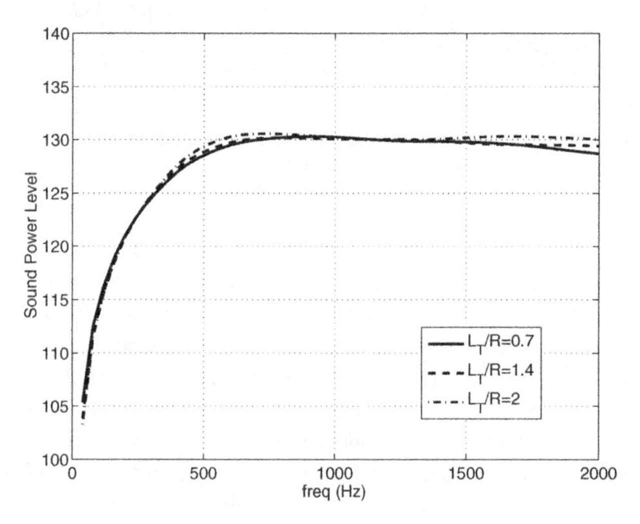

Fig. 4.23 Dependence of the sound power level on the length of the hot region L_T.

tive to the length of the hot region than the radiated sound power. The larger the hot region, the broader the radiation pattern becomes. This effect is shown in Fig. 4.25. The polar plots are calculated at two frequencies for $T_m = 773°$K.

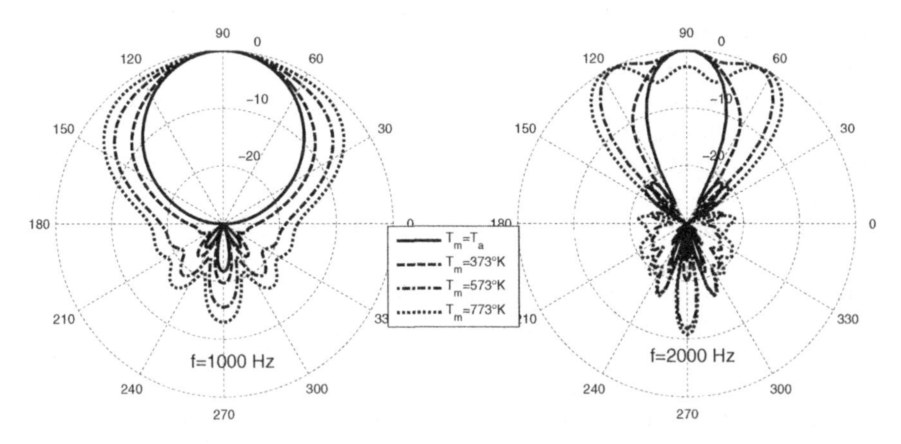

Fig. 4.24 Dependence of the directivity on the temperature T_m.

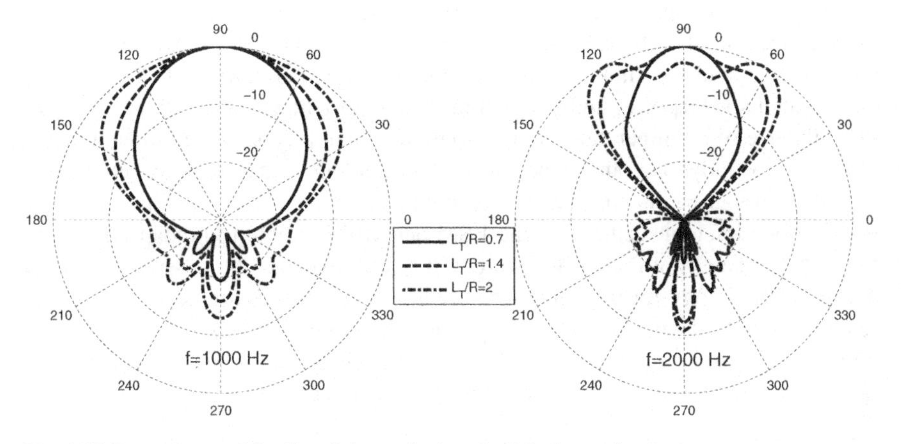

Fig. 4.25 Dependence of the directivity on the length of the hot region L_T.

4.4 Conclusions

In this paper, a hybrid approach for the prediction of the sound radiation of flames, which couples a incompressible Large Eddy Simulation (LES) with methods of computational acoustics, namely the Equivalent Source Method (ESM) and the Boundary Element Method (BEM) is presented. The coupling is performed at a control surface surrounding the flame, where the output data of the LES code are sampled and serve as boundary conditions for the acoustic methods. The location of this control surface as well as its discretization strongly influences the accuracy of the acoustic calculations. To reduce the large variance of the LES data, a procedure is introduced, similar to an ensemble average of the original time series. The influence of a half space boundary on the sound radiation is briefly discussed. A validation of the hybrid approach is made by comparison of the simulated results with measurement data. A reasonable agreement of the radiated sound power of the turbulent flame in the frequency domain is obtained. But still, some aspects need more study. An elaborate spatial filtering technique, which remedy aliasing effects due to a spatial downsampling of the LES data should be developed. Also, passing vortical structures of the flow are expected to induce a high velocity level at the closing disk of the control surface. The elimination of these vortical perturbations could also improve the accuracy of the velocity boundary condition and consequently the accuracy of the sound field solution. Work is in progress for generating such a splitting technique. In the second part of the paper, an extension of the presented hybrid approach has been developed by means of the Dual Reciprocity Boundary Element Method (DRBEM) to handle configurations, where the sound has to propagate through an inhomogeneous medium. The DRBEM has been applied to study the effect of an temperature field at the exit of a generic combustion chamber. The method accounts for the reflection and refraction effects that are expected in this configuration and the simulation shows, that high temperature gradients decrease the radiated sound power and broaden the radiation patterns of the flame. Besides the influence of a temperature field, the DRBEM can be considered to be a useful technique to study also other types of inhomogeneities in the propagation domain. All in all, it can be stated, that the comparatively economical coupling of a LES with surface integral methods as the ESM and BEM is able to predict effectively the sound radiation of flames and the DRBEM still expands the possibilities of the hybrid methodology for the numerical simulation of combustion noise. Further research should be devoted to the investigation of an combination of half space setting with finite impedance boundary condition and the DRBEM for the simulation of the radiated sound of real flames enclosed by combustion chambers.

Acknowledgements The authors gratefully acknowledge the financial support of the German Research Foundation (DFG) through the research unit FOR 486 "Combustion Noise".

References

[1] Amini S, Harris PJ, Wilton DT (1992) Coupled Boundary and Finite Element Methods for the Solution of the Dynamic Fluid-Structure Interaction Problem. No. 77 in Lecture Notes in Engineering, Springer, Berlin

[2] Brick H, Piscoya R, Ochmann M, Költzsch P (2004) Modelling of combustion noise with the boundary element method and equivalent source method. In: Proc. of the 33rd International Congress and Exposition on Noise control -

internoise 2004, Prague, Czech Republic
[3] Bui TP, Ihme M, Meinke M, Schröder W, Pitsch H (2008) Numerical applicability of different sound source formulations to compute combustion noise using acoustic perturbation equations for reacting flows. In: 14th AIAA/CEAS Aeroacoustics Conference (29th AIAA Aeroacoustics Conference), Vancouver, British Columbia Canada, AIAA 2008-2948
[4] Burton A, Miller G (1971) The application of integral equation methods to the numerical solution of some exterior boundary problems. Proc Roy Soc Lond 323:201–220
[5] von Estorff O (2000) Boundary Element Acoustics: Advances and Applications. WIT Press, Southhampton / Boston
[6] Flemming F (2007) Investigation of combustion noise using a LES/CAA hybrid approach. Proc Combust Inst 31:3189–3196
[7] Flemming F (2007) On the simulation of noise emissions by turbulent nonpremixed flames. PhD thesis, Institute for Energy and Powerplant Technology, TU Darmstadt, Germany
[8] Flemming F, Nauert A, Sadicki A, Janicka J, Brick H, Piscoya R, Ochmann M, Költzsch P (2005) A hybrid approach for the evaluation of the radiated noise from a turbulent non-premixed jet flame based on large eddy simulation and equivalent source and boundary element methods. In: Proc. of the Twelfth International Congress on Sound and Vibration ICSV12, Lisbon, Portugal
[9] Holste F (1997) An equivalent source method for calculation of the sound radiated from aircraft engines. J Sound Vib 203:667–695
[10] Hüttl T, Wagner C, Delfs J (eds) (2002) LES for Acoustics, Göttingen, Germany
[11] Kempf A, Flemming F, Janicka J (2005) Investigation of lengthscales, scalar dissipation and flame orientation in a piloted diffusion flame by LES. Proc Combust Inst 30:557–565
[12] Lenze B, Pauls D (1974) Die Entstehung und Bekämpfung von Brennergeräuschen. gwf-gas/erdgas 115(2):48–54
[13] Lorenz I (1971) Verbrennungsgeräusche. gwf-gas/erdgas 112(8):367–375
[14] Lyrintzis AS (2003) Surface integral methods in computational aeroacoustics - from the (CFD) near-field to the (acoustic) far field. Int J Aeroacoust 2(2):95–128
[15] Manoha E, Elias G, Troff B, Sagaut P (1999) Towards the use of boundary element method in computational aeroacoustics. In: 5th AIAA/CEAS Aeroacoustics Conference, Bellevue, WA, USA
[16] Marburg S (2002) Six boundary elements per wavelength. Is that enough? J Comput Acoust 10(1):25–51
[17] Montavon C, Jones I, Szepessy D, Henriksson R, el Hachemi Z, Dequand S, Piccirillo M, Tournour M, Tremblay F (2002) Noise propagation from a cylinder in a cross flow: comparison of SPL from measurements and from a CAA method based on a generalised acoustic analogy. In: IMA Conference on Computational Aeroacoustics, University of Greenwich - Maritime Greenwich Campus

[18] Morino L, Gennaretti M, Iemma U, Salvatore F (1998) Aerodynamics and aeroacoustics of wings and rotors via BEM - unsteady, transonic, and viscous effects. Comput Mech 21(4/5):265–275

[19] Ochmann M (1995) The source simulation technique for acoustic radiation problems. Acustica 81:512–527

[20] Ochmann M (1999) The full-field equations for acoustic radiation and scattering. J Acoust Soc Am 105:2574–2584

[21] Ochmann M (2004) The complex equivalent source method for sound propagation over an impedance plane. J Acoust Soc Am 116(6):3304–3311

[22] Ochmann M, Brick H (2008) Acoustical radiation and scattering above an impedance plane. In: Marburg S, Nolte B (eds) Computational Acoustics of Noise Propagation in Fluids. Finite and Boundary Element Methods, Sringer Verlag, Berlin, pp 459–494

[23] Ochmann M, Mechel F (2002) Analytical and numerical methods in acoustics. In: Mechel F (ed) Formulas of Acoustics, Springer Verlag, Berlin / Heidelberg, pp 930–1005

[24] Oppenheim AV, Schafer RW, Buck JR (2004) Zeitdiskrete Signalverarbeitung, Pearson Studium, München, chap 4

[25] Osetrov AV, Ochmann M (2005) A fast and stable numerical solution for acoustic boundary element method equations combined with the Burton and Miller method for models consisting of constant elements. J Comput Acoust 13(1):1–20

[26] Perrey-Debain E (1999) Analysis of convergence and accuracy of the DRBEM for axisymmetric Helmholtz-type equation. Eng Anal Bound Elem 23:703–711

[27] Perrey-Debain E, Boineau P, Gervais Y (1999) A numerical study of refraction effects in combustion-generated noise. In: Proc. of the Sixth International Congress on Sound and Vibration ICSV 6, Copenhagen, Denmark

[28] Piscoya R, Ochmann M (2007) Sound propagation in a region of hot gas using the DRBEM. In: Proc. of the 14th International Congress on Sound and Vibration ICSV 14, Cairns, Australia

[29] Piscoya R, Brick H, Ochmann M, Költzsch P (2005) Numerical aspects of the Equivalent Source Method applied to combustion noise. In: Proc. of the 12th International Congress on Sound and Vibration ICSV 12, Lisbon, Portugal

[30] Piscoya R, Brick H, Ochmann M, Költzsch P (2008) Equivalent Source Method and Boundary Element Method for calculating combustion noise. Acta Acustica united with Acustica 94(4):514–527

[31] Piscoya R, Brick H, Ochmann M (submitted) Determination of the far field sound radiation from flames using the dual reciprocity boundary element method. Acta Acustica united with Acustica

[32] Polacsek C, Burguburu S (2003) CFD-BEM coupling for computing noise radiated from engine inlets. In: Proceedings Computational Aeroacoustics - Euromech 449, Chamonix, France

[33] Rienstra SW, Hirschberg A (2005) An introduction to acoustics. Tech. rep., Eindhoven University of Technology

[34] Schenck H (1968) Improved integral formulation for acoustic radiation problems. J Acoust Soc Am 44:41–58
[35] Smith TJB, Kilham JK (1963) Noise generation by open turbulent flames. J Acoust Soc Am 35(5):715–724
[36] Welch PD (1967) The use of the fast fourier transform for the estimation of power spectra: A method based on time averaging over short, modified periodograms. IEEE Trans Audio Electroacoust 15(2):70–73
[37] Wrobel LC (2002) Application in Thermo-Fluids and Acoustics, The Boundary Element Method, vol 1. John Wiley & Sons
[38] Wu T (ed) (2000) Boundary Element Acoustics: Fundamentals and Computer Codes. WIT Press, Southampton, Boston
[39] Zorumski WE (1992) Classic theoretical approaches to computational aeroacoustics. In: Hardin J, Hussaini MY (eds) Computational Aeroacoustics, Springer, Berlin, pp 41–49

Chapter 5
Investigation of the Correlation of Entropy Waves and Acoustic Emission in Combustion Chambers

Friedrich Bake, André Fischer, Nancy Kings and Ingo Röhle

Abstract The entropy noise mechanism was experimentally investigated under clearly defined flow and boundary conditions on a dedicated test setup. Previous experimental research on the topic of entropy noise could draw only indirect conclusions on the existence of entropy noise due to the complexity of the physical mechanism. In order to reduce this complexity, a reference test rig has been set up within this work. In this test rig well controlled entropy waves were generated by electrical heating. The noise emission of the entropy waves accelerated in an adjacent nozzle flow was measured accurately and therewith an experimental proof of entropy noise could be accomplished. In addition to this, a parametric study on the quantities relevant for entropy noise was conducted. The results were compared to a one-dimensional theory by Marble & Candel. In a next step investigations on a combustor test rig showed a broadband noise generation mechanism in the frequency range between 1 and 3.2 kHz. The combustor rig was set up with a similar outlet-nozzle geometry like the reference test rig (EWG) and provided therefore outlet-boundary conditions like in real-scale aero-engines (outlet Mach number = 1.0). It was found that this broadband noise has a strong dependency on the nozzle Mach number in the combustor outlet. The summed-up broadband sound pressure level increases exponential with the nozzle Mach number. However, investigations of comparable cold flow conditions did not show this behavior. Since the results of the reference experiment with artificially generated entropy waves did not show this exponential increase with the nozzle Mach number, this leaves the conclusion that this additional noise is generated by the interaction of small-scale fluctuations, e.g. in entropy or vorticity, with the turbulent nozzle flow in the combustion chamber outlet nozzle.

Friedrich Bake · André Fischer · Nancy Kings
German Aerospace Center (DLR), Institute of Propulsion Technology, Department of Engine Acoustics, e-mail: friedrich.bake@dlr.de

Ingo Röhle
German Aerospace Center (DLR), Institute of Propulsion Technology, Turbine Department

A. Schwarz, J. Janicka (eds.), *Combustion Noise*, DOI 10.1007/978-3-642-02038-4_5,

5.1 Introduction

The total noise emitted by a combustion chamber consists of direct and indirect combustion noise. Only the direct combustion noise is directly related to the combustion process. Indirect combustion noise also called entropy noise is related to the acceleration of gas temperature nonuniformities which result from the unsteady combustion processes. Since the nozzle guide vanes (NGV) of the first turbine stage are choked under almost every relevant operating condition of aero-engines, hot spots passing through the nozzle are connected with mass flow variations (monopole sound source) and also with momentum flux variations (dipole sound source). Gas temperature nonuniformities may cause also broadband noise in all turbine stages, since the related density fluctuations cause pressure fluctuations during the acceleration through each turbine stage.

Entropy noise receives increased interest by the aero-engine industry because it may have a major contribution to the total noise emission of combustion systems. With the noise reducing improvements achieved for other aero-engine components, e.g. low noise fan design and jet noise reduction by high bypass ratios, the noise concern in aero-engine developments also includes the combustion noise issue. Especially at helicopter engines, which emit almost no jet noise, the entropy noise seems to be of high importance.

The generation mechanisms and parameter dependencies of entropy noise are still not completely explained. Hence, this work within the framework of a DFG research unit on combustion noise (`http://www.combustion-noise.de`) presents in a first step investigations of entropy noise phenomena on a reference test rig called Entropy Wave Generator (EWG), where the parametric dependencies of entropy noise could be evaluated. In a second step combustion noise investigations have been conducted in a downscaled aero-engine model combustor with a similar outlet-nozzle geometry like the reference test rig. The goal of the combustor experiments is to assess if the combustion noise characteristics can be estimated and explained by the entropy noise mechanism.

5.2 Theoretical Background, Test Specification and Data Analysis

The understanding of instationary fluid flow phenomena in a medium at rest as a superposition of different physical modes of perturbation probably goes back to the beginning of modern research in fluid dynamics. The modes are characterized by their physical properties as entropy, vorticity and acoustic mode of perturbation. Later Chu and Kovasznay [20] analyzed the interaction of these modes. They found that in a medium at rest on average, the interaction would be a second order effect [20]. The entropy mode, which is silent in a constant flow may transfer energy to acoustic and vorticity mode and vice versa by the nonlinear interaction. Furthermore, when considering a flow system for energy conversion from hot or cold gases, the average flow would vary in the order of the mean state of pressure and density and the velocity variation is in the range of the speed of sound. Then, the analysis of Chu and Kovasznay [20] provides the qualitative statement, that the interaction of the order zero mean flow variation with a first order perturbation leads to a first order interaction effect. However, such a variation of the base flow is not explicitly considered in this early theory.

Thus, one of the first analytical investigations of noise generation by accelerated or decelerated entropy waves was published by Morfey [47]. This work was an extension of the Lighthill theory [42] on jet noise by the so-called "excess jet noise" caused by density inhomogeneities in a free jet, e.g. of aero-engines. Following an analytical estimation by Morfey the excess jet noise scales with the sixth power of the jet velocity. Howe [35] picked this extended Lighthill analogy up and formulated the noise generation in inhomogeneous and non-isentropic flows with an acoustic wave operator (see [35] equation 4.14).

Ffowcs Williams and Howe [31] developed 1975 an analytical solution for the sound generation of sharp-fronted or spherical pellet-like entropy fluctuations in a nozzle flow. Applying the Green function Ffowcs Williams & Howe formulated the sound propagation of entropy noise in an adjacent duct as well as under free field conditions. However, these solutions were limited to low Mach number flows. Also for low Mach number flows Lu [43] developed a one-dimensional analytical model for the prediction of entropy noise based on correlation quantities of temperature, pressure and velocity fluctuations.

The noise generation by entropy waves in nozzle and diffuser flows at higher Mach numbers was described by Marble and Candel [44] for compact elements in a one-dimensional theory. Here, the length of the nozzle or diffuser, respectively, have to be small in comparison to the regarded wave length (entropy or acoustic wave) in order to fulfill the compactness assumption. The results of this estimation will be compared with the experimental and numerical results acquired at the Entropy Wave Generator test rig (EWG).

Cumpsty & Marble [23, 24] refined and applied this one-dimensional theory on an unreeled turbine stage. In a quasi-two-dimensional (axial and azimuthal) system the turbine stage was modeled as an in axial direction infinite thin discontinuity plane where the static pressure as well as the amplitude and direction of the flow velocity is changed. One result of these investigations was a strong increase of entropy noise generation with an increase of the pressure drop over a turbine stage. Furthermore, in [22] the generation of pressure, vorticity and entropy waves in a flow with fluctuating heat release is described. That allowed to compare the amplitude of directly generated noise to the one of entropy noise in a simplified turbine stage. As a result of this analytic estimation was the indirect entropy noise dominating the direct combustion noise. First comparisons of the total sound power of several aero-engines showed a good agreement to the results of these prediction method especially for operating conditions with low jet noise contribution [23].

In general, the interaction of turbulence and combustion is identified by Strahle in a review article [55] as the main source of sound generation in combustion systems. Furthermore, at this stage the contribution of entropy noise to the total noise emission was not known mainly due to the lack of experimental work in this field.

In the seventies also the first numerical approaches concerning entropy noise arose. Mathews et al [46] compared on behalf of the Federal Aviation Administration, U.S. Department of Transportation, different prediction tools for combustion noise in aero-engines. Following this study for turbojet engines of the manufacturer Pratt & Whitney the direct combustion noise was dominating the entropy noise.

Using the method of characteristics Bloy [17] calculated numerically the pressure pulses produced by accelerated entropy waves.

Referring to combustion test rigs with open outlet conditions, i.e. without acceleration in the combustor outlet, Strahle and Muthukrishnan [56] developed in a numerical-empirical way a correlation for the total combustion sound power. However, due the restriction to open combustion chamber outlet conditions the indirect noise phenomena have been neglected.

Direct numerical simulations of turbulent combustion flows by Tanahashi et al [57] resolved the sound generated by accelerated entropy waves as the main contribution to the total noise radiation.

In the field of experimental entropy noise research only little work was published yet like Strahle [55] already mentioned. At the California Institute of Technology (Caltech) a similar test rig like the Entropy Wave Generator test rig (EWG) presented in this paper was investigated [59, 18, 19]. But in these experiments the amplitude of the induced temperature fluctuation was with approx. 1 K very low. Furthermore, due to technical restrictions at this time a post-processing of the acquired data in the time domain was not possible.

By means of coherence analysis of different sensor signals from the inside and the outside of the combustion chamber Muthukrishnan et al [49] determined the separation of the different combustion noises sources on a test rig for aero-engine combustors. The results showed a dominating broadband entropy noise contribution to the total noise spectrum. Similar experiments are described by Guedel and Farrando [34] on a helicopter engine from the manufacturer Turbomeca. Guedel identifies using a three-signal coherence technique a mainly low frequency sound source domain located between combustion chamber and low pressure turbine. Recently, microphone-array measurements of a GE aero-engine for regional aircraft (CF34-10E) in an open air test bed have been published by Martinez [45]. Comparing different acoustic damping materials for the hot stream liner a significant part of the total noise emission related to combustion noise was detected. However, the allocation of this noise origin to direct or indirect combustion noise generation mechanisms could not be determined.

During the last 20 years a lot of research was conducted in order to reveal the role of entropy waves and entropy noise with respect to the feedback mechanisms of thermoacoustic instabilities. In this context Keller et al [37] found in a linear stability analysis that entropy noise can be one of the possible feedback mechanisms. But Keller [36] also showed that the entropy noise phenomenon does not cover the entire instability region for combustion oscillations. Similar formulated Dowling [26, 27,

28] indirect combustion noise as one source term in an acoustic energy equation for combustion systems.

Using a linear stability analysis Lieuwen [40, 41] described the interaction of sound pressure waves with the flame front as the main reason for combustion instabilities in lean premixed flames. In a review paper [39] about the state of the art concerning combustion-acoustic interaction Lieuwen refers to an existing demand for experimental investigations in this field.

With respect to combustion oscillations Polifke et al [52] presented analytically the possibly constructive or destructive interference of direct combustion noise and entropy noise. But following Sattelmayer [53] the indirect combustion noise has no noticeable destabilizing effect on the stability of a simplified combustor model due to the high dispersion rate of the convecting entropy waves. In a mathematical approach Ali and Hunter [1] formulate on the other hand a possible resonant interaction between sound waves and quasi spatially fixed entropy waves.

A numerical study by Dowling [25] did explicitly exclude the entropy noise mechanism since the oscillation behavior of configurations only with open combustion chamber outlets have been taken into account. Zhu et al [58] specified as a result of a flow simulation (CFD) of a choked combustion chamber outlet system with a subsequent stability analysis entropy noise as the dominant feedback mechanism for the combustion instability of the regarded configuration.

Experimental research to evaluate the importance of entropy noise on combustion oscillations was conducted by Eckstein [30, 29] by the variation of the outlet condition (open and choked nozzle) on a combustor test rig. The investigation showed on the one hand that combustion oscillation can be also very strong in the absence of entropy noise (configuration with an open combustion chamber outlet) but on the other hand in case of the choked outlet nozzle a clear portion of entropy noise in the emitted noise spectrum exists.

Just recently the results of a numerical study by Leyko et al [38] have been published, which predict the relevance of indirect combustion noise compared to direct combustion noise to be a factor 10 larger at aero-engines.

To sum up, a clear and distinct experimental proof of the existence of entropy noise was not carried out up to now and a comprehensive parameter study on entropy noise for validation purpose of numerical and analytical models is still missing. This was the motivation of the work presented here, where the complexity of the entropy noise generation process is reduced and therewith a doubtless experimental proof of entropy noise can be provided.

5.2.1 Test Specification and Data Analysis

The experiments have been conducted at two different test rigs - at the Entropy Wave Generator and the Model Combustor Rig. In the non-reactive reference test rig, called Entropy Wave Generator (EWG) the sound emission of artificially induced entropy waves in an accelerated tube flow can be investigated. The idea of this set-up is to test and optimize detection methods for entropy noise and to study the parameter dependencies of the entropy noise generation mechanism. In addition the EWG allows with its well-defined boundary conditions the validation of numerical simulations and the comparison with theoretical considerations. The received knowledge at the EWG can be transmitted and applied to the complex combustor test rig. In this Combustor Test Rig the noise emitted by accelerated entropy perturbations in case of a reactive combustor flow can be analyzed.

5.2.1.1 Entropy Wave Generator Test Rig (EWG)

The setup of the EWG, shown in Fig. 5.1, consists basically of a straight tube flow with a heating module and a nozzle where the flow is accelerated. The flow, supplied by the laboratory compressed-air system through a mass flow controller, enters the setup into a settling chamber with a honeycomb flow straightener. From the settling chamber the flow is conducted via a bell mouth intake into the first tube section with a length of ≈ 250 mm and a diameter of 30 mm.

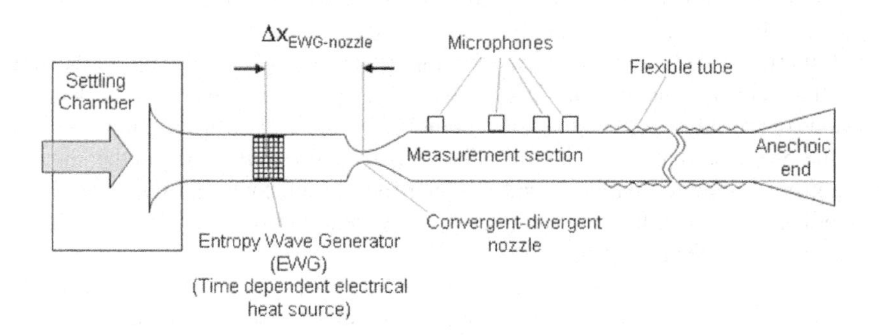

Fig. 5.1 Sketch of the **E**ntropy **W**ave **G**enerator test rig (EWG); Tube section $\Delta X_{EWG\text{-}nozzle}$ is variable, corresponding to different propagation lengths of entropy waves.

In the middle of this tube section the heating module is located. It is composed of six rings each of them with ten heating wires stretched over the cross section. The wires made of platinum have a diameter of 25 μm and summed up over the six rings an effective wire length of ≈ 1260 mm. Using custom-built electronic delay circuit it is possible to heat electrically either the six wire rings at the same time or to operate the six rings delayed corresponding to the convection time of the flow. Therewith, different edge shapes of the induced entropy waves can be produced.

After the heating module the flow is strongly accelerated through a convergent-divergent nozzle structure. Depending on the mass flux the Mach number in the nozzle throat (diameter $= 7.5$ mm) can be adjusted up to $Ma = 1$. Adjacent to the divergent nozzle part follows another tube section with a length of ≈ 1020 mm and a

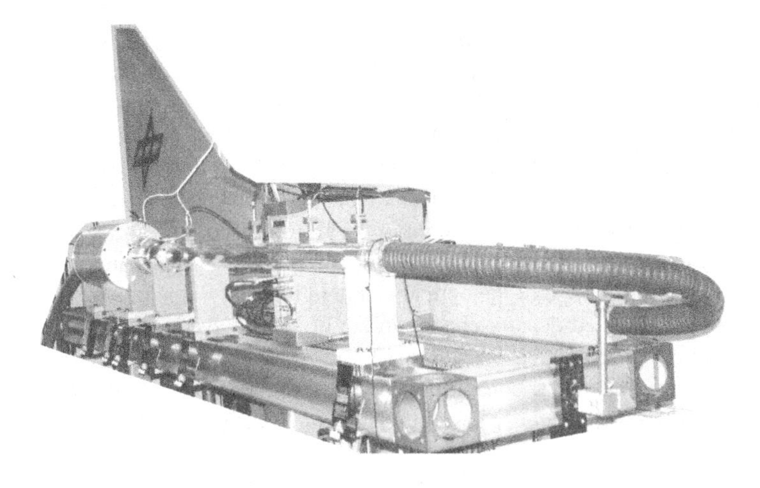

Fig. 5.2 Photo of the Entropy Wave Generator test rig (EWG).

diameter of 40 mm. In this tube section four microphones are wall-flush mounted at different axial positions. Setting the origin of the coordinate system in the plane of the most downstream heating module ring on the center axis of the tube and with the direction of the x-axis along the tube axis in downstream direction, the microphone positions are $x_{AGR1} = 456$, $x_{AGR2} = 836$, $x_{AGR3} = 1081$ and $x_{AGR4} = 1256$ mm. According to this coordinate system the nozzle throat is located at $x_{nozzle} = 105.5$ mm.

The microphone tube section is connected to a flexible tube of ≈ 980 mm and an adapter section (from round to square cross section) of ≈ 280 mm. The test rig ends with an anechoic termination with an inner square cross section (40 mm x 40 mm). A photo of the EWG test rig is shown in Fig. 5.2.

Reference Test Cases

Due to the modular setup of the EWG test rig the influence of a variety of different parameters on entropy noise generation could be investigated.

For the comparison with numerical simulations two reference cases have been selected and documented in detail. The two cases differ in the nozzle Mach number and in the amount and control of the electrically induced heating energy. Furthermore, a different entropy wave detection and determination method was applied.

In case 1 the mass flow rate was set to 42 kg/h with a corresponding nozzle Mach number of 1.0. The entropy wave was generated by the simultaneous pulse shape heating of four heating module rings (no. 3 to 6) for a pulse duration of 100 ms. This pulse excitation was repeated once per second to enable a phase averaging over 300 pulse events. The entropy wave was detected downstream of the heating module at the axial position of $x = 34$ mm by a bare wire thermocouple with a wire diameter of $1/1000$ inch. The average temperature fluctuation amplitude was ≈ 9.1 K.

The mass flow rate in case 2 was set to 37 kg/h which corresponds to a subcritical nozzle Mach number of $Ma_{\text{nozzle}} = 0.7$. In contrast to case 1 the rings of the heating module (no. 1 to 6) were heated one after the other with a delay according to the flow velocity in the tube and the distance between each heating ring. Therefore, a certain spatial sharpness and resolution of the induced entropy wave could be achieved. In case 2 the temperature pulse was measured by applying a vibrometer and evaluating the change of the optical path length through the tube resulting from the temperature caused density change in the flow. The averaged amplitude of the temperature fluctuation in case 2 was ≈ 13.4 K.

All relevant parameters concerning the flow, the excitation and the geometric conditions of the two reference cases are listed in Table 5.1.

Parameter	**Case 1**	**Case 2**
Mass flow rate	42 kg/h	37 kg/h
Nozzle Mach number	1.0	0.7
Pressure(settling chamber; between honeycomb flow straightener and tube inlet; against ambient pressure)	11.17 kPa	4.34 kPa
Pressure(nozzle; against ambient pressure)	-52.71 kPa	-32.65 kPa
Pressure(ambient)	1008 hPa	1013 hPa
Mean flow velocity in the tube (upstream of the nozzle; bulk velocity determined by pressure, mass flow rate and tube cross section)	12.18 m/s	11.39 m/s
Pulse duration	100 ms	100 ms
Pulse distance	1 s	1 s
Pulse voltage (averaged over entire pulse)	67.5 V	62.4 V
Pulse current (averaged over entire pulse)	2.1 A	3.1 A
Heating power (electrical)	143.7 W	192.7 W
Temperature increase ΔT (measured by R-Typ-Thermocouple (1/1000inch) at $x = 34$ mm for case 1 and by vibrometer at $x = 47.5$ mm for case 2	9.1 K	13.4 K
Heating power (determined with ΔT)	106.8 W	138.2 W
Heated wire rings (No.1 is the most upstream wire ring; No.6 is the closest wire ring to the nozzle (also position of $x = 0$))	No. 3 to 6 simul-taneously	No. 1 to 6 with delay according to flow velocity
x-position of the nozzle throat (x_{nozzle})	105.5 mm	105.5 mm

Table 5.1 Parameters concerning the flow, the excitation and the geometric conditions of the EWG reference test cases

5.2.1.2 Combustor Test Rig

In a next step the noise characteristics of a combustion system has been analyzed. Here, the combustion chamber was terminated with an outlet nozzle geometry very similar to the EWG reference test rig which replicates the flow conditions in the combustor outlet of a real scale aero-engine. The scope of these investigations was to identify the major noise sources by comparing the noise characteristics to the parametric results of the reference test rig.

The combustor is designed to replicate a fuel-air-mixing characteristic of a full-scale gas turbine combustor while still permitting analysis by experimental means. The thermal power can be varied from 5 kW to 20 kW at different air ratios from $\lambda = 0.8$ to $\lambda = 1.8$. These air ratios correspond to equivalence ratios from $\phi = 1.25$ to $\phi = 0.55$ applying the definition $\phi = 1/\lambda$. The axially symmetric test rig consists of three sections: the combustion chamber, the convergent-divergent outlet nozzle, and the exhaust duct. A swirl generating dual air-flow nozzle is used to drive the combustion zone with non-preheated air. Methane gas is injected as fuel through an annular slot between the air streams. The combustion chamber itself is made of a fused quartz glass or steel cylinder with 100 mm inner diameter.

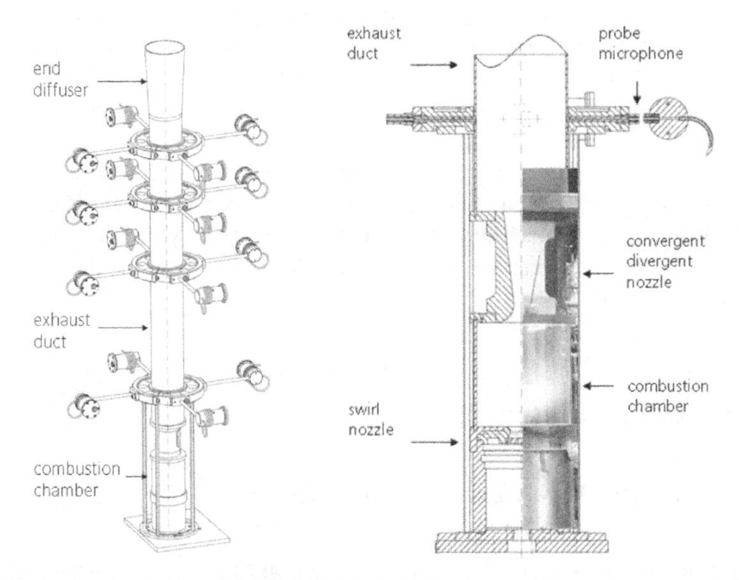

Fig. 5.3 Isometric view (left) and sketch (right) including a picture of the combustor test facility.

A convergent-divergent nozzle with a throat diameter of 7.5 mm that emulates the acceleration through the first turbine stage terminates the combustion chamber (see Fig. 5.3) The outlet nozzle (exit diameter: 40 mm) is attached to an exhaust duct with the same diameter as the combustion chamber. In order to reduce the impedance jump at the exhaust outlet, an end diffuser perforated with holes of 2 mm diameter and with increasing perforation density toward the exit is installed.

Sound pressure measurements in combustion environments are quite demanding with respect to the acoustic equipment. High temperatures, up to 2000 K, and high corrosive exhaust gases preclude the usage of classical microphone set-ups. To prevent sensor destruction a probe microphone configuration is used. Due to the spatial separation of the measurement location at the combustion chamber wall or exhaust duct and the microphone itself, common $1/4$ inch microphones can be used. A steel tube of 2 mm inner diameter connects the exhaust duct wall and the microphone. For impedance matching and for minimizing standing wave effects the probe tube is extended according to the principle of a semi-infinite acoustic duct. The microphone itself is perpendicular and wall-mounted inside the cylindrical chamber or exhaust duct. From the rear end, the probe tube is purged with cooling air with a well-controlled small flow rate. The air purging prevents damage of the microphone diaphragm by corrosive combustion products. Of course, the inherent phase shift in the collected data due to the propagation delay through the probe tube has to be corrected afterwards. Since the probe tube is finite, small reflection effects remain in the transfer function of the probe microphone.

The combustion chamber can be equipped with up to three probe microphones at different axial positions. On the exhaust duct system sixteen microphones can be installed at four axial and four circumferential positions. From the acoustic time series, the downstream and upstream propagating acoustic waves can be determined using an in-house processing code for mode analysis. Thus, the total sound power emitted by the combustion system can be determined. The frequency range, which is of importance for the investigations extents up to the cut-off frequency of the first higher mode. Therefore, only plane acoustic waves have to be taken into account. The first higher mode is an azimuthal wave at approximately 3.2 kHz, depending, on the exhaust gas temperature. Steel and quartz glass cylinders were chosen as different combustion chamber materials in order to compare the different thermal radiation conditions.

5.3 Results and Discussion

5.3.1 Entropy Wave Generator Test Rig (EWG)

The results of the two reference cases is shown in Fig. 5.4. The dashed red line in the upper half of the plots displays the time signal of the temperature fluctuation measured in case 1 with the fast thermocouple and in case 2 with the vibrometer (see also Table 5.1).

In both cases the temperature pulse with a duration of 100 ms is observable. However, the edge steepness in case 1 is decreased due to the time constant of the thermocouple. The solid black line in the lower half of the figure represents the response pressure signal of the fourth microphone in the measurement tube section downstream of the nozzle.

In both cases the pressure signal shows a positive pressure pulse by the time when the temperature pulse reaches the nozzle. This pulse is followed by a certain oscillating behavior presumably due to acoustic reflections in the test setup. At the

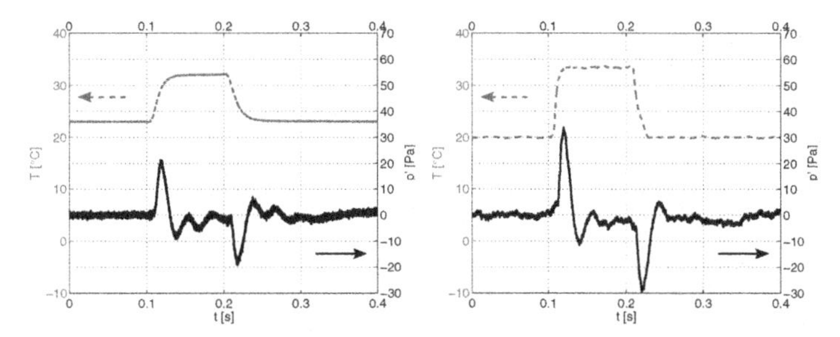

Fig. 5.4 Reference test cases 1 (left) and 2 (right): Flow temperature signal (dashed line, measured by thermocouple (case 1) or by vibrometer (case 2)) and pressure fluctuation signal at the fourth microphone position x_{AGR4} (solid line).

time when the trailing edge of the temperature pulse reaches the nozzle a negative pressure pulse is generated and detected by the microphones.

In order to estimate the necessary space and time resolution required for the numerical simulations, the spectral content of the acoustic pressure signals in the measurement tube was evaluated. Fig. 5.5 illustrates in result the power spectral density of the microphone signal at position x_{AGR4} for both reference cases 1 (solid black line) and 2 (dashed red line). It is obvious that most of the energetic spectral content of the acoustic pressure signals occurs below ≈ 100 Hz.

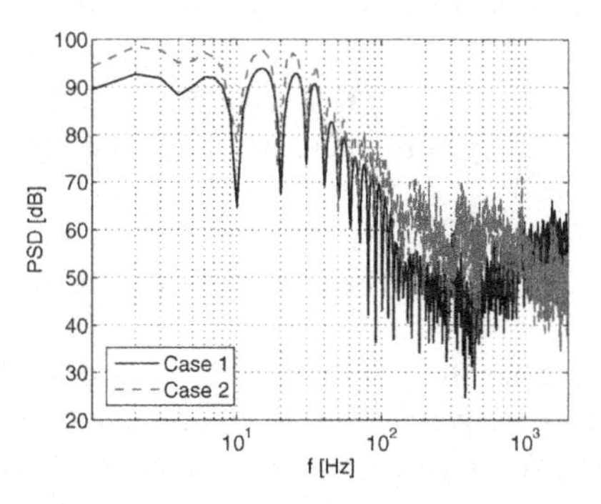

Fig. 5.5 Spectral content of the acoustic pressure signal at microphone position x_{AGR4} for both reference test cases 1 (solid black line) and 2 (dashed red line).

Besides the reference cases in the previous section the dependency of entropy noise on different parameters has been investigated. Two of the main parameters are here, on the one hand, the amplitude of the accelerated temperature fluctuation and on the other hand the Mach number in the nozzle throat.

In result Fig. 5.6 (left) shows the amplitude of the generated sound pressure pulse as a function of the amplitude of the temperature fluctuation for two different nozzle Mach numbers $Ma_{\mathrm{nozzle}} = 0.15$ and $Ma_{\mathrm{nozzle}} = 1$. For both nozzle Mach numbers the amplitude of the generated entropy noise increases linearly with the amplitude of the temperature pulse. This confirms the expected behavior found in the literature by Marble and Candel [44].

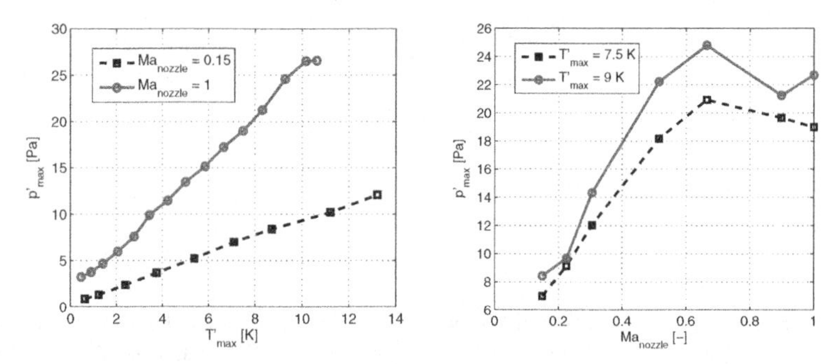

Fig. 5.6 Entropy noise over entropy wave amplitude (left) and nozzle Mach number (right): Amplitude of the generated sound pressure pulse as a function of the amplitude of the temperature fluctuation for two different nozzle Mach numbers $Ma_{\text{nozzle}} = 0.15$ and $Ma_{\text{nozzle}} = 1$ (left) and as a function of the nozzle Mach number for two different amplitudes of the temperature fluctuation $\Delta T = 7.5$ K and $\Delta T = 9$ K (right).

Fig. 5.6 (right) displays the amplitude of the generated sound pressure pulse as a function of the nozzle Mach number for two different amplitudes of the temperature fluctuation $\Delta T = 7.5$ K and $\Delta T = 9$ K. Here, the relationship is not linear as in Fig. 5.6 (left). In the nozzle Mach number regime between $Ma_{\text{nozzle}} = 0.15$ and $Ma_{\text{nozzle}} = 0.5$ the entropy noise amplitude shows a strong increase with the Mach number. However, above a nozzle Mach number of $Ma_{\text{nozzle}} \approx 0.5$ the sound pressure amplitude saturates and even decreases slightly for higher Mach numbers up to $Ma_{\text{nozzle}} = 1$. This behavior applies to both amplitudes of the temperature fluctuation for $\Delta T = 7.5$ K and $\Delta T = 9$ K, although for the temperature pulse of $\Delta T = 9$ K the entropy noise increases again slightly for the last operating point at $Ma_{\text{nozzle}} = 1$.

5.3.1.1 Comparison With Theoretical Prediction

The generation of entropy noise is described, in principle, for either a nozzle or a diffuser separately in the one-dimensional theory of Marble & Candel [44]. Since in the EWG a convergent-divergent nozzle was used, a combination of the theoretical expressions concerning the nozzle and the diffuser part of the set-up is necessary.

The downstream propagating entropy fluctuation generates a pressure pulse when it is accelerated in the convergent nozzle. Furthermore, the deceleration of the entropy wave in the diffuser generates a downstream and an upstream propagating pressure wave. The upstream propagating pressure wave is partially reflected in the nozzle throat. The total downstream propagating sound pressure wave can be compared to the experimental data measured by the microphones in the duct section downstream of the nozzle.

The comparison of the experimental microphone data with the theory of Marble & Candel [44] is shown in Fig. 5.7 displaying the entropy sound pressure amplitude normalized by the total pressure and divided by the normalized relative temperature

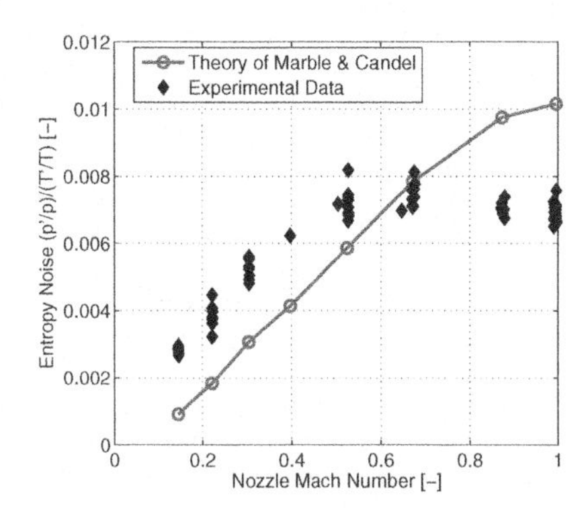

Fig. 5.7 Comparison of experimental data with theoretical prediction; Normalized entropy sound pressure over nozzle Mach number.

perturbation over the nozzle Mach number. The different temperature amplitudes investigated in the parametric study for each nozzle Mach number result in several marker points for a certain Mach number.

In the low nozzle Mach number region up to $Ma \approx 0.5$ the theoretical results are slightly lower than the acquired data whereas for higher nozzle Mach numbers between 0.5 and 0.7 the experimental data and the predicted values show a good agreement. For the choked nozzle ($Ma = 1$), the measured entropy noise amplitudes are lower than the theoretically predicted ones. The differences are not fully understood yet but a possible root of the deviation could be the one dimensional concept of the theory and the assumed compactness of the nozzle. The theory implies that the nozzle length has to be much smaller than the wavelengths of the entropy and the sound waves. This assumption may not be valid anymore, especially in case of the entropy wave which has a very short wavelength due to its low propagation speed. Furthermore, any radial velocity components and therewith correlated acceleration or deceleration occurring in the experimental nozzle flow are not included in the one-dimensional theory.

Another aspect which had not been taken into account for this comparison is the possible influence of acoustic reflection effects presumably occurring at the inlet and outlet boundaries of the experimental setup. Recent numerical investigations by Muehlbauer et al. [48] showed that these reflections cannot be neglected and should be implemented in the future.

5.3.2 Combustor Test Rig

During the experiments on the combustor test rig differences in the sound pressure levels between combustion chamber and exhaust duct especially at high Mach number outlet conditions have been observed.

Fig. 5.8 (left) compares the sound pressure levels of two microphones mounted in the combustion chamber and the exhaust duct at 15 kW thermal power and an equiv-

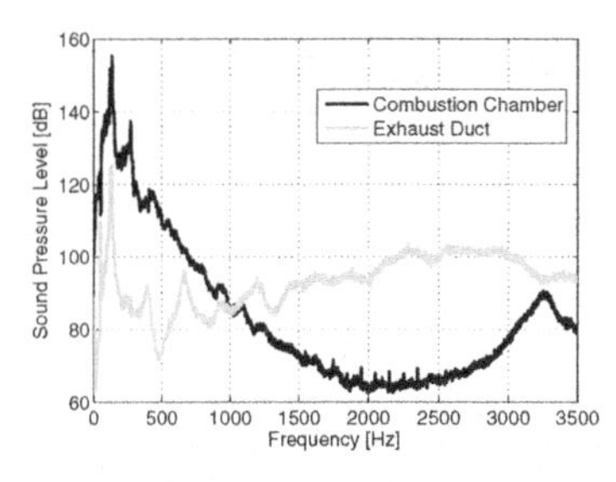

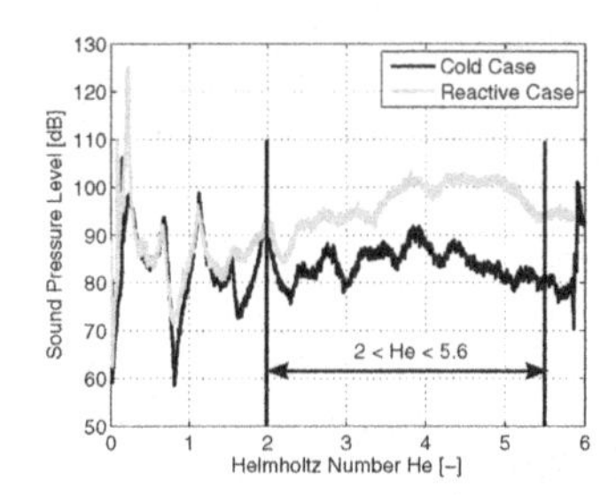

Fig. 5.8 Comparison of sound pressure levels in combustion chamber and exhaust duct at $\lambda = 1.8$ ($\phi = 0.55$) and 15 kW (left) and sound pressure levels in the exhaust duct at same Mach number outlet condition ($Ma = 1$) for reactive and isothermal conditions (right).

alence ratio of $\phi = 0.55$. It can be seen that waves of the dominating frequencies propagate from combustion chamber to the exhaust duct via the nozzle. Especially in the high frequency region (higher than 1 kHz) an increased sound pressure level in the exhaust duct is remarkable - differences up to 30 dB can be recognized. The combustion chamber is much noisier than the exhaust duct in the low frequency region because lower frequencies are dominated by direct combustion noise. Since only the outlet nozzle is installed between combustion chamber and exhaust pipe an additional noise generating mechanism can be assumed inside this nozzle producing sound in the higher frequency region.

To exclude turbulent flow noise as a possible origin for this additional noise the acoustic field of a comparable cold flow was investigated. For this kind of nozzle flows the throat Mach number is a significant parameter for turbulent flow noise and allows the comparison of hot and cold flow conditions. Nonreactive/cold conditions are established by an air flow through the combustion chamber without methane and therefore without combustion. For identical microphone positions in the exhaust duct and the same nozzle Mach numbers but different conditions in the combustion chamber broadband noise is lower in amplitude for cold conditions (see Fig. 5.8 (right)).

For a better comparison the spectra (cold and hot case) are plotted versus the Helmholtz number He instead of the frequency. This Helmholtz number was calculated with the respective speed of sound and a unit length of 1 m.

In order to achieve a better quantification of this broadband noise phenomenon the SPL's were summed up for Helmholtz numbers from 2 to 5.6 and plotted in Fig. 5.9 to show the dependency on the nozzle throat Mach number. In both images the summed up sound pressure levels in the combustion chamber and in the exhaust duct for cold respectively reactive conditions are determined and plotted versus the nozzle Mach number.

Different curves in the reactive case result from different thermal power, e.g. the first region of $Ma = 0.1$ to nearly 0.4 relates to the 5 kW case followed by the 10 kW case for $Ma = 0.45$ to 0.85. Overlapping Mach numbers can be observed in the region of $Ma = 0.6$ to 0.85. Here the 10 kW power consumption for lean combustion produces outlet conditions in the nozzle similar to those of the rich

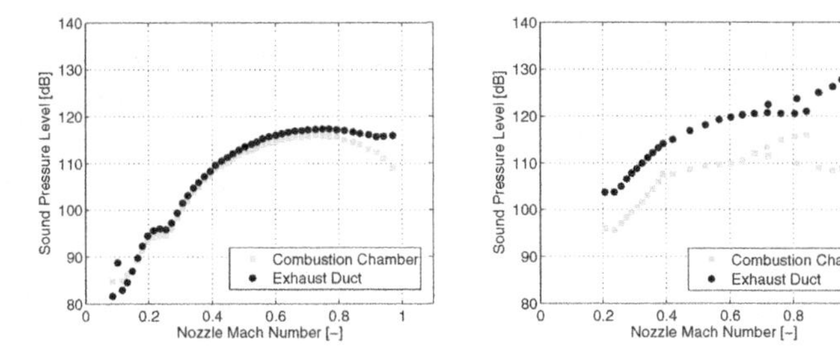

Fig. 5.9 Summed-up sound pressure levels in the combustion chamber and the exhaust duct versus Mach number for cold flow (left) and reactive flow (right) conditions.

combustion at 15 kW power consumption. For both conditions the summed up SPL's in the exhaust duct are always higher than the SPL's in the combustion chamber for Helmholtz numbers between 2 to 5.6.

A better illustration of the additional noise in the exhaust duct is given by Fig. 5.10, which displays the ratio between the sound pressure levels in exhaust duct and the combustion chamber for cold and reactive conditions. It shows a significantly higher increase in sound pressure in the exhaust duct at higher Mach numbers, up to a ratio of 20 dB. The differences between cold and reactive conditions may be explained by noise generation due to the strong acceleration of small scale inhomogeneities, e.g. in temperature or vorticity, in the combustion chamber outlet nozzle. Another explanation taken into account was the noise generation of the shear layer at the nozzle exit into the exhaust duct. The mixing of temperature fluctuations in this area generates the so called excess jet noise (see [47]), but by estimating the nozzle outlet flow velocity in the order of 30 m/s at the highest operating point, this excess jet noise generation mechanism is not strong enough to produce the additional noise of 20 dB measured in the exhaust duct.

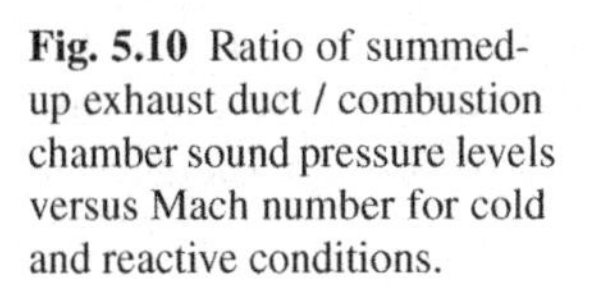

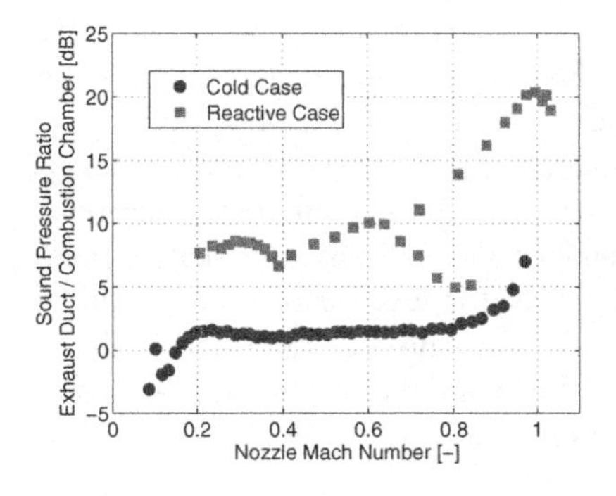

Fig. 5.10 Ratio of summed-up exhaust duct / combustion chamber sound pressure levels versus Mach number for cold and reactive conditions.

The results of summed-up sound pressure levels may be distorted by reflections or interference noise caused by the flow. A better quantity to evaluate the noise source mechanism is therefore the emitted sound power instead of the sound pressure. The sound power can be derived by decomposing the sound field into forward and backward traveling waves (see also [32]). Considering only the forward traveling plane waves in the exhaust duct the emitted sound power can be calculated. Simultaneously the influence of the reflected waves is eliminated. To overcome the turbulence noise caused by the flow the three microphone method [21] was applied to calculate the acoustic spectra.

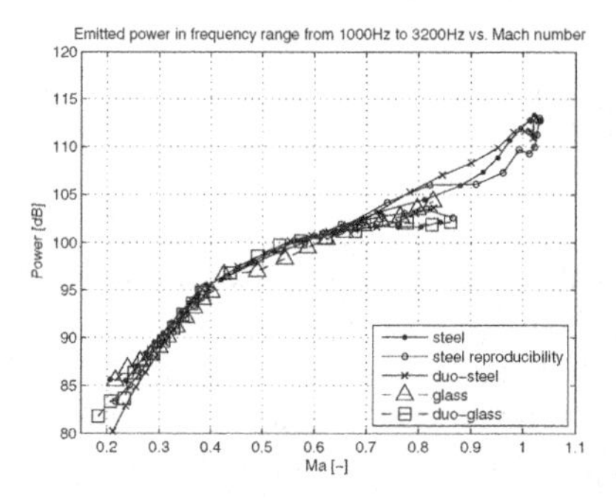

Fig. 5.11 Emitted sound power (1-3.2 kHz) in the exhaust duct versus Mach number.

Fig. 5.11 shows the dependency of the summed-up emitted sound power in the exhaust duct investigating different combustor materials and lengths at different Mach numbers. It displays a similar increase of the sound power compared to the sound pressure level. It seems that neither different combustor wall temperatures and radiation conditions, provided by different materials (steel/glass), nor different travel lengths of temperature inhomogeneities (doubled combustion chamber length allows a better mixing than the standard length) change the acoustic output power of the combustion system in this frequency range. An increase in the emitted sound power with increasing Mach number is detectable for all different combustion chamber setups.

This behavior was not expected with the assumption that the noise generating entropy and vorticity fluctuations have their origin in the primary reaction zone only. In this case doubling the length of the combustion chamber and therewith increasing the residence time of the perturbations significantly would have to reduce the noise generation due to dissipation effects on the convecting perturbations.

The fact that the emitted sound power does not change considerably with the combustor length indicates that the entropy and vorticity fluctuations result from shear layer interaction of the flow with the colder walls in the combustion chamber or the outlet nozzle. Here, the doubled combustor length would in fact increase the dissipation but it also would produce additional boundary layer turbulence in entropy and vorticity.

5.4 Conclusions

Experimental research was performed on a dedicated test setup in order to investigate the entropy noise mechanism under clearly defined flow and boundary conditions. Previous experimental research to the topic of entropy noise could draw only indirect conclusions on the existence of entropy noise due to the complexity of the physical mechanism.

In order to reduce this complexity, a reference test rig with clearly defined boundary and flow conditions has been set up within this work. In this test rig well controlled entropy waves were generated by electrical heating. The noise emission of the entropy waves when they were accelerated in an adjacent nozzle flow was accurately measured and therewith the experimental proof of entropy noise could be accomplished. In addition to this, a parametric study on the quantities relevant for entropy noise was conducted. The results of these investigations were compared to a one-dimensional theory by Marble & Candel, showing the same order of magnitude and a similar same trend for an increasing nozzle Mach number. The discrepancies can be explained by multi-dimensional effects which are neglected in the theory and possible acoustic reflections in the test rig.

Furthermore, a laboratory combustor test rig was built and investigated with the following features:

1. Combustion chamber outlet-boundary conditions like in real-scale aero-engines (outlet Mach number = 1.0),
2. the possibility of accurately measuring the acoustic properties of the combustor, especially the total emitted sound power, and
3. an optical access into the combustion chamber for laser diagnostics.

The investigations on this combustor test rig provided manifestations on a very dominant contribution of entropy noise on the total emitted sound power of the system. An unknown broadband noise generation mechanism in the frequency range between 1 and 3.2 kHz ($2 < He < 5.6$) was found which showed a strong dependency on the nozzle Mach number in the combustor outlet. The summed-up broadband sound pressure level increases exponential with the nozzle Mach number. However, investigations of comparable cold flow conditions did not show this behavior. Since the results of the reference experiment with synthetically generated entropy waves did not show this exponential increase with the nozzle Mach number, this leaves the conclusion, that this additional noise is generated by the interaction of small-scale fluctuations, e.g. in entropy or vorticity, with the turbulent nozzle flow in the combustion chamber outlet nozzle. In addition, the flow field in the combustor is dominated by a strong swirl component due to the swirl stabilization of the reaction zone whereas in the EWG test rig a straight tube flow was present.

This broadband entropy or vortex noise has also been observed in aero-engines in the interaction between combustion chamber and high pressure turbine and is called "core noise". This "core noise" has a strong relevance especially for turboshaft engines, since there is hardly any jet noise on the exhaust side of the engine.

Prospective investigations should focus on this broadband noise generation mechanism with the significant challenge of detecting and measuring the small-scale entropy and vorticity fluctuations also called entropy and vorticity turbulence.

Acknowledgment

The authors gratefully acknowledge the financial support by the German Research Foundation (DFG) through the research unit FOR 486 "Combustion Noise". We also would like to thank Prof. Dr. Ulf Michel (DLR) and Dr. Christoph Hirsch (TU Munich) for the profitable and fruitful discussions.

References

[1] Ali G, Hunter JK (2000) The resonant interaction of sound waves with a large amplitude entropy wave. SIAM J Appl Math 61(1):131–148
[2] Bake F (2007) Experimental investigation of the fundamental entropy noise mechanism in aero-engines. In: 8th ONERA-DLR Aerospace Symposium, ODAS 2007, Göttingen, Germany
[3] Bake F, Michel U, Röhle I, Liu M, Aigner M, Noll B, Richter C, Thiele F (2004) Untersuchung zur Entstehung und Bedeutung von Entropielärm an einer Modellbrennkammer. In: 11. Workshop Physikalische Akustik, Bad Honnef, Germany
[4] Bake F, Michel U, Röhle I (2005) Entropieschall - Experimenteller Nachweis und Beitrag zum Fluglärm. In: 69. Jahrestagung der Deutschen Physikalischen Gesellschaft - Fachverband: Akustik, Berlin, Germany
[5] Bake F, Michel U, Röhle I, Richter C, Thiele F, Liu M, Noll B (2005) Indirect Combustion Noise Generation in Gas Turbines. In: 11th AIAA/CEAS Aeroacoustics Conference, Monterey, CA, 2005-2830
[6] Bake F, Michel U, Röhle I (2006) Investigation of entropy noise in aero-engine combustors. In: ASME Turbo Expo 2006, ASME, Barcelona, Spain, GT2006-90093
[7] Bake F, Fischer A, Kings N, Röhle I (2007) Experimental investigation of the fundamental entropy noise mechanism in aero-engines. In: 11th CEAS-ASC Workshop & 2nd Scientific Workshop of X3-Noise: Experimental and Numerical Analysis and Prediction of Combustion Noise, Instituto Superior Tecnico, Lisbon, Portugal
[8] Bake F, Michel U, Röhle I (2007) Entropieschall - Eine Parameterstudie zur Entstehung von indirektem Verbrennungsschall. In: DAGA 2007, DEGA, Stuttgart, Germany, DAGA2007/435

[9] Bake F, Michel U, Röhle I (2007) Experimental investigation of the fundamental entropy noise mechanism in aero-engines. In: 13th AIAA/CEAS Aeroacoustics Conference, Rome, Italy, 2007-3694

[10] Bake F, Michel U, Röhle I (2007) Fundamental mechanism of entropy noise in aero-engines: Experimental investigation. In: ASME Turbo Expo 2007, ASME, Montreal, Canada, GT2007-27300

[11] Bake F, Michel U, Röhle I (2007) Investigation of entropy noise in aero-engine combustors. Journal of Engineering for Gas Turbines and Power 129(2):370–376

[12] Bake F, Kings N, Fischer A, Röhle I (2008) Noise generation by accelerated flow inhomogenities – indirect combustion noise. In: International Conference on Jets, Wakes and Separated Flows, ICJWSF-2008, Technische Universität Berlin, Berlin, Germany

[13] Bake F, Kings N, Röhle I (2008) Fundamental mechanism of entropy noise in aero-engines: Experimental investigation. Journal of Engineering for Gas Turbines and Power 130(1):011,202–1–011,202–6

[14] Bake F, Kings N, Fischer A, Röhle I (2009) Experimental investigation of the entropy noise mechanism in aero-engines. International Journal of Aeroacoustics 8(1 and 2):125–142

[15] Bake F, Kings N, Fischer A, Röhle I (2009) Indirect combustion noise: Investigations of noise generated by the acceleration of flow inhomogeneities. Acta Acustica united with Acustica Submitted for publication

[16] Bake F, Richter C, Mühlbauer B, Kings N, Röhle I, Thiele F, Noll B (2009) The entropy wave generator (EWG): A reference case on entropy noise. J Sound Vibration Submitted for publication

[17] Bloy AW (1979) The pressure waves produced by the convection of temperature disturbances in high subsonic nozzle flows. J Fluid Mech 94:465–475, part 3

[18] Bohn MS (1976) Noise produced by the interaction of acoustic waves and entropy waves with high-speed nozzle flows. PhD thesis, California Institute of Technology, Pasadena, California, USA

[19] Bohn MS (1977) Response of a subsonic nozzle to acoustic and entropy disturbances. J Sound Vibration 52(2):283–297

[20] Chu BT, Kovasznay LSG (1958) Non-linear interactions in a viscous heat-conducting compressible gas. J Fluid Mech 3:494–514, part 5

[21] Chung JY (1977) Rejection of flow noise using coherence function method. Journal of the Acoustical Society of America 62:388–395

[22] Cumpsty NA (1979) Jet engine combustion noise: Pressure, entropy and vorticity perturbations produced by unsteady combustion or heat addition. J Sound Vibration 66(4):527–544

[23] Cumpsty NA, Marble FE (1977) Core noise from gas turbine exhausts. J Sound Vibration 54(2):297–309

[24] Cumpsty NA, Marble FE (1977) The interaction of entropy fluctuations with turbine blade rows; a mechanism of turbojet engine noise. Proc R Soc Lond A (357):323–344

[25] Dowling AP (1995) The calculation of thermoacoustic oscillations. J Sound Vibration 180(4):557–581
[26] Dowling AP (1996) Acoustics of unstable flows. In: Tatsumi T, Watanabe E, Kambe T (eds) Theoretical and Applied Mechanics, Elsevier, Amsterdam, pp 171–186
[27] Dowling AP (1997) Combustion noise and active control. In: VKI Lecture, VKI
[28] Dowling AP, Hubbard S (2000) Instability in lean premixed combustors. In: Proc. Instn. Mech. Engrs., Vol 214 Part A, IMechE
[29] Eckstein J (2004) On the mechanisms of combustion driven low-frequency oscillations in aero-engines. Dr.-Ing. Dissertation, Technische Universität München, München, Germany
[30] Eckstein J, Freitag E, Hirsch C, Sattelmayer T (2004) Experimental study on the role of entropy waves in low-frequency oscillations for a diffusion burner. In: ASME Turbo Expo 2004, ASME, Vienna, Austria, GT2004-54163
[31] Ffowcs Williams JE, Howe MS (1975) The generation of sound by density inhomogeneities in low Mach number nozzle flows. J Fluid Mech 70:605–622, part 3
[32] Fischer A, Bake F, Röhle I (2008) Broadband entropy noise phenomena in a gas turbine combustor. In: ASME Turbo Expo 2008, ASME, Berlin, Germany, GT2008-50263
[33] Flemming F, Olbricht C, Wegner B, Sadiki A, Janicka J, Bake F, Michel U, Lehmann B, Röhle I (2005) Analysis of Unsteady Motion with Respect to Noise Sources in a Gas Turbine Combustor: Isothermal Flow Case. Flow, Turb Combust 75(1-4):3–27
[34] Guedel A, Farrando A (1986) Experimental study of turboshaft engine core noise. J Aircraft 23(10):763–767
[35] Howe MS (1975) Contributions to the theory of aerodynamic sound, with application to excess jet noise and the theory of the flute. J Fluid Mech 71:625–673, part 4
[36] Keller JJ (1995) Thermoacoustic oscillations in combustion chambers of gas turbines. AIAA Journal 33(12):2280–2287
[37] Keller JJ, Egli W, Hellat J (1985) Thermally induced low-frequency oscillations. Journal of Applied Mathematics and Physics (ZAMP) 36:250–274
[38] Leyko M, Nicoud F, Poinsot T (2007) Comparison of indirect and direct combustion noise in aircraft engines. In: 11th CEAS-ASC Workshop & 2nd Scientific Workshop of X3-Noise: Experimental and Numerical Analysis and Prediction of Combustion Noise, Instituto Superior Tecnico, Lisbon, Portugal
[39] Lieuwen T (2003) Modeling premixed combustion-acoustic wave interactions: A review. Journal of Propulsion and Power 19(5):765–781
[40] Lieuwen T, Zinn BT (1999) Theoretical investigation of unsteady flow interactions with a planar flame. In: 37th AIAA Aerospace Sciences Meeting and Exhibit, AIAA, Reno, NV, USA, AIAA-99-0324
[41] Lieuwen T, Torres H, Johnson C, Zinn BT (1999) A mechanism of combustion instability in lean premixed gas turbine combustors. In: International Gas

Turbine & Aeroengine Congress & Exhibition, ASME, Indianapolis, Indiana, USA, 99-GT-3

[42] Lighthill MJ (1952) On sound generated aerodynamically. Proc R Soc Lond A (211):564–587

[43] Lu HY (1977) An analytical model for entropy noise of subsonic nozzle flow. In: AIAA 4th Aeroacoustic Conference, AIAA, 77-1366

[44] Marble FE, Candel SM (1977) Acoustic disturbances from gas non-uniformities convected through a nozzle. J Sound Vibration 55(2):225–243

[45] Martinez MM (2006) Determination of combustor noise from a modern regional aircraft turbofan engine. In: 12th AIAA/CEAS Aeroacoustics Conference, Cambridge, Massachusetts, 2006-2676

[46] Mathews DC, Rekos Jr NF, Nagel RT (1977) Combustion noise investigations. Technical Report FAA RD-77-3, Pratt & Whitney Aircraft Group, United Technologies Corporation, East Hartford, Connecticut, USA

[47] Morfey CL (1973) Amplification of aerodynamic noise by convected flow inhomogeneities. J Sound Vibration 31(4):391–397

[48] Mühlbauer B, Noll B, Aigner M (2008) Numerical investigation of entropy noise and its acoustic sources in aero-engines. In: ASME Turbo Expo 2008, ASME, Berlin, Germany, GT2008-50321

[49] Muthukrishnan M, Strahle WC, Neale DH (1978) Separation of Hydrodynamic, Entropy, and Combustion Noise in a Gas Turbine Combustor. AIAA Journal 16(4):320–327

[50] Olbricht C, Flemming F, Sadiki A, Janicka J, Bake F, UMichel, Röhle I (2005) A study of noise generation by turbulent flow instabilities in a gas turbine model combustor. In: ASME Turbo Expo 2005, ASME, Reno-Tahoe, NV, GT2005-69029

[51] Olbricht C, Hahn F, Kühne J, Sadiki A, Janicka J, Bake F, Röhle I (2007) Flow and mixing in a model GT combustor investigated by LES and Monte-Carlo filtered PDF methods. In: ASME Turbo Expo 2007, ASME, Montreal, Canada, GT2007-27270

[52] Polifke W, Paschereit CO, Döbbeling K (2001) Constructive and destructive interference of acoustic and entropy waves in a premixed combustor with a choked exit. The International Journal of Acoustics and Vibration 6(3):135–146

[53] Sattelmayer T (2000) Influence of the combustor aerodynamics on combustion instabilities from equivalence ratio fluctuations. In: ASME Turbo Expo 2000, ASME, Munich, Germany, 2000-GT-0082

[54] Schemel C, Thiele F, Bake F, Lehmann B, Michel U (2004) Sound Generation in the Outlet Section of Gas Turbine Combustion Chambers. In: 10th AIAA/CEAS Aeroacoustics Conference, Manchester, UK, 2004-2929

[55] Strahle WC (1978) Combustion noise. Prog Energy Combust Sci 4(3):157–176, a

[56] Strahle WC, Muthukrishnan M (1980) Correlation of combustor rig sound power data and theoretical basis of results. AIAA Journal 18(3):269–274, article No. 79-0587R

[57] Tanahashi M, Tsukinari S, Saitoh T, Miyauchi T, Choi G, Ikame M, Kishi T, Harumi K, Hiraoka K (2001) On the sound generation and its controls in turbulent combustion field
[58] Zhu M, Dowling AP, Bray KNC (2000) Self excited oscillations in combustors with spray atomisers. In: ASME Turbo Expo 2000, ASME, Munich, Germany, 2000-GT-108
[59] Zukoski EE, Auerbach JM (1976) Experiments concerning the response of supersonic nozzles to fluctuating inlet conditions. Journal of Engineering for Power (75-GT-40):60–63, ASME

Chapter 6
Influence of boundary conditions on the noise emission of turbulent premixed swirl flames

Fabian Weyermann, Christoph Hirsch and Thomas Sattelmayer

Abstract The main focus of subproject 6 was the development of time resolved experimental techniques for the analysis of the noise formation of turbulent premixed swirling flames. With these techniques the connection between the spatially and temporally fluctuating heat release in the reaction zone and the emitted noise spectrum could be investigated. In particular previous theoretical and experimental work could be extended such that a comprehensive picture of the noise-generating parameters in the flame was obtained.
This understanding was cast into a new method [12, 36], which allows the prediction of the fluctuating heat release spectrum from local mean values of turbulence and heat release. With this method it is possible to estimate the acoustic power spectral density of premixed swirl flames with good accuracy. Furthermore, the basic influences of turbulence intensity and length scale, fuel and mixture composition on the generated acoustical frequency spectrum appear naturally in this analysis, which allows to specifically design and optimize combustion systems with respect to noise emission.
The model was validated globally and in parts [36, 44] by using experimental mean data of turbulence and mean heat release as inputs and comparing the predicted model result with the corresponding measured quantity. It was shown successfully that the model could be used to calculate combustion noise on the basis of CFD results [11].
Further pursuing the title objective of this project, it was found that the model responded correctly to significant changes of boundary conditions. The basic theoretical analysis of a noise source in a confined acoustical system showed how to insert a CFD-based noise source spectrum into a thermo acoustical network code in order to calculate noise emission from a combustor. With this approach measured acoustic pressure spectra were reproduced with very good comparison.

Lehrstuhl für Thermodynamik, Technische Universität München, e-mail: sattelmayer@td.mw.tum.de

A. Schwarz, J. Janicka (eds.), *Combustion Noise*, DOI 10.1007/978-3-642-02038-4_6,

6.1 Introduction

Over the last years turbulent combustion noise has become a focus of interest in industrial gas turbines, jet engines and domestic water heaters in order to optimize designs for durability and minimal sound emission. The fundamental connection between the turbulent reactive flow field and the generation of sound is known since the pioneering work of Lighthill [18] in 1952. There its mathematical description emerged as a by- product of the analysis of aerodynamically generated noise which most of the early theoretical and experimental work was focused on [24, 26].
Almost a decade later the first studies on combustion generated noise were published [3, 30, 31]. While most authors in the field of combustion noise agree that the coherent heat release structures postulated by Strahle [31] determine the overall noise emission of the investigated flames the spectral behavior was hardly considered [3, 19, 31, 39]. In the early 90s Clavin [5] proposed a connection between the turbulent flow field and the noise emission of turbulent flames, that exhibited a reasonable amplitude decay according to an exponential power law but could not provide a quantitative closure for sound pressure levels. Boineau et al. [1] analyzed turbulent diffusion flames with spatial coherence functions arriving at a model for the sound spectrum which was able to reproduce the measured spectral behavior.
Klein and Kok [14, 13] proposed a formal closure for calculating sound pressure from turbulent diffusion flames based on local mean turbulence, mean reaction rate density and an assumed model spectrum for heat release fluctuation. However, it relied on the specification of an empirical coherence length scale to calibrate the model. Also the space time mapping proved to be non-universal. The model for turbulent premixed flames proposed by Hirsch et al. [12] was shown to provide quantitative spectra of acoustic power based on local mean turbulence and heat release rate density without adjustments. Their theory provides a local spatial model spectrum of heat release rate fluctuation from mean quantities and the mapping of the spatial spectrum to the temporal spectrum needed for the calculation of the sound pressure frequency spectrum. With this model the thermo-acoustic source term of the wave equation (as shown in e.g. Crighton et al. [6]) can be evaluated allowing the calculation of the turbulent combustion noise spectrum for an acoustic system, e.g. by a Finite Element method or acoustical network analysis. For free space an analytical solution exists, which is used below in the validation experiments. Recently a new approach has been published by Mühlbauer et al. [20] that uses an assumed shape spatio-temporal correlation function which is fed with local turbulence data and the distribution of the temperature variance from a RANS calculation. Using stochastic forcing temporal acoustic source terms can be calculated which feed an aero - acoustics code to calculate the sound propagation with good comparison to experimental data.
In this final report first the theory and methods developed and applied in this project are presented. Then the experimental setup and methods used to validate the noise source model developed in the project are outlined. In the results section the comparison of experimental and model data are given.

6.2 Theory and Methods

In turbulent reacting flows of low Mach number the main source of sound is the flame. It has been shown previously [3, 6] that for these cases other sources like flow-induced noise can be neglected. The flame generates sound only due to fluctuating heat-release, i.e. laminar steady flames generate no sound. A fluctuating heat release generates fluctuations of mass density and therefore volume fluctuations, which act as an acoustic monopole-source. The source-term that specifies this relationship is found for example in [6]. It appears on the right hand side of the wave-equation, which describes the propagation of sound:

$$\frac{1}{\rho_0 c_0^2}\frac{\partial^2 p'}{\partial t^2} - \nabla\cdot\left(\nabla\frac{1}{\rho}p'\right) = \frac{\partial}{\partial t}\left(\frac{\rho_0}{\rho}\frac{\gamma-1}{\rho c^2}\dot{q}'\right) . \tag{6.1}$$

Equation (6.1) can be further simplified, if γ is assumed as a constant, which is a good approximation in many cases. Doing so, density ρ and speed of sound c can be replaced by constant field values:

$$\rho c^2 = \gamma p_0 = \rho_0 c_0^2 . \tag{6.2}$$

The wave-equation (6.1) becomes:

$$\frac{1}{\rho_0 c_0^2}\frac{\partial^2 p'}{\partial t^2} - \nabla\cdot\left(\nabla\frac{1}{\rho}p'\right) = \frac{\partial}{\partial t}\left(\frac{\rho_0}{\rho}\frac{\gamma-1}{\rho c^2}\dot{q}'\right) \approx \frac{\partial}{\partial t}\left(\frac{\gamma-1}{c_0^2}\dot{q}'\right) . \tag{6.3}$$

As can be seen in equation (6.3) the time derivative of the local heat-release fluctuation acts as source. To solve this equation the locally and temporally resolved heat-release density needs to be known. If that is given, the time dependent solution to equation (6.3) can be calculated and the spectrum of acoustic power at some location of interest can be obtained which typically is the final objective of a noise calculation.
In the next section a particular solution of the linear wave equation is considered. It allows to carry out the procedure analytically, thereby connecting the acoustic power spectrum directly with the power spectrum of the fluctuating heat release.

6.2.1 Calculation of the acoustic power spectrum

In the case of a monopole-source in the free field, equation (6.3) can be solved using the Green-function [13]. The pressure-fluctuation at the observer location $\mathbf{x_B}$ yields:

$$p'(\mathbf{x_B},t) = \frac{1}{4\pi(\mathbf{x_S}-\mathbf{x_B})}\frac{\gamma-1}{c_0^2}\frac{\partial}{\partial t}\int_{V_S} q'(\mathbf{x_S},t-\tau_B)\,d\mathbf{x_S} . \tag{6.4}$$

Here $\mathbf{x_S} - \mathbf{x_B}$ is the distance between the position of the local heat-release $\mathbf{x_S}$ and the observer $\mathbf{x_B}$ and can be replaced by a radius $\mathbf{r_B} = \mathbf{x_S} - \mathbf{x_B}$ in the case of a monopole-source [33]. A flame consists of N sources, which is illustrated in figure 6.1 by the positions $\mathbf{x_{S1}}, \ldots, \mathbf{x_{SN}}$. If the flame is acoustically compact and the distance between the flame and the observer is large, $\mathbf{r_B}$ can be assumed constant for all sources. Then the delay time τ_B which the sound emitted by the different sources needs to reach the observer is also constant. It can be derived by the distance to the observer and the sound-speed $\tau_B = r_B/c$. The acoustic power is the correlation of sound pressure

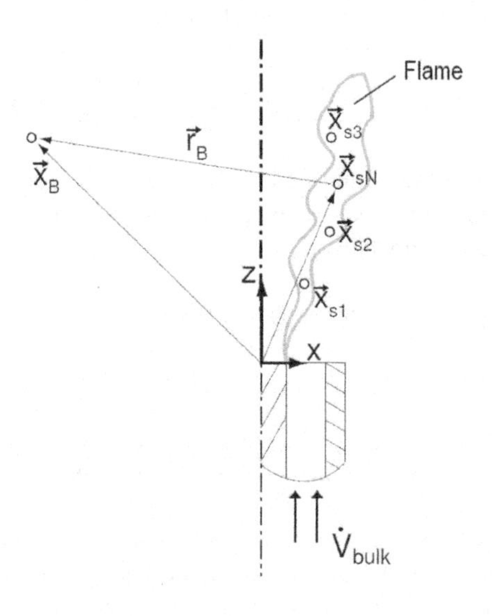

Fig. 6.1 Sketch of a turbulent flame with N sources and the location of the observer **O**, [25]

and sound particle velocity, which is integrated over the Area $A = 4\pi r^2$. In the free field the sound particle velocity u' can be replaced by p' using the impedance $Z = \rho_0 c_0 = p'/u'$:

$$P_{ac} = \int_A \overline{p'u'} d\mathbf{x} = \int_A \frac{\overline{p'^2}}{\rho_0 c_0} d\mathbf{x} = \frac{\overline{p'^2}}{\rho_0 c_0} \cdot A \,. \tag{6.5}$$

Inserting equation (6.4) into equation (6.5) we get the relation between the heat-release as source and the integral acoustic power:

$$P_{ac} = \frac{1}{4\pi\rho_0 c_0} \left(\frac{\gamma - 1}{c_0^2}\right)^2 \overline{\int\int \left(\frac{\partial}{\partial t_1} q'(\mathbf{x_{S1}}, t_1)\right) \left(\frac{\partial}{\partial t_2} q'(\mathbf{x_{S2}}, t_2)\right) d\mathbf{x_2} d\mathbf{x_1}} \,. \tag{6.6}$$

Because of the conservation of energy, P_{ac} does not depend on the radius r_B. Equation (6.6) specifies an important property of the noise formation by turbulent flames:

The fluctuating heat release in the flame must be coherent to some extent. Because the flame is modulated to a high degree by the turbulent flow Strahle [31] proposed to use of cross correlation functions to evaluate the mean correlation in equation (6.6). In figure 6.2 the local separation of two sources at position x_{S1} and x_{S2} is shown.

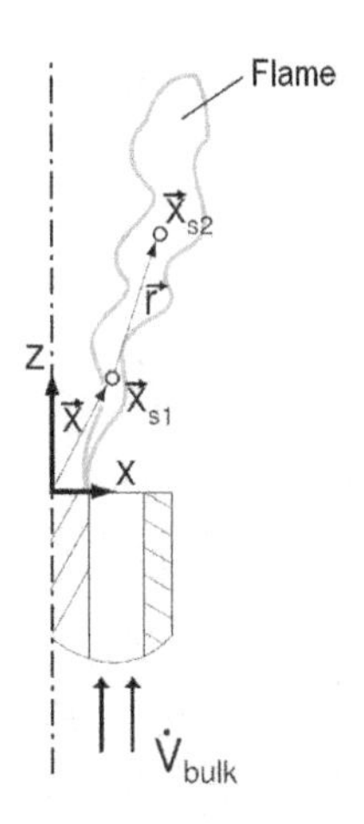

Fig. 6.2 Separation of two sources in a flame

The distance between the two sources is $\mathbf{r}$ in space and τ in time. The relation can be described by the autocorrelation R_{qq} of the heat-release fluctuations:

$$R_{qq}(\mathbf{x_{S1}}, t_1, \mathbf{r}, \tau) = \overline{q'(\mathbf{x_{S1}}, t_1)\, q'(\mathbf{x_{S1}} + \mathbf{r}, t_1 + \tau)} \tag{6.7}$$

The correlation of the sources within the double-integral in equation (6.6) can be written [27]:

$$\overline{\left(\frac{\partial}{\partial t_1} q'(\mathbf{x_{S1}}, t_1)\right)\left(\frac{\partial}{\partial t_2} q'(\mathbf{x_{S2}}, t_2)\right)} = \frac{\partial^2 R_{qq}(\mathbf{x_{S1}}, t_1, \mathbf{r}, \tau)}{\partial t \partial \tau} - \frac{\partial^2 R_{qq}(\mathbf{x_{S1}}, t_1, \mathbf{r}, \tau)}{\partial \tau^2}\,. \tag{6.8}$$

For steady flows the first term on the right hand side is vanishes. The correlation of the heat-release sources, which is described by R_{qq}, is normalized by the root mean square deviation q'_{rms} of the sources and subsequently replaced by a coherence-function $\Gamma(\mathbf{r}, \tau)$ [1]. Here $q'_{rms} = q'_{rms,S1} \approx q'_{rms,S2}$ is assumed. Putting equation (6.8) into equation (6.6) yields:

$$P_{ac} = \frac{1}{4\pi\rho_0 c_0}\left(\frac{\gamma - 1}{c_0^2}\right)^2 \frac{\partial^2}{\partial \tau^2} \int\int q'^2_{rms} \Gamma(\mathbf{r}, \tau) d\mathbf{r} d\mathbf{x}\,. \tag{6.9}$$

For simplification $\mathbf{x} = \mathbf{x_{S1}}$ is written. The coherence-function $\Gamma(\mathbf{r}, \tau)$ considers both, spatial and temporal correlation. In the following the local correlation is replaced by a characteristic volume. Therefore the function $\Gamma(\mathbf{r},\tau)$ is integrated over $\mathbf{r}$ for $\tau = 0$:

$$V_{coh} = \int_r \Gamma(\mathbf{r},0) d\mathbf{r} . \tag{6.10}$$

This integration gives the size of a virtual statistical volume V_{coh}, in which the heat-release is fully coherent. Equation (6.9) describes the behavior of the sound pressure in the time-domain. For determining the spectral character a Fourier-Transformation (FT) of equation (6.9) is done. The Fourier-transformation of $\Gamma(0,\tau)$ for $\mathbf{r} = 0$ corresponds to a power-density spectrum. Using the variance of the heat-release yields the spectral heat-release:

$$\chi_{qq}(2\pi f) = FT\left(q'^2_{rms} \cdot \Gamma(0,\tau)\right) . \tag{6.11}$$

The derivative $\partial/\partial\tau$ in the time-domain equals a multiplication by $2\pi f$ in the frequency-domain. To get the sound power spectrum in units of W/Hz a factor of 2π has to be introduced:

$$P_{ac}(f) = \frac{2\pi}{4\pi\rho_0 c_0}\left(\frac{\gamma-1}{c_0^2}\right)^2 \int_{V_{fl}} (2\pi f)^2 \chi_{qq}(2\pi f) V_{coh}\, d\mathbf{x} \quad [W/Hz] . \tag{6.12}$$

The overall acoustic power is obtained by integration over all frequencies:

$$P_{ac,tot} = \int_0^\infty P_{ac}(f) df . \tag{6.13}$$

6.2.2 *Modeling of the spectral heat-release*

Assuming sufficiently strong turbulence the volumetric mean heat-release rate $\bar{q}$ can be expressed using the variance $\bar{c}'^2$ of the progress-variable and the time-scale of the dissipation:

$$\bar{q} = \rho_0 Y_F H_u C_D \frac{\varepsilon}{k} \bar{c}'^2 . \tag{6.14}$$

Here ρ_0 denotes the density of the mixture, Y_F the fuel-mass fraction, H_u the lower heating value, C_D a constant, ε the turbulent dissipation and k the turbulent kinetic energy. Using equation (6.14) the spectrum of the heat release fluctuations can be derived from the spectrum of the progress variable $E_{\bar{c}'^2}$:

$$E_q = \rho_0 Y_F H_u C_D \frac{\varepsilon}{k} E_{\bar{c}'^2} . \tag{6.15}$$

Under the assumption, that $\bar{c}'^2$ behaves like a passive scalar, we can use a model-spectrum from turbulence-theory for its characterization. Using the approach of Tennekes and Lumley [34] we get for E_q:

$$E_q = \bar{q}\frac{C_S C_D}{\alpha}\alpha\frac{\varepsilon^{2/3}}{k}\kappa^{-5/3}\exp\left(-\frac{3}{2}\left(\pi\beta\alpha^{1/2}(\kappa l_t)^{-4/3}+\alpha(\kappa\eta_{c^2})^{4/3}\right)\right) \quad . \tag{6.16}$$

However $\tilde{c}'^2$ is no passive scalar. Its active behavior appears in two effects, a spectral cut-off and a finite rate chemistry influence on the spectral amplitude.

6.2.2.1 Active scalar treatment

Spectral cut-off:

An eddy of the turbulent flow field is able to disturb the flame front only if its circumferential velocity v_n' is higher than the burning velocity of the flame front. If the circumferential velocity is lower than the burning velocity, the flame-front propagates through the eddy too fast and the eddy has no time to wrinkle the flame-front. For the smallest eddies, which are able to impact the flame, must hold:

$$v_n' = s_L \; . \tag{6.17}$$

For the length-scale of these smallest eddies we get:

$$l_G = \frac{s_L^3}{\varepsilon}, \tag{6.18}$$

which is called the Gibson-Length. It represents a lower cut-off scale for the interaction of turbulent eddies with a flame front. A second effect, which influences the spectral cut-off at high wave-numbers are stretching and dissipative effects. They appear at a length-scale given by the Corrsin-length [21]:

$$L_C = \left(\frac{a^3}{\varepsilon}\right)^{1/4} \quad . \tag{6.19}$$

The cut-off-length η_{c^2} of the noise-model is determined by the effective length:

$$\eta_{c^2} = \max\left(c_G L_G, L_C\right) \; . \tag{6.20}$$

The constant c_G was determined empirically and has a value of $c_G = 3.0$

Scaling of the spectral amplitude:

If an infinitely fast combustion process is assumed every fluctuation of the turbulent flow will instantly result in a fluctuation of the heat release. Conversely, if the chemical reaction is very slow, the heat-release rate will not respond to the turbulent fluctuations any more. This results in a quiet flame. Therefore it appears reasonable to assume that the ratio between the turbulent and the chemical time-scale rules the amplitude of the heat-release fluctuations. This ratio is known as the turbulent Damkoehler-Number Da_t:

$$Da_t = \frac{t_{flow}}{t_{chem}} \ . \tag{6.21}$$

If we use the life-time of the integral eddies as the characteristic time of the flow $t_{flow} = l_t / v'_{rms}$ and the time, which the laminar flame needs to cross the distance of the flame $t_{chem} = \delta_L / s_L$, as chemical time-scale, we get:

$$Da = \left(\frac{l_t}{\delta_L}\right)\left(\frac{s_L}{v'}\right) \ . \tag{6.22}$$

The first factor is the ratio of the characteristic lengths of turbulent eddies and laminar flame front thickness and the second a reciprocal ratio of turbulence-intensity and flame speed. If the lengths are fixed, the Damkoehler-Number behaves inversely to the turbulence-intensity. If the Damkoehler-Number is high the heat release is controlled by the turbulent mixing-process. In this case large intermittency between unburnt and burnt gas occurs in the flame zone and the variance of progress variable is large. For small Damkoehler-Numbers heat release rate is controlled by chemistry. In the flame zone the variance of progress variable will be small.
The scaling-function C_S in equation (6.16) is introduced to consider this Damkoehler-Number dependence on the amplitude of variance. For the Damkoehler-Number $Da_t \to \infty$ the variance of the progress variable reaches its maximum of a fully intermittent distribution $\bar{c}'^2 = \bar{c}\,(1-\bar{c})$. In the case of a slow reaction, $\bar{c}'^2$ decreases and therefore also the emitted noise. To consider these limits the scaling function C_S is constructed as:

$$C_S = \frac{\alpha}{C_D} \frac{\bar{c}'^2\,(Da_t)}{\bar{c}'^2\,(Da_t \to \infty)} \ . \tag{6.23}$$

To obtain an estimator of $\bar{c}'^2$ the comparison is made between equation (6.14) and the combustion model proposed by Schmidt et all [28, 29]:

$$\tilde{q} = 4.96 \frac{\varepsilon}{k} \tilde{c}\,(1-\tilde{c}) \cdot \tag{6.24}$$

$$\left(\frac{s_l}{\sqrt{2/3k}} + (1 + Da_t^{-2})^{-0.25} \right)^2 \cdot \rho_0 \, Y_{F,0} \, H_u$$

$$Da_t = \frac{0.09 \cdot k \cdot s_l^2}{\varepsilon \cdot C_c^2 a_0} \ . \tag{6.25}$$

Inserting this modeled heat-release into equation (6.14) an expression for the variance $\bar{c}'^2$ of the progress-variable can be derived. Together with the requirement for C_S (equation 6.23) the scaling-function results as:

$$C_S = \frac{\alpha}{C_D}\left(\frac{\frac{s_l}{\sqrt{2/3k}} + \left(1 + Da_t^{-2}\right)^{-1/4}}{\frac{s_l}{\sqrt{2/3k}} + 1}\right)^2 , \tag{6.26}$$

which holds the asymptotic requirement $C_S \to 0$ for $Da_t \to 0$ as shown in Fig. 6.3.

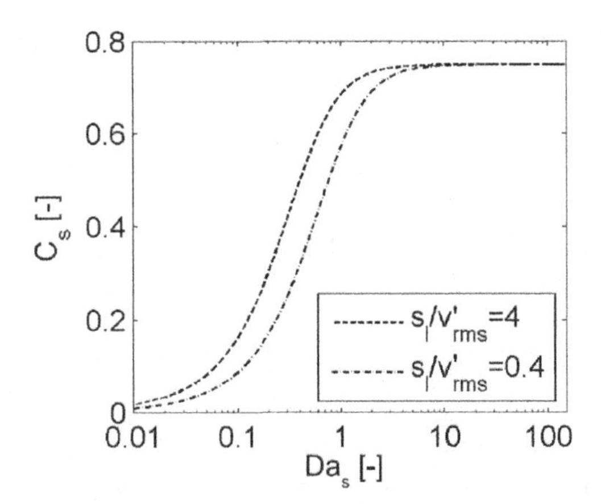

Fig. 6.3 Amplitude scaling-function C_S

6.2.2.2 Frequency-wave-number mapping

For equation (6.12) the temporal correlation spectrum $\chi_{qq}(f)$ is needed, which shall be represented by:

$$\tilde{\chi}_{qq}(\omega) = \tilde{\chi}_q(\omega) \cdot \tilde{\chi}_q^*(\omega) = \tilde{\chi}_q^2(\omega). \tag{6.27}$$

To this point the spectral heat-release $E_q(\kappa)$ was modeled in the wave-number-domain and a transformation is needed to transfer it into the frequency-domain to provide $\tilde{\chi}_q(\omega)$. For homogeneous turbulence a mapping between the spatial and temporal spectra is proposed in Tennekes and Lumley [34]. Based on their development the conservation of the spectral energy for corresponding wavenumber - frequency pairs is assumed.

$$\kappa E_q(\kappa) = \omega \tilde{\chi}_q(\omega) \tag{6.28}$$

$$\omega = \alpha^{1/2} \varepsilon^{1/3} \kappa^{2/3}. \tag{6.29}$$

In the original work the dissipation rate ε is given by

$$\varepsilon = \frac{l_t^2}{\tau_t^3} , \tag{6.30}$$

where l_t is the turbulent integral length scale and $\tau_t = l_t/u'$ is the characteristic turbulent dissipation time-scale. In the case of a premixed flame, the characteristic scalar dissipation time scale τ_c is the time, which a turbulence-structure needs to cross the flame brush thickness:

$$\tau_c = C_\tau \frac{\delta_t}{s_t} \,. \tag{6.31}$$

Here s_t is the turbulent burning velocity and δ_t is the turbulent flame brush thickness which were modeled using again the Schmid model [28, 29].

$$\tau_c = C_\tau \cdot \frac{\delta_t}{s_t} = C_\tau \cdot \frac{l_t}{u'} \cdot \frac{a_0/(s_l \cdot l_t) + (1 + Da_t^{-2})^{0.25}}{s_l/u' + (1 + Da_t^{-2})^{-0.25}} \tag{6.32}$$

Here $C_\tau = 0.5$ is a model constant which is fixed by the requirement that $\tau_c \to \tau_t$ for $Da_t \to \infty$. The mapping (6.28) finally becomes:

$$\kappa(2\pi f) = \frac{(2\pi f \tau_c)^{3/2}}{\delta_t \alpha^{3/4}} \,. \tag{6.33}$$

and so:

$$(2\pi f)^2 \chi_{qq}(f) = (2\pi f \chi_q(2\pi f))^2 = (\kappa(2\pi f) E_q(\kappa))^2 \tag{6.34}$$

For quantitative comparison of the model spectra which are based on $\omega = 2\pi f$ i.e. [rad/s] with experiments which are typically based on frequency [Hz] the proper normalization has to be considered. It can be derived from considering that the total energy-content should be equal in $\chi(\omega)$ and $\chi(f)$ spectra:

$$\int_0^\infty \chi_q(f)\,df = \int_0^\infty \tilde{\chi}_q(\omega)\,d\omega = 2\pi \int_0^\infty \tilde{\chi}_q(2\pi f)\,df \,. \tag{6.35}$$

Thus on substitution of equation (6.28) into equation (6.12) a factor of $(2\pi)^2$ must be inserted to obtain frequency based acoustic power spectra.

6.2.3 Coherence volume

As shown above the cross correlation of heat release fluctuation was simplified by introducing the coherence volume. It describes the spatial domain in which heat release fluctuation is in phase, such that the total integral over the flame domain does not cancel. Early it was accepted, that the size of the coherent sources is closely connected to the turbulent length scale of heat release l_q which itself would be a function of the turbulence length scale l_t and chemical reaction influence.

$$V_{coh} \sim l_q^3 \,. \tag{6.36}$$

This led to many different approaches. Strahle [32] was the first who introduced a combined model, based on turbulence and chemical variables:

$$V_{coh} = C_{coh} \delta_L^{3-n} l_t^n \,. \tag{6.37}$$

The constant C_{coh} is not defined, and $0 \le n \le 2$. An important contribution was given by Boineau et al. [2], who determined the coherence volume experimentally in turbulent diffusion flames. They derived an expression for the spatial distribution of the correlation of heat release, which is similar to the velocity correlation given e.g. in Hinze [10]:

$$\Gamma(r) = e^{-\frac{\pi}{4}\left(\frac{r}{l_q}\right)^2} \,. \tag{6.38}$$

With this the coherence volume is calculated by integrating equation (6.38):

$$V_{coh} = \int_0^{2\pi}\int_0^{\pi}\int_0^{\infty} e^{-\frac{\pi}{4}\left(\frac{r}{l_q}\right)^2} r^2 \sin(\phi)\, dr\, d\phi\, d\vartheta \,. \tag{6.39}$$

This yields:

$$V_{coh} = 4\pi \frac{\sqrt{\pi}}{4\left(\sqrt{\frac{\pi}{4l_q^2}}\right)^3} = 4^{3/2} l_q^3 \,. \tag{6.40}$$

To consider the influence of chemical effects, the turbulent flame thickness $\delta_t = \delta_L + l_t \left(1 + Da_t^{-2}\right)^{1/4}$ [29], which connects turbulent and chemical scales, is used as the length-scale of the heat release $l_q = \delta_t$. The coherent volume becomes:

$$V_{coh} = 8 \cdot \delta_t^3 \,. \tag{6.41}$$

6.2.4 Acoustic power spectrum of an unconfined flame

Given the above model the acoustic power of an unconfined flame equals:

$$P_{ac}(f) = \frac{(2\pi)^3}{4\pi\rho_0 c_0}\left(\frac{\gamma-1}{c_0^2}\right)^2 \int_{V_{fl}} (\kappa(2\pi f) E_q(\kappa))^2 8 \cdot \delta_t^3 dx \quad [\mathrm{W/Hz}]. \tag{6.42}$$

6.2.5 Simulation of confined flames

For the simulation of the sound-emission of an enclosed flame, the whole acoustic system consisting of a plenum, combustion chamber and exhaust-tract has to be considered. For the computation we assume the fluctuations are harmonic in time:

$$p'(x,t) = \hat{p}(x)\,e^{i\omega t}\,. \tag{6.43}$$

Here $i = \sqrt{-1}$ and " ^ " denotes the complex amplitude. Putting this into equation (6.3), we get the so called Helmholtz-Equation:

$$\nabla \cdot \left(-\frac{1}{\rho}\nabla\hat{p}\right) - \frac{\omega^2}{\rho c^2}\hat{p} = \frac{\gamma - 1}{\rho c_0^2} i\omega\hat{q}\,. \tag{6.44}$$

The fluctuating heat-release $\hat{q}$ acts as a acoustic monopole-source [6]. It is equivalent to an oscillating sphere, which extends and shrinks harmonically and thereby causes small velocity fluctuations of the ambient air. Thus a flame generates sound particle velocity $\hat{u}$. Here the question arises, how much acoustic power is performed thereby by a flame. For the acoustic power which crosses a surface A the following equation holds:

$$P_{ac} = \int_A \overline{I \cdot \mathbf{n}}\, d\mathbf{x}\,. \tag{6.45}$$

$\mathbf{n}$ is the normal vector of the Area A and I the acoustic impedance, which is defined as:

$$I = \frac{\omega}{2\pi} \int_0^{2\pi/\omega} \mathrm{Re}\left(\hat{p}e^{i\omega t}\right) \mathrm{Re}\left((\hat{u}\cdot\mathbf{n})\,e^{i\omega t}\right)\, dt\,. \tag{6.46}$$

Putting equation (6.46) into equation (6.45) yields:

$$P_{ac} = \int_A \frac{\omega}{2\pi} \int_0^{2\pi/\omega} \mathrm{Re}\left(\hat{p}e^{i\omega t}\right) \mathrm{Re}\left((\hat{u}\cdot\mathbf{n})\,e^{i\omega t}\right) dt d\mathbf{x} \tag{6.47}$$

$$= \int_A \frac{1}{2}\mathrm{Re}\left(\hat{p}^*\hat{u}\cdot\mathbf{n}\right) d\mathbf{x}\,. \tag{6.48}$$

Using the impedance

$$Z(x,\omega) = \frac{\hat{p}(x,\omega)}{\hat{u}(x,\omega)\cdot\mathbf{n}(x)}\,, \tag{6.49}$$

the acoustic power is:

$$P_{ac} = \int_A \frac{1}{2}\mathrm{Re}(Z)\,|\hat{u}\cdot\mathbf{n}|^2\, d\mathbf{x}\,. \tag{6.50}$$

This means, that the real part of the acoustic impedance (resistance) determines the power, which an acoustic velocity source (flame) can emit. The impedance depends strongly on the geometry of the combustion system. That is the reason, why a proper modeling of the system is very important for the calculation of the noise of confined flames.

6.2.5.1 Acoustic network-models

Because the acoustic wave-lengths are large in comparison to the dimensions of a typical combustion chamber, the sound propagation can be assumed as one-dimensional (plane waves). Therefore the acoustics of the combustion-system can be modeled using a so called network-model. The network-method has two advantages: First a very low computational effort, and secondly an easy to do model setup. So this method is very attractive for industrial usage. Another advantage is, that acoustic losses in the system can be considered in simple way. Details of the network-method are obtained e.g. in [7].
The combustion-system is divided into its component parts (ducts, area-changes, the flame,...). If the acoustics in each part is linear and one-dimensional, every part can be described by a linear transformation T. This transformation connects the acoustic pressure and velocity on either side of the element. Two nodes i and i+1 are related through a complex (2x2)-matrix:

$$\begin{pmatrix} \frac{p'}{\rho c} \\ u' \end{pmatrix}_{i+1} = \begin{pmatrix} T_{11} & T_{12} \\ T_{21} & T_{22} \end{pmatrix} \begin{pmatrix} \frac{p'}{\rho c} \\ u' \end{pmatrix}_{i} . \tag{6.51}$$

Assembling all the transformation-matrices of the systems elements, a linear equation system for the complex nodal pressures and velocities is obtained, which describes the systems acoustics. In Fig. 6.4 a network-model of a combustion-system is shown. It consists of a plenum, a burner and a combustion chamber. These three main parts are modeled by ducts D for which analytical models exist. Changing duct properties like Mach-Number, area or density is accomplished using generalized area changes AC. These are derived from linearized Bernoulli and continuity equations. At the beginning and at the end the system is bounded by a reflecting boundary R where a known boundary impedance is assumed. The excitation by the flame is introduced by the source-element S.

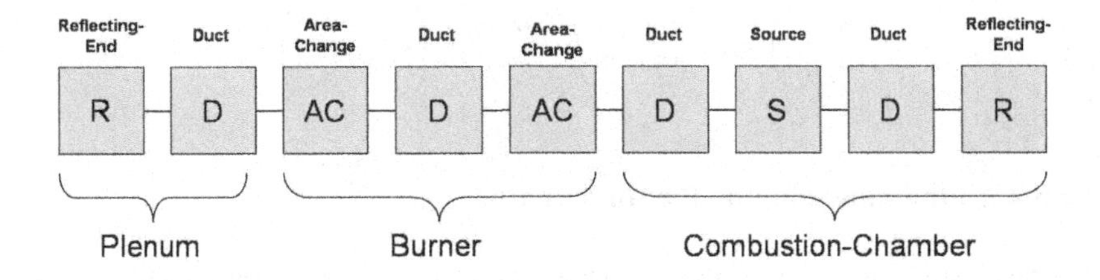

Fig. 6.4 Simple network-model

Noise-source implementation:

For the implementation of the noise-source linearized Rankine-Hugoniot-relations [23] are used. For very low Mach-Number flows $Ma << 1$ they become :

$$p'_{i+1} = p'_i \tag{6.52}$$

$$u'_{i+1} = u'_i + \left(\frac{T_{i+1}}{T_i} - 1\right) u_i \left(\frac{Q'}{Q}\right) , \tag{6.53}$$

whereby Q is the heat-addition per unit area, $[Q] = \frac{W}{m^2}$. The flame front is considered as a discontinuity of negligible thickness, which adds heat to the flow.

For the temperature-term in equation (6.53) holds:

$$\frac{T_{i+1}}{T_i} - 1 = \frac{\gamma - 1}{\rho_i c_i^2} \frac{Q}{u_i} . \tag{6.54}$$

Putting this equation into the Rankine-Hugoniot-equation (6.53) yields:

$$u'_{i+1} = u'_i + \left(\frac{\gamma - 1}{\rho_i c_i^2}\right) \frac{Q}{u_i} u_i \frac{Q'}{Q} = u'_i + \left(\frac{\gamma - 1}{\rho_i c_i^2}\right) Q' . \tag{6.55}$$

We define

$$Q'_{tot} = Q'A \qquad [W] , \tag{6.56}$$

which is the overall heat-release fluctuation of the flame. The spectrum of Q'_{tot} can be calculated using the modeled heat release spectrum E_q from equation (6.16):

$$Q'_{tot}(f) = \sqrt{QQ} = \sqrt{\int_{V_{fl}} (2\pi)^3 \left(\frac{\kappa E_q(\kappa)}{f}\right)^2 V_{coh}\, dx} \quad [W/Hz] . \tag{6.57}$$

The network-element of the flame-noise becomes:

$$\begin{pmatrix} p' \\ u' \end{pmatrix}_{i+1} = \begin{pmatrix} 1 & 0 \\ 0 & 1 \end{pmatrix} \begin{pmatrix} p' \\ u' \end{pmatrix}_i + \begin{pmatrix} 0 \\ \frac{\gamma-1}{\rho_i c_i^2} \frac{Q'_{tot}}{A} \end{pmatrix} . \tag{6.58}$$

6.2.5.2 Sound-emission into the surrounding

The level of noise in the combustion chamber itself is not of interest in most cases. More important is the amount of sound which can leave the system and causes the sound emission. Most of the sound is reflected at the exhaust-gases exit and stays in the system. Only a small part leaves and contributes to the outside noise-level.
Levine and Schwinger examined in their work [17] the sound-emission of an open tube into the surrounding. They presented an impedance boundary condition valid for a tube radiating to the free field. For low wave-numbers $\kappa = \omega/c$ ($\kappa r < 0.5$, r:

exit-radius) this impedance simplifies:

$$Z = \rho c \left(\frac{\kappa^2 r^2}{4} + i \cdot 0.6\, \kappa r \right) . \qquad (6.59)$$

By calculating the reflexion-coefficient from this impedance, it can be seen, that the reflexion-coefficient is higher than 96% in the whole frequency range (Fig. 6.5). This means, that not even four percent of the noise actually leaves the system and contribute to the audible sound.

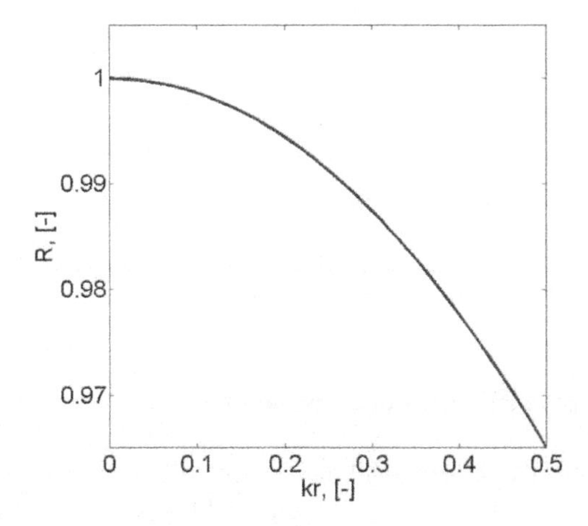

Fig. 6.5 Reflection-coefficient R of an unflanged open duct

6.2.6 Experimental setup and measurement techniques

6.2.6.1 Experimental setup

For the experiments the TD1-burner was used, which was designed at the chair. This burner is a modular swirl-burner shown in Fig. 6.6. For the experiments the outer diameter D of the annular burner exit was $D = 40mm$, the inner diameter $d = 16mm$. For the measurements either pure natural gas (methane-content 98%) was used, or a mixture of natural gas and hydrogen. Upstream of the burner a static mixer provided perfectly premixed reactants. The burner has a power range of $P_{th} = 10 \ldots 120kW$; this correspondes to velocities of $u = 3.3 \ldots 39.4\ m/s$. The air-ratio can be varied in a range of $\lambda = 0.8 \ldots 1.52$. The swirl can be altered by partial closing the slots.

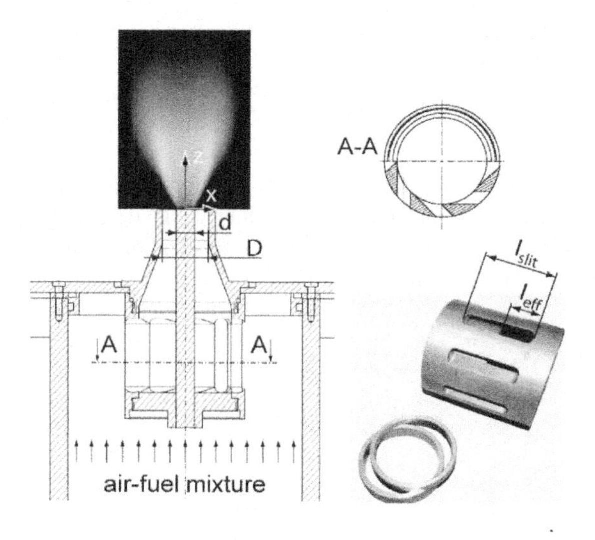

Fig. 6.6 TD1 swirl-burner for premixed operation. Tangential slots deflect the air-fuel-mixture whereby the swirl is generated. A typical natural-gas flame is also displayed

6.2.6.2 Simultaneous PIV-LIF-measurements

For the validation of the noise-model data of the turbulent flow field and the heat-release rate density is required. The velocity-field was investigated using particle image velocimetry (PIV). Simultaneously laser induced fluorescence (LIF) provided planar information of the flame front position and therefore of the progress-variable. A simultaneous PIV-LIF-measurement was necessary to validate the scaling-function C_S (6.26), which depends on the interaction of the flow field with the flame [44]. The LIF-technique was improved from a single shot system with statistical evaluation methods to a real time system with high temporal resolution [37]. The high repetition rate (1 kHz) allows the acquisition of information up to 500 Hz, which is the frequency range that contributes most to the acoustic energy of flame noise. The emitted acoustic power was measured by an intensity probe. The entire measurement setup is depicted in Fig. 6.7.

6.2.6.3 Improved method for the determination of the heat-release-distribution

The technique applied to quantify the average heat release rate density uses the chemiluminescence, which is a direct measure for the heat release in adiabatic, perfectly premixed flames. However, it is known, the chemiluminescence signal depends strongly on the equivalence-ratio. Since in this project unconfined flames had to be investigated, the flame unavoidable entrains air from the ambient; so the

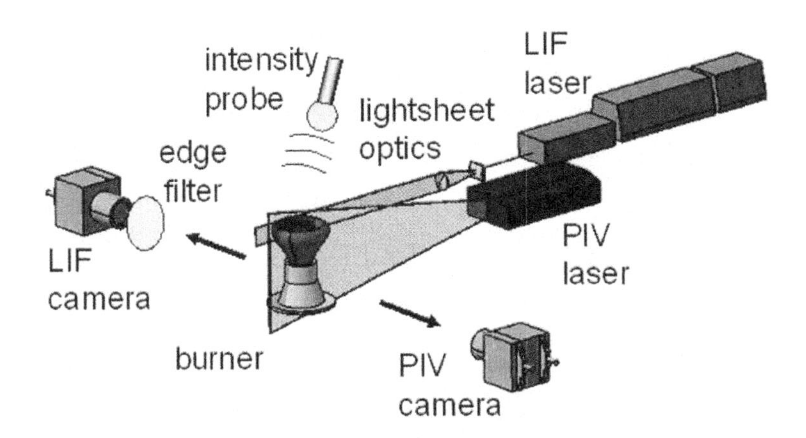

Fig. 6.7 Sketch of the experimental setup for the simultaneous PIV-LIF-measurements

equivalence-ratio alters and the direct correlation between chemiluminescence and heat-release is lost. As the exact knowledge of the mean heat release density is essential for the calculation of the combustion-noise, the influence of the entrainment of ambient air must be considered. In the following the method used in this project is presented, which solves this problem by correcting the chemiluminescence-data using the equivalence-ratio distribution in the flame.
The method is based on the work of Haber [9], who applied it for the investigation of laminar flames. His method was enhanced, so that it can be applied also for the investigation of turbulent flames. The method uses the fact, that the ratio of OH^*- and CH^*-chemiluminescence directly depends on the local flame temperature and thus depends on the equivalence-ratio. However the dependence between the chemiluminescence-ratio and the equivalence-ratio is not a priori known, but it has to be determined for an adiabatic flame with known constant equivalence-ratio. Therefore the flame of the TD1-burner was confined with a quartz-cylinder. Now, with the known correlation between the OH^*/CH^*-ratio and the equivalence-ratio, the variation of the equivalence-ratio in the unconfined case can be determined. In Fig. 6.8 the equivalence-ratio Φ is shown over the z-coordinate. From this data the correction-factor for the chemiluminescence can be calculated.

For the identification of the local heat-release the OH^*-chemiluminescence is recorded by an intensified camera. The data is then corrected by the calculated correction-factor. In Fig. 6.9 the heat-release distribution of the side view before (left) and after (right) correction is shown. In Fig. 6.10 the same data was deconvoluted to provide the local mean heat release rate density. It can be seen, that the considered flame volume increases due to the correction. For a given total thermal

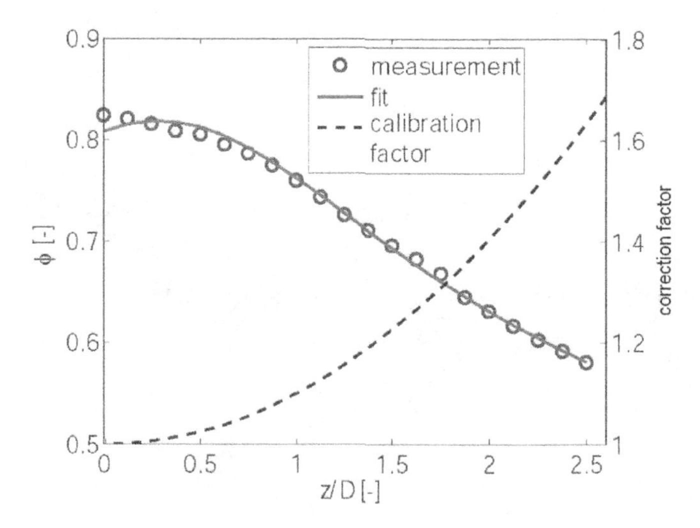

Fig. 6.8 Spatial resolved equivalence-ratio and derived correction-factor plotted over the burner z-axes

power this will decrease the heat release rate density. Now the noise-model (6.42)

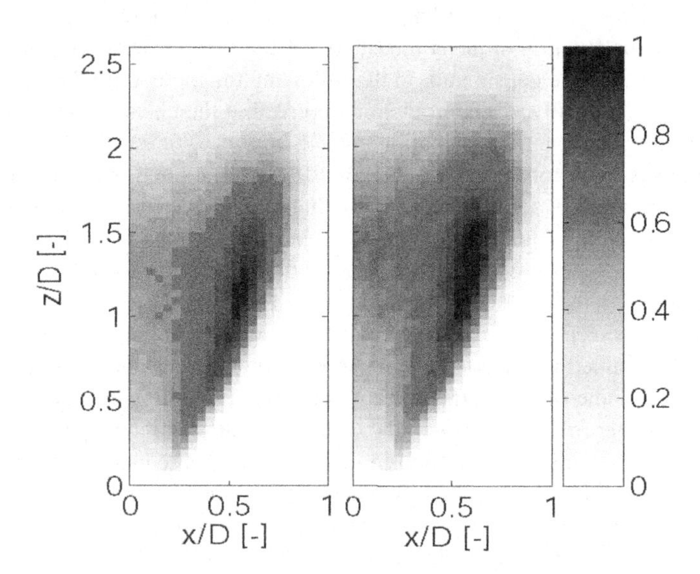

Fig. 6.9 Comparison between measured (left) and corrected (right) OH^*-intensity

can be used to calculate the power-spectra of the emitted combustion noise. The results are shown in figure 6.11. In case (a) for the calculation the raw data of the

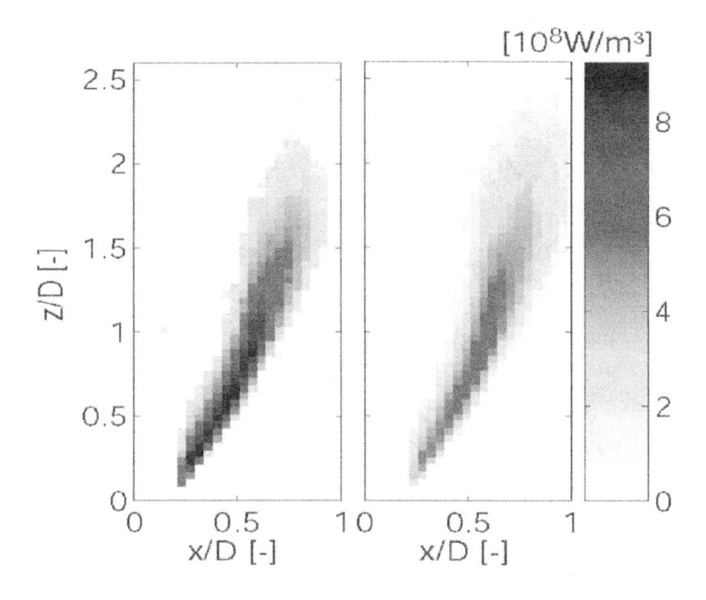

Fig. 6.10 Deconvoluted heat-release distribution, derived from the raw-data (left) and from the corrected data (right) of the OH^*-intensity

OH^*-intensity was used, in case (b) the intensity-corrected values. Both are compared to a spectrum which was measured by a sound-intensity probe. Case (a) overestimates both, the peak-frequency and the amplitude. Case (b) shows a very good agreement with the measured spectrum. The effects of the correction can be understood with the model. The increase of flame length with the correction has increased also the coherence length scales and thus the coherence volume. This compensates the quadratic decrease of acoustic power with heat release rate density. The increasing coherence length due to the correction also shifts the peak frequency to smaller values.

6.2.6.4 Investigation of adiabatic unconfined flames

To evaluate combustion noise without negative influences of the confinement, unconfined freely burning flames are often considered. They provide good optical access, which is crucial for laser optical studies but suffer from entrainment of cold ambient gases, which generates equivalence and temperature changes not present in the confined case. For this reason, it is desirable to protect the flame from the surrounding atmosphere without confining the flame mechanically. This can be achieved, if an inert gas shielding is provided protecting the flow field of the investigated burner from the ambient air. Ideally, the shielding gas should have the same temperature and composition as the combustion products of the investigated

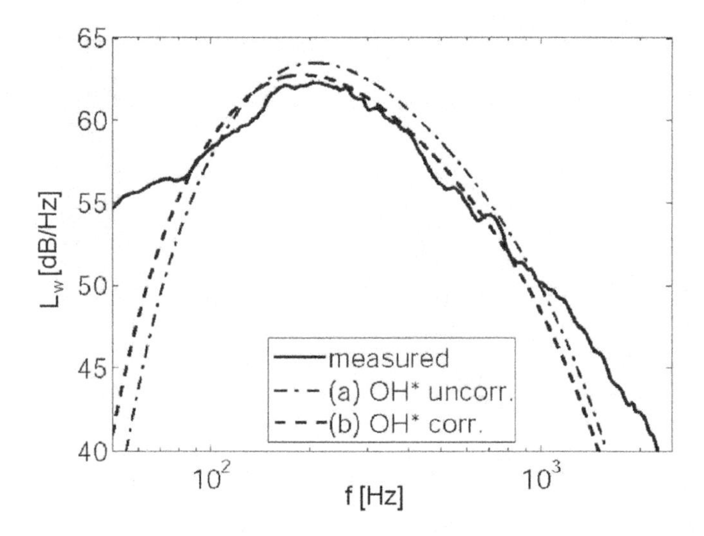

Fig. 6.11 Measured and modeled sound-power-spectrum. The two modeled curves were calculated with and without intensity-correction

flame in order to avoid the formation of gradients. This is achieved if hot exhaust gas is used for shielding the flame under investigation. In the past small laminar flames were used for this purpose [4]. However, the disadvantage of setups with laminar flames stabilized on a perforated plate is the narrow operating range, which does not allow flexible changes of the equivalence ratio. These drawbacks lead to the development of a better suited concept. The LSTM (University Erlangen, Germany) developed a burner with combustion in a porous ceramic matrix [35]. The flame stabilizes in the ceramic filling downstream of the preheating zone. The strong thermal transport against the main flow direction in the porous material leads to high volumetric power density and the formation of a fully combusted exhaust gas, which is free of super equilibrium-radicals and in particular chemically excited OH* and CH*. A burner of this type was adopted for the shielding of a perfectly premixed swirl burner (TD1 burner) (Fig. 6.12).

6.3 Results and Analysis

6.3.1 Validation of the noise-model for unconfined flames

For the validation of the noise-model several parameter-studies were accomplished. The thermal power, the equivalence-ratio and the fuel-composition were varied. The turbulence parameters and the acoustic power were measured, in addition CH^*-chemiluminescence-pictures were recorded. The measured flow-data were used as

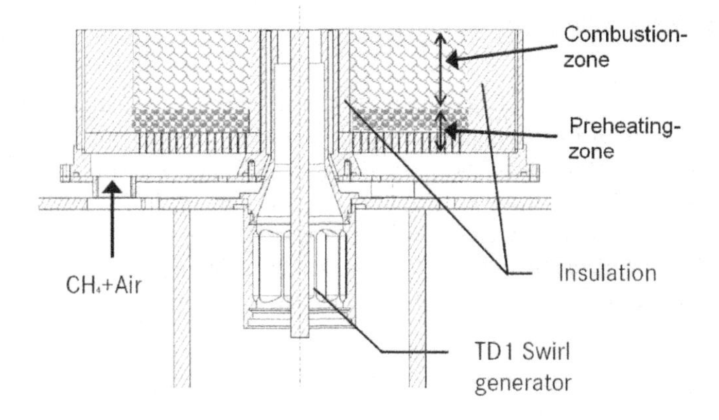

Fig. 6.12 Schematic sketch of the porous burner [41]

input parameter of the noise-model and by applying the free field solution (6.42), the emitted sound-power was calculated. The results were compared to acoustic intensity measurements. Comprehensive comparison is found in [36, 44]. As an example of the results, Fig. 6.13 shows measured and modeled spectra for thermal powers $P_{th} = 40,\ 60,\ 90\ kW$ and an equivalence-ratio $\Phi = 0.8$. The model captures the measured behavior of the spectra very well. Most notably the frequency-offset and the changing of the amplitude with increasing thermal power is reproduced. This validates the correctness of the frequency-wave-number mapping (6.33).

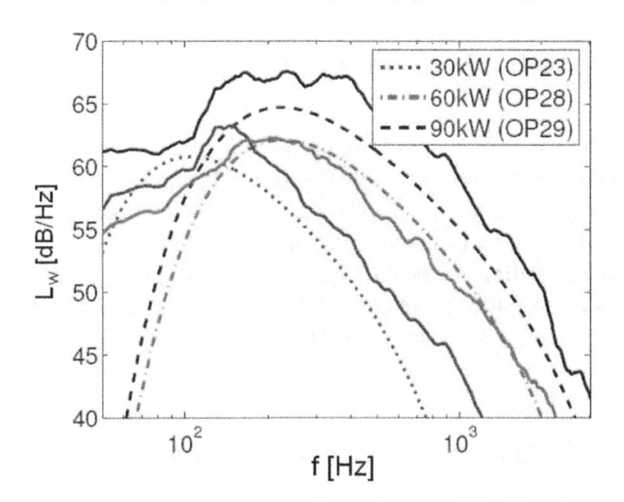

Fig. 6.13 Acoustic power-spectrum-density of a thermal power of $P_{th} = 30, 60, 90\ kW$, natural gas, $\Phi = 0.8$

6.3.2 Adiabatic flames

To remove the effects of cold air entrainment, the experimental setup with the porous burner co-flow described in chapter 6.2.6.4 was used.
In Fig. 6.14 a comparison between the diabatic (left) and adiabatic (right) experiment is shown. The pictures were acquired by recording the PIV-particle scattering. Unlike the cold co-flow, the adiabatic flame is also burning in the outer shear-layers, what can be recognized by their sharp-cut appearance. Without hot co-flow the outer shear entrains cold air from the ambient; thereby the flame is quenched in this region.

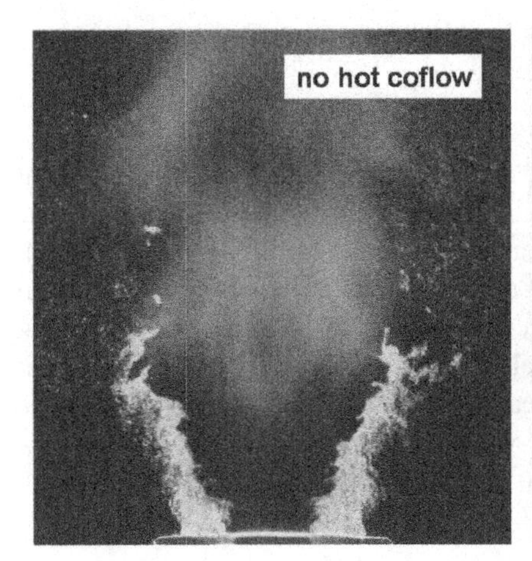

Fig. 6.14 Light sheed of a premixed swirl-flame with and without hot co-flow

Applying the noise model (6.42) to this adiabatic flame gives the result plotted in Fig. 6.15. The modeled spectrum is displayed by the dashed line, the measured one by the solid line. The model-prediction is in good agreement with the experimental data. Notably the spectral decay at higher frequencies fits very well. The difference of approximately 10 dB at the peak-frequency is caused by the porous-burner, which also generates noise, which is not covered by the model. The results clearly indicate, that the noise-model is also valid in the case of an adiabatic flame.

6.3.3 Unconfined flames, modeling based on CFD-data

The ultimate goal of noise modeling is the prediction of sound spectra based on simulation only. Therefore CFD simulations of the experimental setup were performed to assess the capability of the noise model in this context [11]. Figure 6.16 shows

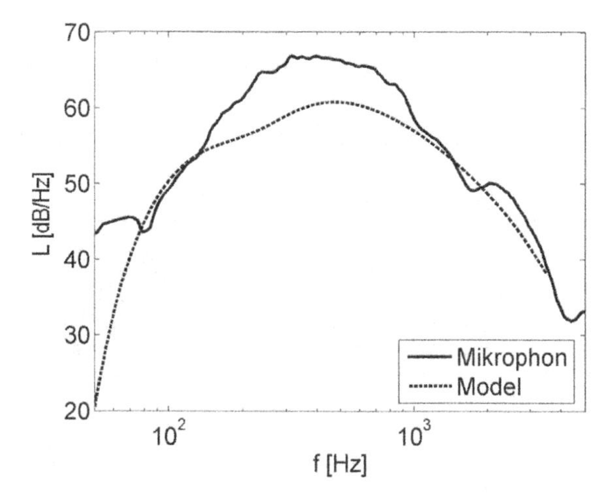

Fig. 6.15 Power spectral density of the noise of an adiabatic confined flame

the comparison of acoustic power-density spectra from measurement (marked lines) and CFD post processing with the developed model (unmarked lines) for two different swirl numbers. The input data (mean heat-release, turbulent kinetic energy, turbulent dissipation rate) was obtained from a RANS-simulation. If we consider, that the measurements are disturbed by the room acoustics, the curves are in very good agreement. The changing of the amplitude and the shifting of the frequency-peak because of the two different swirl-numbers are reproduced fully correct by the model. Likewise the spectral decline at high frequencies. This behavior can be perfectly explained by the model. The flame without swirl ($S = 0$) is much longer than the swirl flame ($S = 0.8$). Therefore the turbulent length- and time-scales are larger and the frequencies are lower. The turbulent velocity-fluctuations become smaller than the laminar flame speed, whereby the heat-release fluctuations and therefore the noise-amplitude become higher. This is also captured by the model, by the increasing of the amplitude-scaling factor C_S.

6.3.4 Sound emission from a complex combustion system into the environment

The characteristics of the noise-emission of a confined flame differs markedly from unconfined ones. This is because of the strong influence of the acoustical properties of the surrounding system (resonance, damping, emission,...).
Using the acoustic network model described in chapter 6.2.5.1 the sound emission of a commercial heater combustor was calculated using CFD input data. A comparison of a measured spectrum and a computed one is given in Fig. 6.17. As the experi-

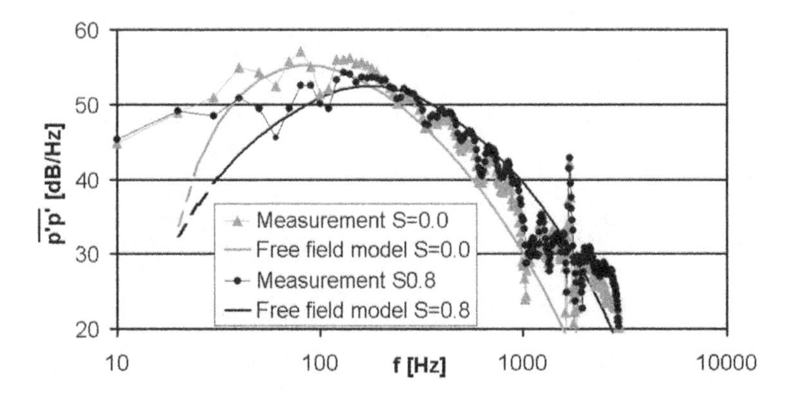

Fig. 6.16 Measured power-spectral-density compared to the spectrum, derived from CFD-data, pictured for two different swirl-numbers

mental data was obtained from a commercial heating system only normalized values are shown. The spectrum was computed using the model in equation (6.16). The required data of the reacting flow was achieved by a RANS-simulation. The acoustic behavior of the burner was determined employing a network-model (see Fig. 6.4). At the exhaust gas exit the impedance-boundary (6.59) is used. The microphone was positioned outside of the burner at a certain distance from the exhaust-gas exit. The sound pressure level at the microphone position was obtained by applying the free field solution to the sound emitted by the exhaust-gas exit. Big differences between measurement and simulation appear only at very low frequencies. The noise measured in this frequency range does not originate from combustion, but is probably induced by the flow as the comparison with cold sound spectra suggests. This sound source is not covered by the model. In the range of higher wave-numbers the calculated spectrum agrees well with the measured one. The peak-frequency shows a good agreement and in particular the spectral descent fits very well. The resonance peaks in the model data are not seen in the experiment. Therefore a better modeling of the acoustic system may be needed.

6.4 Conclusions

In this project the formation of combustion noise in premixed turbulent flames has been investigated. A comprehensive experimental campaign that was sided with theoretical development has resulted in the development of a noise source model. This model was validated in detail and with changing the boundary conditions of the experiment. The model explains the observed effects comprehensively and allows to draw conclusions on the governing parameters of noise spectra. The effect of enclo-

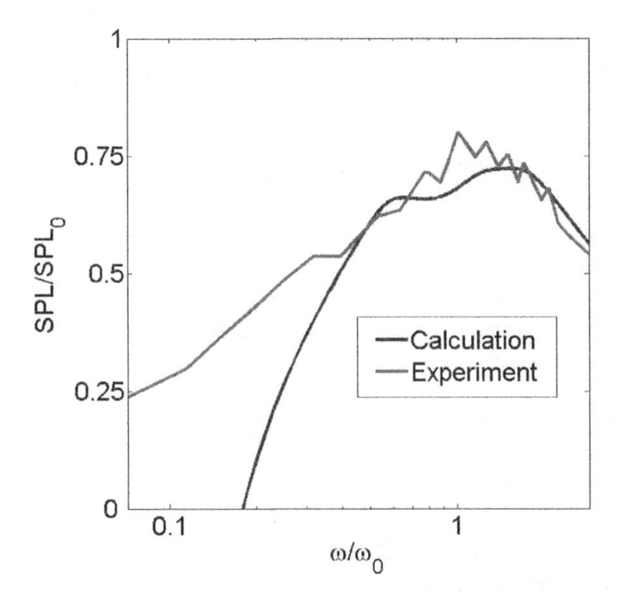

Fig. 6.17 Comparison of the measured and the computed sound-pressure-spectrum of an enclosed flame. Additionally the modeled spectrum of the unconfined flame is shown.

sure was addressed using acoustic network modelling. It was demonstrated that the sound emission of a commercial heater combuster could be reproduced very well on the basis of reactive RANS CFD data and the developed combustion noise model. Further work is needed to extend the current theory to non-premixed flames.

Acknowledgements The authors gratefully acknowledge the financial support by the German Research Council (DFG) through the Research Unit FOR486 "Combustion Noise".

References

[1] Boineau P, Gervais Y, Morice V (1996) An aerothermoacoustic model for computation of sound radiated by turbulent flames. In: International Congress on Noise Control Engineering, Liverpool, Proc. Internoise 96
[2] Boineau P, Gervais Y, Toquard M (1997) Spatio-frequential optical measurements of acoustic sources in a turbulent flame. In: International Congress on Noise Control Engineering, Budapest, Hungary, Proc. Internoise 97
[3] Bragg SL (1963) Combustion noise. J of the Institute of Fuel 36:12–16
[4] Cabra R, Hamno Y, Chen J, Dibble R, Acosta F, Holve D (2000) Ensemble diffraction measurements of spray combustion in a novel vitialed coflow tur-

bulent jet flame burner, NASA/CR 2000-210466. National Aeronautics and Space Administration, Glenn Research Center

[5] Clavin P, Siggia ED (1991) Turbulent premixed flames and sound generation. Combustion Science and Technology 78:147–155

[6] Crighton D, Dowling A, Ffowcs Williams J, Heckl M, Leppington F (1992) Modern Methods in Analytical Acoustics. Springer-Verlag Berlin Heidelberg New York

[7] Fischer A (2004) Hybride, thermoakustische Charakterisierung von Drallbrennern. PhD thesis, Technische Universität München, Germany

[8] Flemming F, Sadiki A, Janicka J, Wäsle J, Winkler A, Sattelmayer T (2005) Large eddy simulation and particle image velocimetry of an isothermal swirling flow. In: 69th Annual Meeting of the German Physical Society (DPG), Vol. 69

[9] Haber LC (2000) An investigation into the origin, measurement and application of chemiluminescent light emissions from premixed flames. Master's thesis, Virginia Polytechnic Institute and State University, Blacksburg, Virginia, U.S.A.

[10] Hinze JO (1975) Turbulence, 2nd edn. Mc Graw-Hill, ISBN 0-07-029037-7

[11] Hirsch C, Winkler A, Wäsle J, Sattelmayer T (2006) Calculating the turbulent noise source of premixed swirl flames from time mean reactive RANS variables. In: 13th International Congress on Sound and Vibration

[12] Hirsch C, Wäsle J, Winkler A, Sattelmayer T (2007) A spectral model for the sound pressure from turbulent premixed combustion. In: 31st Symp (Int.) on Combustion, pp 1435–1441

[13] Klein S (2000) On the acoustics of turbulent non-premixed flames. PhD thesis, University of Twente

[14] Klein SA, Kok JBW (1999) Sound generation by turbulent non-premixed flames. Combustion Science and Technology 149/1-6:267–295

[15] Lauer M, Sattelmayer T (2007) Luftzahlmessung in einer turbulenten Drallflamme auf Basis spektral aufgelöster Chemilumineszenz. In: VDI-Berichte 1988, pp 735–741

[16] Lauer M, Sattelmayer T (2008) Heat release calculation in a turbulent swirl flame from laser and chemiluminescence measurements. In: 14th Int Symp on Applications of Laser Techniques to Fluid Mechanics

[17] Levine H, Schwinger J (1948) On the radiation of sound from an unflanged circular pipe. Physical Review 73:383–406

[18] Lighthill MJ (1952) On sound generated aerodynamically. Proc Royal Soc London, Series a 211:564–587

[19] Mahan JR, Jones J (1984) Recovery of burner acoustic source structure from far-field sound spectra. AIAA Journal 22:631–637

[20] Mühlbauer B, Ewert R, Kornow O, Noll B, Delfs J, Aigner M (2008) Simulation of combustion noise using CAA with stochastic sound sources from RANS. In: 14th AIAA/CEAS Aeroacoutics Conference, Vancouver

[21] Peters N (2000) Turbulent Combustion. Cambridge University Press, ISBN 0-521-66082-3

[22] Pfadler S, Leipertz A, Dinkelacker F, Wäsle J, Winkler A, Sattelmayer T (2006) Two-dimensional direct measurement of the turbulent flux in turbulent premixed swirl flames. In: 31st Symp (Int.) on Combustion
[23] Polifke W, Paschereit C, Döbbeling K (1999) Suppression of combustion instabilities through destructive interference of acoustic and entropy waves. In: 6th International Congress on Sound and Vibration, Copenhagen
[24] Proudman I (1952) The generation of noise by isotropic turbulence. In: Proceedings of the Royal Society of London, Series A. Mathematical and Physical Sciences, vol 214, pp 119–132
[25] Rajaram R, Lieuwen T (2003) Parametric studies of acoustic radiation from premixed flames. Combustion Science and Technology 175(12):29
[26] Ribner HS (1969) Quadrupole correlation governing the pattern of jet noise. J of Fluid Mech 38, part 1:1–24
[27] Rotta J (1972) Turbulente Strömungen. Teubner Verlag
[28] Schmid HP (1995) Ein Verbrennungsmodell zur Beschreibung der Wärmefreisetzung von vorgemischten turbulenten Flammen. PhD thesis, Fakultät für Chemieingenieurwesen, TH Karlsruhe
[29] Schmid HP, Habisreuther P (1998) A model for calculating heat release in premixed turbulent flames. Combustion and Flame 113:79–91
[30] Smith TJB, Kilham JK (1963) Noise generation by open turbulent flames. Journal of the Acoustic Society of America 35(5):715–724
[31] Strahle WC (1971) On combustion generated noise. J of Fluid Mech 49(2):399–414
[32] Strahle WC (1978) Combustion noise. Prog Energy Combust Sci 4:157–176
[33] Strahle WC (1982) Estimation of some correlations in a premixed reactive turbulent flow. Combustion Science and Technology 29:243–260
[34] Tennekes H, Lumley JL (1972) A First Course in Turbulence, 11th edn. The MIT Press, ISBN 0-262-200198
[35] Trimis D, Durst F, Piekenäcker O, Piekenäcker K (1997) Porous medium combustor versus combustion systems with free flames. In: Proceedings 2nd International Symposium on Heat Transfer Enhancement and Energy
[36] Wäsle H (2007) Vorhersage der Lärmemission turbulenter Vormischflammen. PhD thesis, Technische Universität München
[37] Wäsle J, Winkler A, Weinmüller C, Sattelmayer T (2004) Real time measurement techniques for turbulent flame noise. In: 2nd International Workshop (SFB568), Heidelberg, Deutschland, vol 2, pp 111–118
[38] Wäsle J, Winkler A, Sattelmayer T (2005) Experimentelle Untersuchungen zum akustischen Umsetzungsgrad in turbulenten Drallflammen. In: VDI Thermodynamik Kolloquium
[39] Wäsle J, Winkler A, Sattelmayer T (2005) Influence of the combustion mode on acoustic spectra of open turbulent swirl flames. In: 12th International Congress on Sound and Vibration
[40] Wäsle J, Winkler A, Sattelmayer T (2005) Spatial coherence of the heat release fluctuations in turbulent jet and swirl flames. Flow Turbulence and Combustion 75:29–50

[41] Wäsle J, Winkler A, Rössle E, Sattelmayer T (2006) Development of an annular porous burner for the investigation of adiabtic unconfined flames. In: 13th Int. Symp. on Applications of Laser Techniques to Fluid Mechanics
[42] Wäsle J, Winkler A, Lauer M, Sattelmayer T (2007) Combustion noise modeling using chemiluminescence data as indicator for the heat release distribution. In: ECM Proceedings
[43] Weyermann F, Hirsch C, Sattelmayer T (2008) Numerische Simulation der Schallabstrahlung von eingeschlossenen turbulenten Flammen. In: Fortschitte der Akustik 2008, DEGA
[44] Winkler A (2007) Validierung eines Modells zur Vorhersage turbulenten Verbrennungslärms. PhD thesis, Technische Universität München
[45] Winkler A, Wäsle J, Sattelmayer T (2004) Investigation of combustion noise by real time laser measurement techniques. CFA/DAGA Gemeinschaftstagung, Strassburg
[46] Winkler A, Wäsle J, Sattelmayer T (2004) Laserinduzierte Fluoreszenz in Echtzeit zur Bestimmung des Flammenlärms. Tagungsbericht GALA, Karlsruhe
[47] Winkler A, Wäsle J, Sattelmayer T (2005) Experimental investigations on the acoustic efficiency of premixed swirl stabilized flames. In: 11th AIAA/CEAS Aeroacoustics Conference, AIAA-2005-2908
[48] Winkler A, Wäsle J, Hirsch C, Sattelmayer T (2006) Peak frequency scaling of combustion noise from premixed flames. In: 13th International Congress on Sound and Vibration

Chapter 7
Theoretical and Numerical Analysis of Broadband Combustion Noise

Thanh Phong Bui and Wolfgang Schröder

Abstract Combustion noise and sound source mechanisms of turbulent reacting flows are investigated theoretically and numerically. Two major targets have been addressed in this work. To deepen the understanding about the sound source mechanisms involved in the noise generation process of turbulent reacting flows, an appropriate system of equations is required, which is based on an acoustic analogy concept. For this purpose, the acoustic perturbation equations for reacting flows (APE-RF) have been derived by reformulating the conservation equations for multi-component turbulent reacting flows. Furthermore, numerical methods have been developed to predict combustion generated noise in an efficient and accurate manner. A hybrid computational fluid dynamics / computational aeracoustics (CFD/CAA) approach is employed. In the first step of the hybrid analysis the reactive flow field is modeled via large-eddy simulations (LES) or direct numerical simulations (DNS) followed by the acoustic simulation in the second step using the APE-RF system. The acoustic impact of the various source mechanisms are discussed and analyzed in detail, while special emphasis is placed on the requirements in terms of numerical aeroacoustics. Numerical results compared with experimental data are shown to evidence the applicability of the proposed hybrid CFD / CAA method.

7.1 Introduction

In the very beginning of combustion noise research, theoretical considerations [5] and experimental observations [61] indicated turbulent flames to be low frequency radiators of essentially monopole type. For this reason, the far-field pressure radiated by turbulent flames has been related to a change of volume increase due to combustion which acts as a monopole. From these findings, and after LIGHTHILL

T. Ph. Bui and W. Schröder
Institute of Aerodynamics, RWTH Aachen University Wüllnerstr. 5a, e-mail: p.bui@aia.rwth-aachen.de

A. Schwarz, J. Janicka (eds.), *Combustion Noise*, DOI 10.1007/978-3-642-02038-4_7,

published his brilliant acoustic analogy [47, 48], KOTAKE & HATTA [41] derived in 1965 an inhomogeneous wave equation to account for reacting flow effects. Since those days, many attempts have been undertaken to clarify the origin of combustion noise [70, 34, 53, 26] and to develop a theory to predict the thermoacoustic efficiency of turbulent flames [56, 39, 1, 16]. Motivated by experimental results by PRICE ET AL. [52], in which the far-field acoustic pressure could be correlated to the time derivative of the methylidyne (CH) emission intensity, STRAHLE [63] showed in a study, that there is a direct relation between the acoustic pressure and the volume integral of the time derivative of the global reaction rate. This theory predicts the radiated acoustic field being isotropic for sound with a wavelength $\lambda \gg L$, where L denotes the characteristic flame length scale. However, weak directivity effects have also been reported by SMITH & KILHAM [61], in which they measured a preferred direction of the radiated sound field, which was between 40° and 80° for the flame tested. Theoretical and numerical studies [63, 29, 35] about the acoustic spectrum and directivity effects of turbulent flames also were based on estimating the acoustic output of the different contributions in LIGHTHILL's tensor. In combustion noise research for flows at low MACH numbers, however, the effect of unsteady heat release was considered the most dominant source contribution and therefore, the investigations mainly pursued analyzing the acoustic output of that mechanism [63, 38, 29, 15]. The dominant contribution of the unsteady heat release to the acoustic pressure field led in the research field of flame diagnostics to the idea of formulating an inverse problem in which the unsteady heat release distribution is determined by measuring the acoustic pressure field in open [54] and in confined [65, 46] systems. In the literature, there are only a few papers in which reacting flows are simulated using a fully compressible formulation, i.e., in which the acoustic wave propagation is taken into account. To solve the problem of compressible multicomponent reacting flows, the transport equations of each species have to be considered. This, however, restricts the chemical reaction schemes to a very limited number of subreactions due to the resulting computational costs [58]. Since many applications within the combustion community can be considered low MACH number problems, numerous articles deal with combustion simulations using a low MACH number approximation, where detailed chemical reactions can be used at reasonable computational costs. However, the disadvantage of this low MACH number approximation from an acoustic point of view is that no sound propagation can be directly computed in the flow analysis. In other words, the possible impact of the acoustic field on the flow and the combustion problem, respectively, is neglected. A second step has to be performed to investigate the acoustic field. This approach has been followed in the present two-step method. That is, a large-eddy simulation (LES) based on incompressible NAVIER-STOKES equations is used to provide the unsteady flow data to determine the source terms of the acoustic perturbation equations.

The primary objective of this work is to develop a theoretical basis to investigate combustion noise in a hybrid computational fluid dynamics and computational aeroacoustics (CFD/CAA) framework. As mentioned above, combustion noise investigations were mainly based on inhomogeneous wave equations using an ordinary wave operator [35, 63, 71, 38]. Like Lighthill's acoustic analogy, those theories were derived by rearranging the conservation equations of mass and momentum and the first and second law of thermodynamics to obtain an inhomogeneous wave equation. This operator describes wave propagation in a uniform mean flow and the right-hand side contains all non-linear effects including linear mechanisms like sound refraction and convection. In the present work, we follow the idea by Strahle [62] and extend an existing set of equations, i.e., the acoustic perturbation equations (APE) [21], to reacting flows. The homogeneous APE formulation [21] is the basis to derive the APE system for reacting flows (APE-RF), since the left-hand side of the APE system describes wave propagation in non-uniform and irrotational mean flows with varying mean speed of sound, i.e., wave refraction and convection are taken into account. The stability of the system to arbitrary mean flows was proven by EWERT & SCHRÖDER [21]. CHIU & SUMMERFIELD [15] pointed out that the reactive flow field of an open jet flame can be divided into three parts, namely the outer region, where the speed of sound is uniform, an intermediate region, in which the mean speed of sound varies in space, and the reaction zone, where non-linear reacting processes occur. According to this analysis, the inhomogeneous APE-RF equations are only solved in the reaction zone where the acoustic sources are non-zero, and the wave propagation in the intermediate and outer region can be described using the homogeneous system. The overall numerical approach to simulate combustion noise is a two-step method, the first step of which is based on an LES, followed by the computational aeroacoustics (CAA) simulation to compute the acoustic field. This hybrid LES/CAA approach is similar to that used in airframe noise analyses [21, 22]. However, in this study the description of the acoustic field is extended to reacting flow effects, which results in the APE-RF system. The reader is reminded that when chemical reactions are considered, the application of such a hybrid approach is more essential since the disparity of the characteristic hydrodynamic and acoustic length scales is even more pronounced than in the non-reacting case.

The investigation of broadband combustion noise possesses the following structure. For the sake of completeness, a description of a wave equation based theory to compute combustion noise is given in the first part of section 7.2. In the second part of section 7.2, the acoustic perturbation equations extended to reacting flows (APE-RF) are derived. The sources especially on the RHS of the APE-RF pressure-density relation are related to sound producing mechanisms in turbulent reacting flows. Furthermore, simplified source formulation are presented, which are most suitable in conjunction with a hybrid LES/APE-RF approach. Besides the physical mechanisms inherently contained on the RHS of the APE-RF system, a local form of RAYLEIGH's criterion for wave amplification can be derived from the APE-RF system. The structure of the two-step approach is outlined in section 7.3. To show the capability of the APE-RF system, the acoustic field of the so-called H3 flame is simulated using a simplified source formulation ($q_e = -\bar{c}^2\bar{\rho}/\rho D\rho/Dt$). The results are presented in section 7.4.1. The source terms, especially the thermoacoustic sources, which are related with multicomponent reacting flows are investigated in section

7.4.2 using DNS data. Additional simplified source formulations for the APE-RF system and their numerical applicability are studied using LES data from the DLR-A flame in section 7.4.3. In section 7.5 the summary is given and conclusions of this work are drawn.

7.2 Aeroacoustic theories to compute combustion generated noise

The aeroacoustics research field deals with the noise generation due to aerodynamic/flow effects. LIGHTHILL'S acoustic analogy [47, 48] could be seen as the first attempt to quantify noise generation due to aerodynamic effects. The key idea was to reformulate the equations governing the fluid motion into one single scalar equation, i.e., an inhomogeneous wave equation for the density. Using an additional equation describing the relation between the pressure and the density, the wave equation can be rewritten in terms of the pressure. From classical acoustics the homogeneous wave equation was known to describe wave propagation in an uniform medium at rest, while an acoustic field can be forced by fluctuating contributions, i.e., source terms on the right-hand side (RHS) of the wave equation. The RHS of LIGHTHILL'S wave equation now contains expressions resulting from the reformulation of the fluid governing equations, which have been consequently interpreted as noise generating mechanisms related to flow effects *analogously* to classical acoustics. From this starting point, aeroacoustic investigations for non-reacting flows, e.g., jet noise [27, 32, 2, 28], rotor noise [6, 24], airframe noise [17, 23, 60], as well as combustion noise have been performed.

7.2.1 Acoustic analogies based on a scalar wave equation

First attempts to quantify the thermoacoustic efficiency from open turbulent flames can be attributed to BRAGG's [5] work in 1963. However, the first article, which appeared in the literature, in which LIGHTHILL'S acoustic analogy has been extended to reacting flows has been published by KOTAKE & HATTA [41] in 1965. Additionally to the governing equations for mass and momentum, they included the first law of thermodynamics in their derivation to yield an inhomogeneous wave equations for the pressure with sources describing the effect of reacting flows. Since these days, many derivatives of an inhomogeneous wave equation have been published [15, 64, 18, 71, 38] to account for reacting flow effects and trying to derive scaling laws for combustion generated noise.

Since the publication of LIGHTHILL'S acoustic analogy, knowledge about how to interpret the sources of the RHS of the inhomogeneous wave equation has grown substantially. For instance, it has been pointed out by DOAK [19] and FFOWCS WILLIAMS [25] that besides the "real" acoustic source mechanisms, the RHS also includes effects like wave refraction, i.e., pseudo source terms. Since the homogeneous wave equation is valid to describe wave propagation in a uniform mean flow only, the otherwise erroneous description of wave propagation in a non-uniform mean flow has to be compensated by source terms, which are no sound producing but rather sound field modifying or correction terms. One major goal in the field of aeroacoustic research, which has been followed since then, is to identify and separate the "real" sources from the pseudo sources. An intermediate step to reach this goal has been proposed by e.g. HOWE [32], MÖHRING [50], and others. They were able to separate the refraction and convection effects from the other sources and shifted these effects to the left-hand side of the wave equation. The resulting wave operator then was capable of describing wave propagation in non-uniform mean flows, which leads to a reduction of the source area. However, no explicit integral formulation can be given, since in general no solutions to the associated GREEN's function are known.

EWERT & SCHRÖDER [21] used this concept in the case of the linearized EULER equations to yield a system of acoustic perturbation equations (APE) capable of describing wave propagation in an irrotational non-uniform mean flow.

7.2.2 Acoustic perturbation equations for reacting flows (APE-RF)

Four different APE formulations have been presented in [21]. The APE-1 system represents the fundamental system of acoustic perturbation equations. This system has been derived by reformulating the continuity and NAVIER-STOKES equations into a system with independent variables $(h,u)^T$

$$\frac{\partial h}{\partial t} + c_\infty^2 (\nabla \cdot u) = q \tag{7.1}$$

$$\frac{\partial u}{\partial t} + \nabla h = f \tag{7.2}$$

with

$$q = -u \cdot \nabla h - \left(c^2 - c_\infty^2\right)(\nabla \cdot u) + \frac{c^2}{R}\frac{Ds}{Dt} \tag{7.3}$$

$$f = -(u \cdot \nabla)\,u + \frac{\nabla \cdot \tau}{\rho} + T\nabla s, \tag{7.4}$$

where h and u are the enthalpy and the velocity vector, respectively. The key idea of this rearranging is that the resulting homogeneous system is linear, or in other words, that its coefficients $(c_\infty, 1)$ are constant. This allows for deriving filter matrices, which can be applied to the RHS to yield a source formulation, which in turn excites acoustic modes only. A shifting of the terms describing refraction and convection of acoustic waves from the RHS to the LHS, yields the APE-1 system for the perturbation variables $(p', u^a)^T$. In a next step, they have shown, that this system is able to describe wave propagation in an irrotational non-uniform mean flow. Moreover, the stability for arbitrary mean flows has been proven. The advantage of a filtered source, which does not excite other modes than the acoustic one, however, requires additional computational effort in evaluating those source terms [21]. Once, the benign properties of the homogeneous APE system has been shown and the stability has been proved, they introduced the APE-4 formulation, which incorporates an easy to evaluate source formulation. Therefore, the continuity and NAVIER-STOKES equations are rewritten, such that the LHS structure corresponds to the homogeneous APE-1 system, while all other terms are shifted to the RHS. Note, though the APE-1 and the APE-4 formulations differ in their source formulation, they comprise the same number of modes, namely the acoustic and vorticity modes, while the entropy modes are excluded. The acoustic perturbation equations for reacting flows (APE-RF), which will be discussed in the next subsection, are based on the homogeneous APE system [21]. A set of three equations, such as the APE-2 formulation, is chosen to form the APE-RF system. A comparison with the two-equation formulation, like the APE-4, will be given below. For combustion noise simulations this homogeneous system has been chosen to take advantage of its benign properties to simulate wave propagation, i.e., its validity for irrotational non-uniform mean flows, while excitations of instabilities are prevented. Like the APE-4 formulation, the resulting sources are not filtered. Using the decomposition of the density (ρ), the pressure (p), and the velocity (u) into a mean, denoted by an overbar, and a fluctuating part, denoted by a prime

$$\rho = \overline{\rho} + \rho' \tag{7.5}$$

$$p = \overline{p} + p' \tag{7.6}$$

$$u = \overline{u} + u' \tag{7.7}$$

the homogeneous APE system [21] in a three-equation variant reads

$$\frac{\partial \rho'}{\partial t} + \nabla \cdot \left(\rho' \overline{u} + \overline{\rho} u'\right) = q_c \tag{7.8}$$

$$\frac{\partial u'}{\partial t} + \nabla \left(\overline{u} \cdot u'\right) + \nabla \left(\frac{p'}{\overline{\rho}}\right) = q_m \tag{7.9}$$

$$\frac{\partial p'}{\partial t} - \overline{c}^2 \frac{\partial \rho'}{\partial t} = q_e, \tag{7.10}$$

where q_c, q_m and q_e are zero. To derive the APE-RF system, the governing equations for reacting flows are rearranged such that the left-hand side describes the original homogeneous APE system, whereas the right-hand side (RHS) consists of all remaining terms including the sources related to chemical reactions. This derivation of the APE-RF system, which is based on the conservation of mass, momentum, and energy of multicomponent reacting flows, is given in detail in Bui et al.[13]. Since the main combustion noise source, i.e., the unsteady heat release rate, can be found on the RHS of the pressure-density relation, the different source mechanisms contained in this equation are discussed in the following section 7.2.2.1.

7.2.2.1 Source mechanisms contained in the pressure-density relation

To clarify the different contributions, i.e., the different source mechanisms contained in the RHS of the pressure density relation, the energy equation for reacting flows, i.e.,

$$\frac{D\rho}{Dt} = \frac{1}{c^2}\frac{Dp}{Dt} + \frac{\alpha}{c_p}\left(\sum_{n=1}^{N}\left.\frac{\partial h}{\partial Y_n}\right|_{\rho,p,Y_m}\rho\frac{DY_n}{Dt} + \nabla\cdot q - \frac{\partial u_i}{\partial x_j}\tau_{ij}\right), \tag{7.11}$$

is introduced to express the substantial time derivative of the density, where α is the volumetric expansion coefficient. To derive the required expressions, the equation of species conservation, which reads

$$\rho\frac{DY_n}{Dt} = \omega_n - \nabla\cdot J_n \tag{7.12}$$

is also introduced, with J_n and ω_n representing the flux of species n by diffusion and the production rate per unit volume, respectively. Then the sum within eqn. (7.11) $\sum_{n=1}^{N}\left.\frac{\partial h}{\partial Y_n}\right|_{\rho,p,Y_m}\rho\frac{DY_n}{Dt}$ can be rewritten using the species conservation equation and the chain rule

$$\left.\frac{\partial h}{\partial Y_n}\right|_{T,p,Y_m} = \left.\frac{\partial h}{\partial Y_n}\right|_{\rho,p,Y_m} + \left.\frac{\partial h}{\partial \rho}\right|_{p,Y_n}\left.\frac{\partial \rho}{\partial Y_n}\right|_{T,p,Y_m}. \tag{7.13}$$

Therefore, the term appearing in the pressure-density relation can be divided into three parts

$$\sum_{n=1}^{N}\left.\frac{\partial h}{\partial Y_n}\right|_{\rho,p,Y_m}\rho\frac{DY_n}{Dt} = \underbrace{\sum_{n=1}^{N}\left.\frac{\partial h}{\partial Y_n}\right|_{T,p,Y_m}\omega_n}_{S_{IA}} - \underbrace{\left.\frac{\partial h}{\partial \rho}\right|_{p,Y_n}\left(\sum_{n=1}^{N}\left.\frac{\partial \rho}{\partial Y_n}\right|_{T,p,Y_m}\omega_n\right)}_{S_{IB}}$$
$$- \underbrace{\sum_{n=1}^{N}\left.\frac{\partial h}{\partial Y_n}\right|_{\rho,p,Y_m}\nabla\cdot J_n}_{S_{IC}}. \tag{7.14}$$

The first part S_{IA} is given by the heat release rate per unit volume and the second part S_{IB} describes the volumetric expansion due to non-isomolar combustion, if the average molecular weight is not assumed to be constant, as was shown by TRUFFAUT ET AL. [71]. The third term S_{IC} is determined by the effects due to species diffusion. Additionally to these sources, heat diffusion and viscous effects are represented by $\nabla\cdot q$ and $\frac{\partial u_i}{\partial x_j}\tau_{ij}$ in eqn. (7.11).

In a more physical formulation, the pressure-density relation of the APE-RF system can therefore be written as

$$\frac{\partial p'}{\partial t} - \overline{c}^2 \frac{\partial \rho'}{\partial t} =$$

$$-\overline{c}^2 \left[\left(\frac{\overline{\rho}}{\rho} + \frac{p-\overline{p}}{\rho \overline{c}^2} \right) \left\{ \underbrace{\frac{1}{c^2}\frac{Dp}{Dt} + \frac{\alpha}{c_p}\left(S_{IA} + S_{IB} + S_{IC} + \nabla \cdot q - \frac{\partial u_i}{\partial x_j}\tau_{ij} \right)}_{D\rho/Dt} \right\} \right.$$

$$\left. - \frac{1}{\overline{c}^2}\frac{Dp}{Dt} - \nabla \cdot (u\rho_e) \underbrace{- \frac{\gamma - 1}{\gamma} u \cdot \nabla \overline{\rho} - \frac{p}{\overline{c}^2} u \cdot \left(\frac{\nabla \overline{p}}{\overline{p}} - \frac{\nabla \overline{\rho}}{\overline{\rho}} \right)}_{\overline{\chi}} \right] . \tag{7.15}$$

Additionally to the aforementioned source mechanisms, the remaining sources of eqn. (7.15) describe the non-uniform mean flow effects ($\overline{\chi}$), the excitation of acoustic waves due to combustion at non-constant pressure, and the influence of acceleration of density inhomogeneities ($\nabla \cdot (u\rho_e)$) on the radiated field within a region, where density and sound speed differ from the ambient values.

7.2.3 Summary of the APE-RF formulation

To better understand the form of the APE-RF system, the most important aspects of the APE-RF system are summarized in the following. Rearranging the conservation equations of mass and momentum for a multicomponent fluid, such that the left-hand side result in the original APE system leads to the equations for the perturbation density and velocities of the APE-RF system. Additionally to these two equations, a pressure density relation has to be included to close the system whose RHS is zero, if only isentropic flows are considered. However, in the case of combustion this source term can not be neglected. It is this RHS of the pressure-density relation (7.15), which includes the source dominating the excitation of acoustic waves in combustion noise. By introducing the energy equation for multicomponent reacting flows, several other source mechanisms could be identified besides the effect of unsteady heat release. Hence, the general APE-RF system yields

$$\frac{\partial \rho'}{\partial t} + \nabla \cdot \left(\rho' \overline{u} + \overline{\rho} u' \right) = q_{c,rf} \tag{7.16}$$

$$\frac{\partial u'}{\partial t} + \nabla \left(\overline{u} \cdot u' \right) + \nabla \left(\frac{p'}{\overline{\rho}} \right) = q_{m,rf} \tag{7.17}$$

$$\frac{\partial p'}{\partial t} - \overline{c}^2 \frac{\partial \rho'}{\partial t} = q_{e,rf} \tag{7.18}$$

with the right-hand side sources

$$q_{c,rf} = -\nabla\cdot\left(\rho' u'\right)' \tag{7.19}$$

$$q_{m,rf} = \nabla p\left(\frac{1}{\bar{\rho}}-\frac{1}{\rho}\right)-\frac{1}{\bar{\rho}}\left[\nabla\bar{p}+\left(\frac{\rho}{\bar{\rho}}c^2-\bar{c}^2\right)\nabla\bar{\rho}\right]$$
$$+\left[\frac{\nabla\cdot\tau}{\rho}+\sum_k Y_k f_k-(\bar{u}\cdot\nabla)\bar{u}-(u'\cdot\nabla)u'-(\omega'\times\bar{u})-(\overline{\omega}\times u')\right] \tag{7.20}$$

$$q_{e,rf} = -\bar{c}^2\left[\left(\frac{\bar{\rho}}{\rho}+\frac{p-\bar{p}}{\rho\bar{c}^2}\right)\left\{\frac{1}{c^2}\frac{Dp}{Dt}+\frac{\alpha}{c_p}\left(S_{IA}+S_{IB}+S_{IC}+\nabla\cdot q-\frac{\partial u_i}{\partial x_j}\tau_{ij}\right)\right\}\right.$$
$$\left.-\frac{1}{\bar{c}^2}\frac{Dp}{Dt}-\nabla\cdot(u\rho_e)-\frac{\gamma-1}{\gamma}u\cdot\nabla\bar{\rho}-\frac{p}{\bar{c}^2}u\cdot\left(\frac{\nabla\bar{p}}{\bar{p}}-\frac{\nabla\bar{\rho}}{\bar{\rho}}\right)\right]. \tag{7.21}$$

It has to be emphasized that during the derivation of the APE-RF system no assumptions about non-premixed or premixed flames have been made, such that the APE-RF system is generally valid for multicomponent reacting flows. During derivation, the "excess density" has been introduced, which is defined as

$$\rho_e = (\rho-\bar{\rho})-\frac{p-\bar{p}}{\bar{c}^2}. \tag{7.22}$$

In general, the APE-RF system can be reduced to a two equation system like the APE-4 variant. This would lead to an APE system for reacting flows, which is directly comparable to the APE-4 version, i.e., the homogeneous system possesses only vortical and acoustic modes, while the RHS sources are not filtered. In terms of combustion noise, however, it has to be emphasized that no assumptions or simplifications have been applied during the derivation procedure, such that the APE-RF system is generally valid for simulating noise excited by multicomponent turbulent reacting flows compared to the APE-4 formulation, in which the entropy source term formulations are based on a first order approximation of the first law of thermodynamics. The three equation version, like the APE-2 variant, has been chosen for the APE-RF system to avoid additional derivatives to be evaluated for the source terms and to more clearly relate the different source expressions to physical source mechanisms. However, as has been shown for the APE-2 system [21], the entropy modes are reintroduced into the system.

7.2.4 Source term formulations

7.2.4.1 Simplified formulation I: Substantial time derivative of the density

Eqs. (7.19,7.20,7.21) contain all the source terms or, in other words, all acoustic source mechanisms, which excite the sound field when reacting flows are considered. On the one hand, this set of equations represents the intricate physics involved in combustion generated sound fields, on the other hand, it evidences that the simulation of the sound generation process is extremely cumbersome to compute. For this reason, we now turn to introduce a simpler source term that can be easily determined by an LES solution.

The easier source formulation is compared with the complete system to evidence what mechanisms are neglected and what sources are retained in the simplified APE-RF system. By juxtaposing numerical and experimental data in sec.7.4.1, 7.4.2, and 7.4.3 it will be shown how reliable the concise source formulation is or, in other words, it will be indicated how little the impact of the neglected terms on the sound field of an open non-premixed flame is.

Comparing the pressure spectra between the reacting and non-reacting case of the radiated field of non-premixed and premixed flames it can be assumed that the effect of chemical reactions dominates over vortex induced noise, which generally is the dominant source term in non-reacting jets. For instance, jet noise simulations of non-reacting jets have been performed by GRÖSCHEL ET AL. [28] for single and coaxial jets using LES and the APE system. The major source term was shown to be the perturbed Lamb vector, which appears on the RHS of the APE-momentum equation.

To simulate combustion noise, former studies showed the effect of unsteady heat release is to be included in the investigation. Since the total time derivative of the density contains this major effect and the instantaneous density fluctuation is immediately available from an LES, this term represents the simplified source expression of the RHS of the APE-RF system, while mean flow effects and the acceleration of density inhomogeneities are neglected. Recall from eqn. (7.11) that besides the unsteady heat release, the effects of non-isomolar combustion, heat and species diffusion, and viscous effects are implicitly taken into account. This leads to a reduced RHS of the pressure-density relation

$$q_{e,rf,D\rho} = -\bar{c}^2 \left(\frac{\bar{\rho}}{\rho} + \frac{p-\bar{p}}{\rho \bar{c}^2} \right) \frac{D\rho}{Dt}. \tag{7.23}$$

In the following, the second term being multiplied by a factor of $\frac{p-\bar{p}}{\rho \bar{c}^2}$ will be shown to be negligible in the low MACH number limit. Applying a low MACH number approximation by expanding the flow variables of the flow in a power series of $\varepsilon = \gamma M^2$, the first and second term of the expansion of the pressure yields $p^{(0)}$ and $p^{(1)}$. These terms can be interpreted as a thermodynamic pressure and a hydrodynamic pressure, respectively, as was done by MCMURTRY ET AL. [49] or KLEIN [37]. The lowest order in ε of the conservation equations [49] leads to a relationship, in which the thermodynamic pressure $p^{(0)}$ is spatially uniform. Additionally $p^{(0)}$ is assumed to be constant in time, since an open physical domain is considered here. This means, the only contribution to the pressure fluctuation is due to the hydrodynamic pressure fluctuation. Rewriting eqn. (7.23) and introducing reference quantities yields

$$
\begin{aligned}
q_{e,rf,D\rho} &= -\overline{c}^2 \left(1 + \frac{p-\overline{p}}{\gamma\overline{p}}\right) \frac{\overline{\rho}}{\rho}\frac{D\rho}{Dt} \\
&= -\overline{c}^2 \left(1 + \frac{p-\overline{p}}{\rho_{ref}c_{ref}^2}\frac{\rho_{ref}c_{ref}^2}{\gamma\overline{p}}\right) \frac{\overline{\rho}}{\rho}\frac{D\rho}{Dt}.
\end{aligned} \tag{7.24}
$$

The two terms containing the product $\rho_{ref}c_{ref}^2$ can be estimated by

$$
\frac{p-\overline{p}}{\rho_{ref}c_{ref}^2} \propto \frac{\rho_{ref}u_{ref}^2}{\rho_{ref}c_{ref}^2} = M^2 \tag{7.25}
$$

$$
\frac{\rho_{ref}c_{ref}^2}{\gamma\overline{p}} = \frac{p_{ref}}{\overline{p}}. \tag{7.26}
$$

Since $p_{ref} = \overline{p}$ the second term in eqn. (7.23) scales with M^2 and as such it can be neglected in the low MACH number limit. Thus, the RHS of the pressure-density relation reduces to the very convenient form

$$
q_{e,rf,D\rho} = -\overline{c}^2 \frac{\overline{\rho}}{\rho}\frac{D\rho}{Dt}. \tag{7.27}
$$

7.2.4.2 Simplified formulation II: Partial time derivative of the density

From the definition of the "excess density" eqn. (7.22) it is evident, that neglecting the pressure fluctuations within the source term of the pressure-density relation leads to a source formulation, which is just represented by the partial time derivative of the density multiplied by the square of the mean speed of sound. In the following paragraph, this formulation is again derived. Moreover, the starting point is the pressure-density relation in eqn. (7.21) to clarify the included source mechanisms within the partial time derivative of the density.

If the effect of acceleration of density inhomogeneities $\nabla\cdot(u\rho_e)$ is additionally taken into account besides the effects included in the substantial time derivative of the density, while the pressure variation within the source term is neglected ($\rho_e \approx \rho - \overline{\rho}$), eqn. (7.21) reduces to

$$
\begin{aligned}
q_{e,rf,\partial\rho} &= -\overline{c}^2 \left[\frac{\overline{\rho}}{\rho}\frac{D\rho}{Dt} - \nabla\cdot(u\rho_e) - u\cdot\nabla\overline{\rho}\right] \\
&= -\overline{c}^2 \left[\frac{\overline{\rho}}{\rho}\frac{D\rho}{Dt} - \left(u\cdot\nabla\rho - u\cdot\nabla\overline{\rho} + \rho\nabla\cdot u + \frac{\overline{\rho}}{\rho}\frac{D\rho}{Dt}\right) - u\cdot\nabla\overline{\rho}\right] \\
&= -\overline{c}^2 \frac{\partial\rho}{\partial t},
\end{aligned} \tag{7.28}
$$

where the mass conservation equation has been used to substitute the divergence of the velocity field. Note, the two simplified source formulations, however, requires

CFD input data, in which acoustic components are excluded, i.e, incompressible flow solutions. This is addressed in more detail in section (7.3.1.3).

7.2.5 Rayleigh's criterion for acoustic wave amplification

Since the phenomenon of acoustic-flame interaction is often discussed based on the RAYLEIGH criterion [55], it is worth mentioning that this fundamental criterion is inherently contained in the pressure-density relation of the APE-RF system. This RAYLEIGH criterion reads that the acoustic response is amplified if the pressure fluctuation and the heat release occur in phase. In recent studies, LAVERDANT & THéVENIN [42] and LAVERDANT ET AL. [43] used a so-called local RAYLEIGH'S criterion to investigate the interaction of an acoustic wave with a turbulent flame. This criterion can be derived using the balance equation for the acoustic energy to yield [42]

$$\frac{\gamma-1}{\gamma\rho c^2}\ln\left(\frac{p}{p_0}\right)S_{IA} > 0, \tag{7.29}$$

for the acoustic amplification due to acoustic-flame interaction.
The term $S_{IA} = \sum_{n=1}^{N} \frac{\partial h}{\partial Y_n}\Big|_{T,p,Y_m} \omega_n$ was defined in eqn. (7.14) and represents the heat release rate per unit volume. The same criterion can be directly determined from the APE-RF pressure-density relation. If only the unsteady heat release is considered, the RHS of eqn. (7.21) reduces to

$$\begin{aligned} q_{e,rf} &= -\bar{c}^2\left(\frac{\bar{\rho}}{\rho} + \frac{p-\bar{p}}{\rho\bar{c}^2}\right)\frac{\alpha}{c_p}S_{IA} \\ &= -\bar{c}^2\left(1 + \frac{p-\bar{p}}{\gamma\bar{p}}\right)\frac{\bar{\rho}}{\rho}\frac{\gamma-1}{c^2}S_{IA}. \end{aligned} \tag{7.30}$$

The acoustic-flame coupling is contained in the second term. Hence, an amplification of the acoustic response occurs if

$$\frac{\gamma-1}{\gamma\rho c^2}\frac{p-\bar{p}}{\bar{p}}S_{IA} > 0, \tag{7.31}$$

i.e., if the pressure fluctuation is in phase with the unsteady heat release, as has been stated by RAYLEIGH [55]. A statement about the amplification or attenuation of an acoustic wave due to acoustic-flame interaction can be given using either eqn. (7.29) or (7.31). In this study, however, this acoustic-flame interaction mechanism is neglected, since this source is assumed to be small for unconfined flames.

7.3 Hybrid CFD/APE-RF method to simulate combustion noise

Different strategies can be pursued to simulate combustion noise using a hybrid approach. Generally, hybrid aeroacoustic methods are based on the scale separation. Due to the large difference in hydrodynamic and acoustic length scales, the problem of interest, e.g., jet noise predictions, is divided into two sub-problems in non-reacting flows. Turbulent reacting flows, however, possess an additional range of scales, i.e., the chemical time and length scales. Therefore, a hybrid aeroacoustic approach in terms of combustion noise is to be divided into three sub-levels, namely the chemical, hydrodynamic, and the acoustic range of length and time scales. In this work, the APE-RF system is used to describe the acoustic field excited by open turbulent flames. Unsteady CFD solutions of turbulent reacting flows, e.g., compressible DNS (section 7.4.2) and incompressible variable density LES data (sections 7.4.1, 7.4.3), are used as input data to drive the acoustic equations. In particular, this section deals with several aspects of the hybrid approach used in this work. CFD/CAA interface conditions are highlighted in the first step in terms of their impact on the acoustic solution, when the CFD results are transferred to the CAA mesh. An effect, which is termed "artificial acceleration due to interpolation" and could lead to erroneous results is presented in the following. The use of compressible or incompressible CFD data to evaluate the source terms of interest is addressed in section 7.3.1.3, followed by an analysis of the influence of the variable mean speed of sound. In the last part, the properties of the numerical methods used in the CAA simulations are discussed and the numerical schemes are summarized.

7.3.1 CFD/CAA interface conditions

7.3.1.1 Inner damping zone

For the simulation of trailing edge noise, SCHRÖDER & EWERT [57] have introduced in conjunction with the APE system a formulation on the embedded boundary of the inner source and the outer acoustic wave propagation region, which suppresses spurious noise generated by the artificial boundary. It was shown in [57] that a sudden appearance of a convecting vortical disturbance due to the truncated source region, i.e., the CFD region, generates spurious noise. This spurious noise contaminates the acoustic solution, since its origin is non-physical. The fundamentally derived formulation for those silent embedded boundaries was shown to effectively suppress spurious noise for sources of the momentum equation of the APE system.

In the following, this zonal formulation will be demonstrated to be also applicable to the pressure-density relation and to the modified continuity equation in the APE-RF system. We start by reformulating the APE-RF system as a single scalar wave equation by applying the time derivative to the continuity equation (7.16) and the pressure-density relation (7.18) and by taking the divergence of the momentum equation (7.17). The resulting wave equation then reads

$$\mathscr{L}(p') = \frac{\partial q_c}{\partial t} - \frac{\partial q_e}{\partial t} + \nabla \cdot q_m, \tag{7.32}$$

where $\mathscr{L}(p')$ represents the resulting wave operator and q_c, q_m, and q_e are the sources of the modified continuity equation, the momentum equation and the pressure-density relation, respectively. It was shown in [57] that only those components of the source vector on the RHS of the momentum equation produce spurious noise, which are not parallel to the normal vector of the inner boundary zone. For instance, if the normal vector of this zone is aligned with the x-direction and the sources $q_m = (0, q_{m,y}, 0)^T$ and q_e are considered, eqn. (7.32) reduces to

$$\mathscr{L}(p') = -\frac{\partial q_e}{\partial t} + \frac{\partial q_{m,y}}{\partial y}. \tag{7.33}$$

After reformulating eqn. (7.33)

$$\begin{aligned}
\mathscr{L}(p') &= -\frac{\partial}{\partial t}\left(\frac{\partial}{\partial y}\int^{y} q_e dy'\right) + \frac{\partial q_{m,y}}{\partial y} \\
&= -\frac{\partial}{\partial y}\left(\frac{\partial}{\partial t}\int^{y} q_e dy'\right) + \frac{\partial q_{m,y}}{\partial y}
\end{aligned} \tag{7.34}$$

the term in parentheses can be understood as a momentum source such that the filter being used on the inner zone

$$h(x) = \begin{cases} 0 & \text{for } x \leq -d/2 \\ \frac{1}{2}\left[1 + \sin\left(\frac{\pi x}{d}\right)\right] & \text{for } -d/2 < x < d/2 \\ 1 & \text{for } x \geq d/2, \end{cases} \tag{7.35}$$

with d representing the filter zone width, can also be applied to the first term of eqn. (7.34). The resulting form of eqn. (7.34) on the embedded boundary reads

$$\begin{aligned}
\mathscr{L}(p') &= -\frac{\partial}{\partial y}\left(h(x)\frac{\partial}{\partial t}\int^{y} q_e dy'\right) + \frac{\partial}{\partial y}\left(h(x) q_{m,y}\right) \\
&= -\frac{\partial}{\partial y}\left(\frac{\partial}{\partial t}h(x)\int^{y} q_e dy'\right) + \frac{\partial}{\partial y}\left(h(x) q_{m,y}\right) \\
&= -\frac{\partial}{\partial t}\left[\frac{\partial}{\partial y}\left(h(x)\int^{y} q_e dy'\right)\right] + \frac{\partial}{\partial y}\left(h(x) q_{m,y}\right) \\
&= -\frac{\partial}{\partial t}\left[\frac{\partial h(x)}{\partial y}\int^{y} q_e dy' + h(x) q_e\right] + \frac{\partial}{\partial y}\left(h(x) q_{m,y}\right) \\
&= -\frac{\partial}{\partial t}\left(h(x) q_e\right) + \frac{\partial}{\partial y}\left(h(x) q_{m,y}\right).
\end{aligned} \tag{7.36}$$

It is evident from eqn. (7.36), that the formulation derived in [57] can be applied to the source terms of the pressure-density relation of the APE-RF system. Since the source of the modified continuity equation q_c is also expressed via the time derivative like the source of the pressure-density relation q_e in eqn. (7.32) the filter zone can also be applied to q_c.

To show the impact of the filter zone, a convection of an iso-thermal density distribution with a non-dimensional mean flow of $\overline{u} = (0.1, 0)^T$ is simulated, which can be considered as a convecting entropy spot. Since the local temperature is constant the speed of sound does not change. In an iso-thermal case, the ideal gas law can be rewritten as

$$\frac{\partial p'}{\partial t} - RT\frac{\partial \rho'}{\partial t} = 0. \tag{7.37}$$

Inserting eqn. (7.37) into the source term of the pressure-density relation eqn. (7.18)

$$q_e = -\overline{c}^2 \frac{\partial \rho_e}{\partial t} \tag{7.38}$$

yields the source

$$q_e = -\overline{c}^2 \frac{\partial \rho'}{\partial t}\left(1 - \frac{1}{\gamma}\right). \tag{7.39}$$

The density distribution is assumed to be GAUSSIAN

$$\rho' = A\exp\left(-\ln(2)\frac{(x-x_0)^2 + (y-y_0)^2}{\sigma^2}\right) \tag{7.40}$$

with an amplitude of $A = 0.1$, a half-width of $\sigma = 0.05$, and a moving center point of $x_0 = (x_0 = -0.5 + \overline{u}t, y_0 = 0)^T$. The computational domain of this test problem consists of 300×300 grid points.

Figs. 7.1(a) show the perturbation density field at the non-dimensional time $t = 0.1$. This perturbation was induced by an onset of the source term at $x_0 = -0.5, t = 0$. Figs. 7.1(a) illustrate the density contours resulting from a source term based on an analysis without (left) and with (right) filter zone. This filter zone ranges from $x = -0.5$ through $x = -0.2$, i.e., the width of the filter zone is $d = 0.3$. The perturbation density at a later non-dimensional time of $t = 0.4$, i.e., the entropy spot has been convected to $x = -0.1$, is depicted in Fig. 7.1(b). Fig. 7.1(b) (left) evidences besides the convected perturbation at $x = -0.1$ a density distribution confined to the interior boundary at $x = -0.5$. The distribution in Fig. 7.1(b) (right), which is based on a computation with the embedded filter zone, does not show any disturbances at $x = -0.5$. From the test of the convecting entropy spot being illustrated in Fig. 7.1 it can be concluded that spurious noise especially in the higher frequency range is effectively suppressed by applying a filter zone to the source term q_e similarly to that being used for q_m in [57].

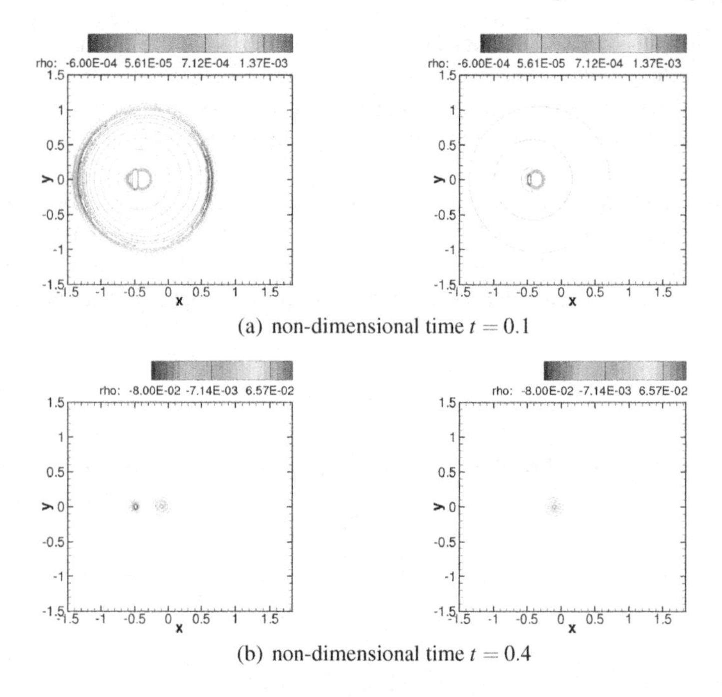

(a) non-dimensional time $t = 0.1$

(b) non-dimensional time $t = 0.4$

Fig. 7.1 Density contours to illustrate spurious sound induced at an inflow boundary located at $x = -0.5$ by a passively convecting iso-thermal entropy spot. Inner zone width: $d = 0$ (left); $d = 0.3$ (right)

7.3.1.2 Artificial acceleration due to interpolation

What exactly is different between $q_{e,rf,D\rho}$ and $q_{e,rf,\partial\rho}$, which could lead to this unexpected behavior? To tackle this problem, a numerical study is performed, in which a GAUSSIAN density spot

$$\rho(x,t) = 0.1 \exp\left(-\ln(2) \frac{(x - x_0(t))^2 + (y - y_0)^2}{0.5^2} \right) \tag{7.41}$$

with

$$x_0(t) = -9.0 + \overline{u}t \tag{7.42}$$

is constantly convected with a uniform mean flow $(\overline{\rho}, \overline{u}, \overline{v}, \overline{p})^T = (1,\ 0.5,\ 0,\ 0.7142)^T$. The difference between those two formulations $q_{e,rf,\partial\rho}$ and $q_{e,rf,D\rho}$ is that the source term $q_{e,rf,\partial\rho}$ incorporates the effect of acceleration of density inhomogeneities. However, in this case, neither $q_{e,rf,D\rho}$ nor $q_{e,rf,\partial\rho}$ excite acoustic responses, since an entropy spot, which is being convected constantly with the mean flow does not produce any noise. This is apparently directly holds for $q_{e,rf,D\rho}$, since by definition, this source term vanishes. Unlike $q_{e,rf,D\rho}$, the source term $q_{e,rf,\partial\rho}$ is non-zero for a constantly convecting entropy spot, or in other words. Using the density distribution in eqn. (7.41), this test case is analytically prescribed with

$$q_{e,rf} = -\overline{c}^2 \frac{\partial \rho}{\partial t} \tag{7.43}$$

on a very fine mesh (Fig. 2(a)). For the simulations on the two coarse grids, the prescribed sources on the very fine mesh are interpolated onto two coarser meshes (Fig. 2(b), 2(c)). Numerically, the entropy spot is initialized at $x = (-9.0, 0.0)$ without using an embedded inner damping zone, such that an initial acoustic response occurs. However, in the following, this entropy spot is just being convected at $\overline{u}$ and should therefore not produce any sound. In the analytically prescribed case in Fig. 2(a) no acoustic disturbance is visible due to the convected spot. At this instant this entropy spot has already been convected 15 units in the positive x-direction.

In the next step, the source terms are sampled with a time increment of $\Delta t = 0.1$ and are interpolated onto two different grids, i.e., the same physical domain is discretized by 63×63 grid points and 112×112 mesh points. Fig. 2(d) shows the source position and the source contours at $t = 30$, while Since the mean flow speed in the x-direction has been set to be constant at $\overline{u} = 0.5$, it appears that the entropy spot on the coarser grids is not only been convected, but also been accelerated. A dipole-like sound field is been produced by this effect, which will be referred to as the "artificial interpolation induced acceleration" effect. The resulting sound fields are shown in Fig. 2(b) and Fig. 2(c). It is clearly demonstrated that this effect is able to generate significant spurious noise. The numerical results obtained from the simulation using the fine mesh and the source formulation $q_{e,rf,\partial\rho}$ show the best agreement with the experimental findings. However, as was shown before, the use of this source formulation requires a preserving of the convection speed during source term interpolation.

7.3.1.3 Compressible versus incompressible

In this subsection, the impact of different source formulations of the pressure density relation in conjunction with input data from either a compressible or an incompressible CFD solution is investigated. It will be shown that input data taken from a compressible flow simulation requires a special source formulation, while incompressible input data allows alternative expressions for the right-hand side of the pressure-density relation.

Consider an entropy fluctuation due to unsteady heat release at constant background pressure and zero mean flow velocities. A constant-pressure process is assumed, which is a classical result of the combustion analysis for flames burning in an open domain. To show the difference in using either compressible or incompressible input data, first, the entropy fluctuation is modeled by a periodic fluctuating source term with GAUSSIAN shape via

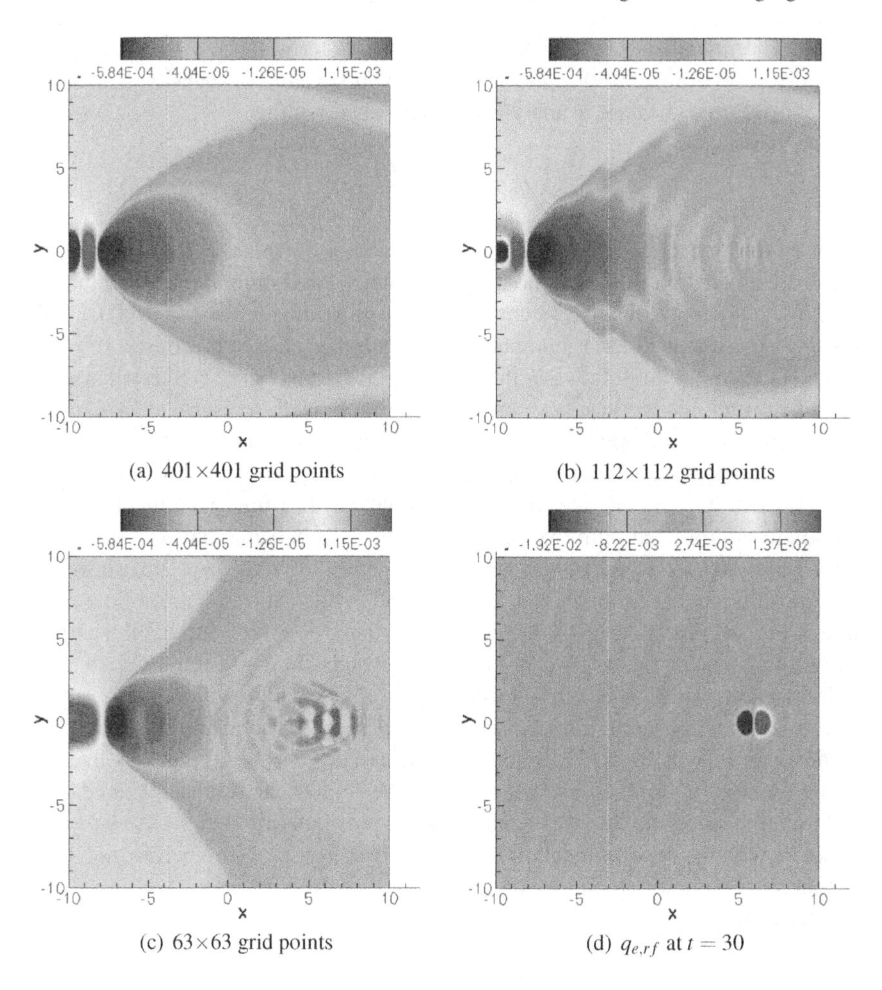

Fig. 7.2 Instantaneous contours of an entropy spot at $t = 30$ convecting with a mean flow speed of $\overline{u} = 0.5$ in the x-direction, on three meshes with different resolutions (a,b,c). At this moment, the entropy spot has been convected to the position $x_0 = (6,\ 0)$ (d).

$$s = s_0(x,y)\cos(\omega t), \tag{7.44}$$

instead of using input data from an incompressible or compressible CFD simulation, while q_c and q_m are zero.

It is evident from eqn. (7.15) that a non-zero RHS of the pressure-density relation indicates an entropy fluctuation in a certain flow region. In the following, it will be demonstrated that for low MACH number flows such as the combustion processes studied in this work the entropy fluctuation can be evaluated equivalently from incomressible or compressible CFD data. In such a case, it is assumed that an unique decomposition of the flow variables into time averaged mean, fluctuating hydrodynamic (h) and acoustic (a) quantities exists

$$\rho = \overline{\rho} + \rho^h + \rho^a \tag{7.45}$$

$$u = \overline{u} + u^h + u^a \tag{7.46}$$

$$p = \overline{p} + p^h + p^a. \tag{7.47}$$

Since the APE-RF system is an exact consequence of the conservation equations governing chemically reacting flows, the entropy source term can be evaluated alternatively from CFD data if the pressure and the density fluctuation are given via

$$\frac{\partial s'}{\partial t} = \frac{c_p}{\overline{\rho}\overline{c}^2}\left(\frac{\partial p'}{\partial t} - \overline{c}^2\frac{\partial \rho'}{\partial t}\right). \tag{7.48}$$

Substituting the aforementioned decomposition into eqn. (7.48) yields

$$\frac{\partial s'}{\partial t} = \frac{c_p}{\overline{\rho}\overline{c}^2}\left[\left(\frac{\partial p^h}{\partial t} - \overline{c}^2\frac{\partial \rho^h}{\partial t}\right) + \underbrace{\left(\frac{\partial p^a}{\partial t} - \overline{c}^2\frac{\partial \rho^a}{\partial t}\right)}_{=0}\right], \tag{7.49}$$

while it is evident that the last two terms in eqn. (7.49) cancel out each other due to the isentropic behavior of the acoustic waves. Furthermore, using eqn. (7.49), it is also demonstrated that in this case the entropy fluctuation can be either evaluated using data taken from an incompressible or compressible CFD solution, or in other words, the entropy fluctuation in this case is governed by the hydrodynamic fluctuations only. At this point, two aspects have to be emphasized. The prescribed entropy fluctuation can be considered being evaluated from an incompressible CFD solution, while the solution of the APE-RF system represents a compressible version of this flow case. The solution of the prescribed entropy fluctuation will be referred to as the solution of "case (1)". Based on the results of "case (1)" two simulations are performed to show on the one hand the error caused by the use of a simplified source formulation in conjunction with compressible CFD input data "case (2)", and on the other hand to show that simplified source formulations have to fulfill a special constraint if compressible input data are taken. In the next simulation "case (2)", the solution of the density fluctuation of "case (1)" is taken as input data to model the source term evaluated from a compressible CFD simulation using the simplified formulation (eqn. (7.28)). The acoustic solution differs from the previous case, since the density fluctuation, which has been fed into the RHS of the pressure-density relation (Fig. 7.3(c)), includes acoustic components, which leads to non-physical acoustic-acoustic excitations. However, if the entropy fluctuation is evaluated in the third case using eqn. (7.48), which is shown in Fig. 7.3(b), i.e., the pressure fluctuations from the compressible solution are also taken into account, the acoustic components of q_e cancel out each other and the entropy fluctuation is again governed by the hydrodynamic density fluctuation only. A comparison of the perturbation pressure time history of each test case is shown in Fig. 7.3(c). For a low

MACH number case, these test cases clearly show that an examination of the impact of different source components using a compressible CFD input data requires a careful formulation of the source terms of interest, such that no unphysical acoustic components are included on the RHS of the APE-RF system. A similar behavior has also been observed by EWERT [20]. He showed that the use of acoustic containing pressure fluctuations as input data in conjunction with the APE-2 formulation leads to unphysical acoustic-acoustic excitations. Moreover, if incompressible input data is available, no special care has to be taken in formulating simplified source terms. In general, each source of interest can be examined separately, since no acoustic information is included by definition and errors due to acoustic-acoustic excitations are avoided.

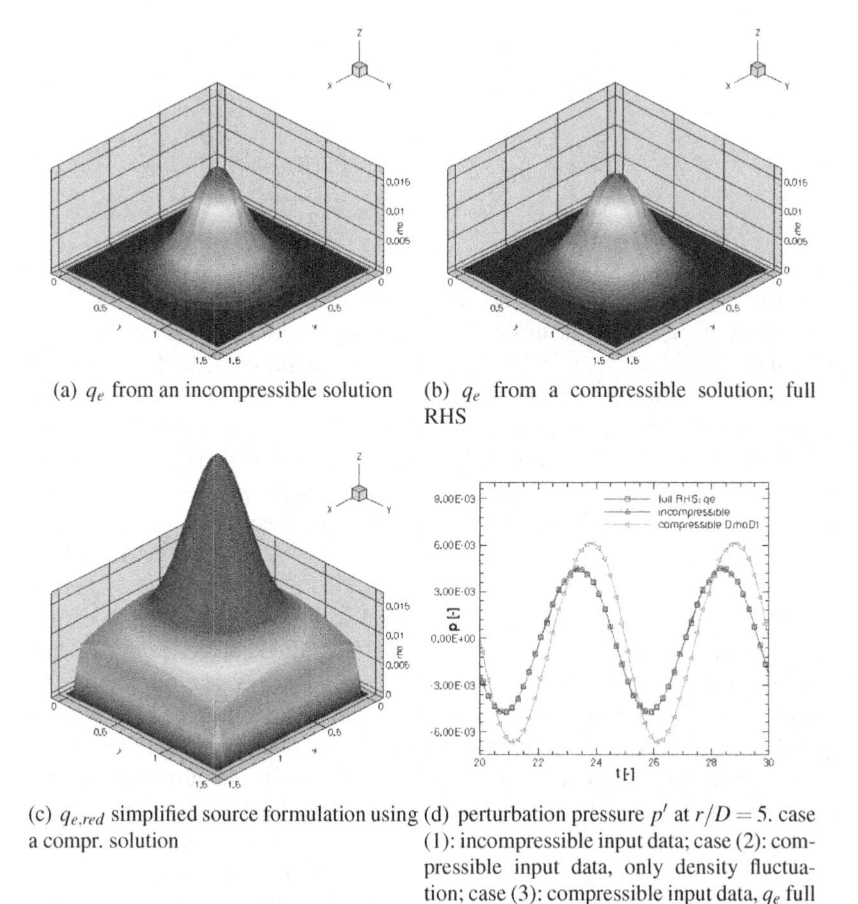

(a) q_e from an incompressible solution

(b) q_e from a compressible solution; full RHS

(c) $q_{e,red}$ simplified source formulation using a compr. solution

(d) perturbation pressure p' at $r/D = 5$. case (1): incompressible input data; case (2): compressible input data, only density fluctuation; case (3): compressible input data, q_e full RHS

Fig. 7.3 (a)-(c): Source term distribution using compressible or incompressible input data and different source formulations.(d): acoustic impact of the different input data and source formulations

7.3.1.4 Influence of the variable mean speed of sound

In this subsection, the need of the inclusion of the variable mean speed of sound in the acoustic simulation will be shown. A numerical study is performed, in which the acoustic field of an artificial heat source is simulated using the mean flow field of the DLR-A flame. Reconsider the RHS of the pressure-density relation

$$q_e = -\bar{c}^2 \frac{\partial \rho'}{\partial t}, \tag{7.50}$$

in which the pressure fluctuation vanishes. In the numerical tests, the fluctuating density distribution is assumed to be GAUSSIAN in space and fluctuating in time, represented by following equation

$$\frac{\partial \rho'}{\partial t} = 0.1 \exp\left(-\ln(2)\frac{(x-x_0)^2+(y-y_0)^2+(z-z_0)^2}{\sigma^2}\right)\cos(\omega t). \tag{7.51}$$

The source is placed at ${x_0}^T = (20,\ 0,\ 0)^T$, since the major sound source from the H3- (section 7.4.1) as well as the DLR-A flame (section 7.4.3) can be found in the region $x/D < 40$. Four different periodic cases are considered, in which the source period is varied ($T = 15$; $T = 30$; $T = 60$; $T = 90$). These periods correspond to STROUHAL numbers of ($St = 0.540$; $St = 0.269$; $St = 0.135$; $St = 0.090$), respectively. Fig. 7.4(a) shows the contours of the mean density. The instantaneous perturbation pressure in the streamwise center plane at the non-dimensional time $t = 300$ is illustrated in Figs. 7.4(b)-(e) for the four different aforementioned cases. From these figures, the influence of the variable mean speed of sound is already noticeable. The refraction effect due to the variable mean speed of sound is more visible from the directivity pattern of each case in Fig. 7.5. To evaluate the directivity patterns, uniformly distributed observer points located on a circle in the x/y-plane with radius $r/D = 40$ and center point $x^T = (20,\ 0,\ 0)^T$ have been used. The ($\Theta = 0°$)-axis in Fig. 7.5 coincides with the flame axis. A "zone of silence", in the region between $0° < \Theta < 20°$, can be observed from Fig. 7.5, which has also been reported from numerical studies in [36]. Moreover, the refraction effect is dependent on the frequency. Two main directivity effects are readily identifiable. First, the preferred direction of acoustic radiation is changed from ($\Theta = 84°, = 98°, = 113°, > 121°$) for ($T = 90, = 60, = 30, = 15$), respectively. The second effect is the change of the shape of the directivity pattern, from a monopole-like pattern to a multipole structure of the excited acoustic field. This behavior, however, has not been observed by IHME ET AL. [36], since they have placed their periodic source term at the jet exit position ($x = (0,\ 0,\ 0)$), i.e., the upstream traveling wave fronts are not refracted due to the uniform mean flow in the upstream area. In the current study, the source term is located at a position ($x = (20,\ 0,\ 0)$) downstream of the jet exit, which allows for waves being refracted while propagating in the upstream direction. In a real flame, the "zone of silence" has not been observed from experiments, since the acoustic driving sources are distributed over a wide spatial area. However, from these find-

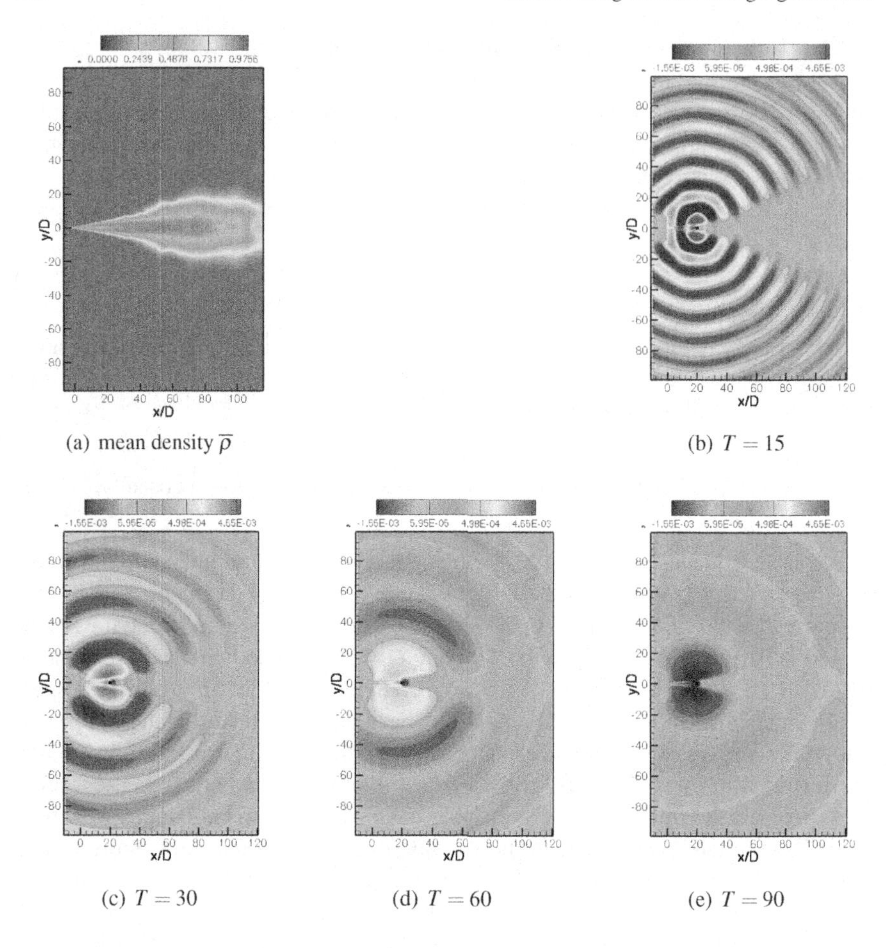

(a) mean density $\overline{\rho}$

(b) $T = 15$

(c) $T = 30$

(d) $T = 60$

(e) $T = 90$

Fig. 7.4 Contours of the mean flow density (a) and the instantaneous perturbation pressure p' at $t = 300$ using a periodic artificial heat source with different angular frequencies $\omega_i = 2\pi/T_i$ and non-uniform mean speed of sound (b-e)

ings, the multipole character of the acoustic field can be related at least partly to the refraction effects as has been theoretically presumed by STRAHLE [63].

7.3.2 Numerical methods used in the CAA

Numerical methods, which deal with aeroacoustic problems have to face a completely different set of challenges than common CFD methods due to several reasons. Many optimized schemes to discretize the partial differential equations, governing the acoustic wave propagation have been developed successfully in the past

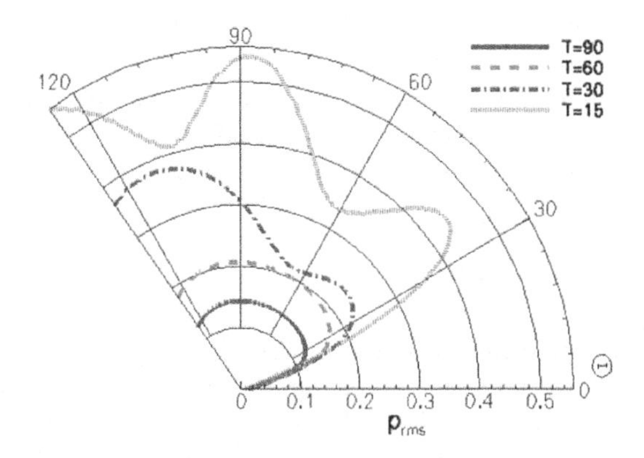

Fig. 7.5 Directivity patterns for different angular frequencies $\omega_i = 2\pi/T_i$

decade [68, 33, 4, 44, 72, 66] to account for the requirements listed above. A sound summary and review about the development and the available numerical schemes for CAA has been given for instance in [67, 45, 20]. The APE-RF system has been solved using the standard 7-point DRP stencil by TAM & WEBB [68] for spatial discretization, while the time integration is done via the alternating 5/6-LDDRK scheme by HU ET AL. [33]. Since the APE system, as well as the APE-RF system does not describe the convection of other modes than the acoustic ones, the radiation boundary condition by TAM & WEBB [68] is sufficient to model the far field boundaries of the computational domain for the acoustic simulation of open turbulent flames. The source terms from an unsteady CFD solution are sampled with a certain time increment. During the acoustic simulation, these source terms have to be interpolated in time, which is done via a quadratic interpolation scheme.

7.4 Results and Analysis: Application of the APE-RF system to open turbulent flames

7.4.1 H3 Flame: A non-premixed open turbulent flame

The acoustic properties of an open turbulent non-premixed flame, the so-called H3 flame, a benchmark flame of the workshop for turbulent non-premixed flames [3], are investigated, to discuss the quality of the APE-RF formulation. The fuel, a mixture of 50/50 vol% H_2/N_2, is injected coaxially through a round nozzle with a diameter of $D = 0.008$ m into ambient air at a temperature of $T = 300$ K and a constant coflow velocity of $U_{coflow} = 0.2$ m/s. The bulk velocity of the fuel is $U_{bulk} = 34.8$ m/s yielding a REYNOLDS number based on D of $Re = 10{,}000$. The stoichiometric mixture of this oxidizer/fuel combination is $Z_{st} = 0.31$.

A detailed description of the numerical specifications can be found in [12]. The results of the aforementioned study [12] are briefly summarized in this section in terms of directivity behavior and acoustic intensity spectrum.

7.4.1.1 Directivity behavior

Observer points with an increment of two degrees are placed on a circle with a radius of $R/D = 45$ in the x-y plane, the origin of which is located at $(x/D, y/D, z/D) = (15, 0, 0)$, to determine the directivity effects. Directivity plots at several values of the STROUHAL number $St = Df/U_{bulk}$, which is based on the bulk jet exit velocity U_{bulk} and the exit diameter D, are shown in figure (7.6). The maximum acoustic energy can be found within the low frequency range (figures 7.6a, b). This result corresponds with theoretical analyses [18, 63, 40] and experimental observations [59, 61], in which low speed turbulent flames are characterized as low frequency acoustic radiators. As discussed before, the total time derivative of the density includes the effects of unsteady heat release, species and heat diffusion as well as viscous effects. However, using the simplified source term formulation $q_{e,rf,D\rho}$ means that the unique contribution of each single effect to the radiated non-isotropic source field cannot be determined. To do so, a more elaborate computation based on the complete $q_{e,rf}$ formulation would be required. A detailed analysis of the acoustic impact of the different source contributions within the RHS of the pressure-density relation is given in section 7.4.2. It has been argued in STRAHLE [63] that the non-uniform mean speed of sound within the flame causes the deviation from the isotropic case via refraction effects. This multipole pattern, which are shown in figures 7.6c-f, can be also related to be caused by either a multipole source distribution or refraction effects. This non-isotropic directivity pattern caused by refraction effects have been evidenced in section 7.3.1.4.

7.4.1.2 Intensity spectrum

The acoustic field of the H3 flame has been measured by PISCOYA ET AL. [7]. The microphones to measure the radial intensity were placed on a circle in the y-z plane with a radius of $r/D = 62.5$ whose center was located at $(x/D, y/D, z/D) = (15, 0, 0)$. Figure (7.7) shows the radial intensity spectra determined from the experimental data and CAA results. The experimental and computational radial intensity spectra are obtained by averaging over all data on the evaluation circle. To get the smallest spectral variance per data point, the Fast Fourier Transformation using a 50% overlapping window technique [51] was applied to the data. The spectrum resulting from the CAA computation shows a good agreement with the experimental data as far as level and spectrum characteristics are concerned. However, in the limited frequency range $0.36 < St < 0.45$ the amplitudes of the radial intensity show

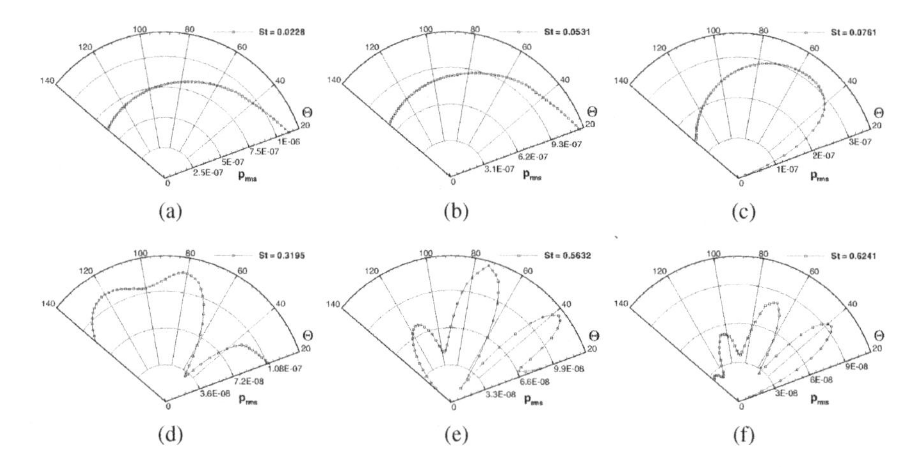

Fig. 7.6 Directivity for different STROUHAL numbers $0.0228 < St = Df/U_{bulk} < 0.6241$. The root mean square value of the non-dimensionalized pressure p_{rms} is plotted as a function of the angle in the streamwise (x-y) center plane. Note, the scaling of the magnitude differs in each subfigure.

deviations. This local discrepancy is conjectured to be caused by not taking into account the complete RHS of the APE-RF system in the CAA analysis, especially the non-uniform mean flow effects, the excitation of acoustic waves due to combustion at non-constant pressure, and the influence of acceleration of density inhomogeneities, but only the simplified source term $q_{e,rf,D\rho}$.

7.4.2 Premixd Methane Flame

A hybrid CFD/CAA approach using large-eddy simulation (LES) and acoustic perturbation equations for reacting flows (APE-RF) has been performed in the previous section to simulate the acoustic field of a non-premixed jet flame. The major source term has been shown to be also represented by the total time derivative of the density. This expression includes the unsteady heat release rate, which is well known to be the most relevant source as to acoustic radiation of open turbulent flames[59, 61, 63]. Using the energy equation for reacting flows, the total time derivative of the density can be separated into several terms which describe the unsteady heat release rate, species and heat diffusion, viscous effects, and a source, which represents the effect of non-isomolar combustion[71]. These source mechanisms are analyzed in this study. For this purpose, the data of a direct numerical simulation (DNS) of a premixed flame has been used. The DNS of the premixed flame is based on computing the full chemical reaction. That is, no look-up tables are taken into account and as such the necessary information for the source terms of the APE-RF formulation is available. However, the total number of grid points and the computational costs for

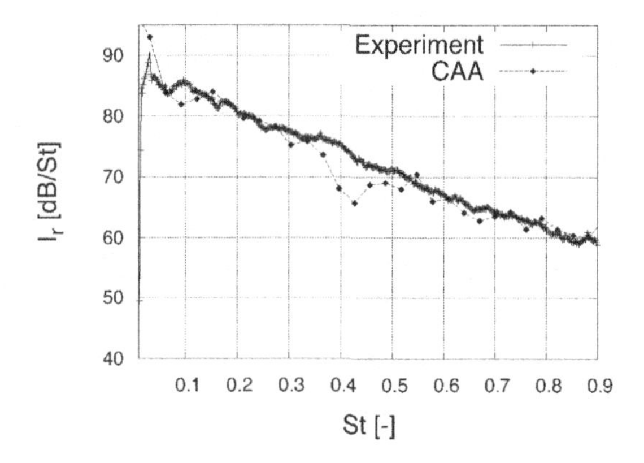

Fig. 7.7 Comparison of the CAA results with experimental data for the radial intensity in [dB/St] as a function of the STROUHAL number, i.e., perturbation acoustic pressure times the radial acoustic velocity component related to the evaluation circle which is defined subsequently. The reference intensity level is $I_{ref} = 10^{-12} W/m^2$ The observer position is located on a circle in the (y/z)-plane whose center point is located at $(x/D, y/D, z/D) = (15,0,0)$ having a radius of $R/D = 62.5$.

the full chemical reaction scheme restrict the DNS to a very focused spatial domain. Since the information of all species is directly provided in the DNS, the aforementioned source terms can be easily evaluated for the CAA analysis. The purpose of the following analysis is to determine the major source terms, to evidence the directivity patterns of the various sources over a wide frequency range and to relate them to the classical pole-like structure in acoustics. The acoustic impact of several source mechanisms are investigated, such as the effect of unsteady heat release, that of heat flux, viscous effects, the effect of non-isomolar combustion, and that of species diffusion. The study shows the unsteady heat release to be the dominant source. All source terms but the heat diffusion term possess a monopole-like structure in the low frequency range. At high frequencies a multipole-like pattern is also determined for the sources due to species diffusion and viscous effects.

7.4.2.1 Flame configuration

The combustion configuration considered in the present study is a turbulent premixed CO/H_2/Air flame. The numerical domain under investigation is a square box of dimension 8x8 mm^2 as shown in Fig. 7.8. A fixed mesh of 251 points is used in each direction. This leads to a spatial resolution of 32 μm, necessary to resolve intermediate radicals. The left-hand boundary condition is a subsonic inlet at an imposed velocity of 1m/s, while on the right-hand boundary a non-reflecting subsonic outlet condition is formulated. A complete reaction scheme with 13 species (CO,

HCO, CH_2O, CO_2, H_2O, O_2, O, H, OH, HO_2, H_2O_2, H_2, N_2) and 67 individual reactions is taken into account. First, the corresponding premixed laminar flame is computed for a one-dimensional flow along the x-direction and the inlet velocity is adapted to keep the flame front in the center of the domain. Then, the obtained steady solution is transposed to a 2-D flow, with fresh gases on the left side and burnt gases on the right side. An isotropic 2-D turbulent velocity field is superposed using a von KÁRMÁN spectrum coupled with PAO correction for near-dissipation scales [69, 30, 31].

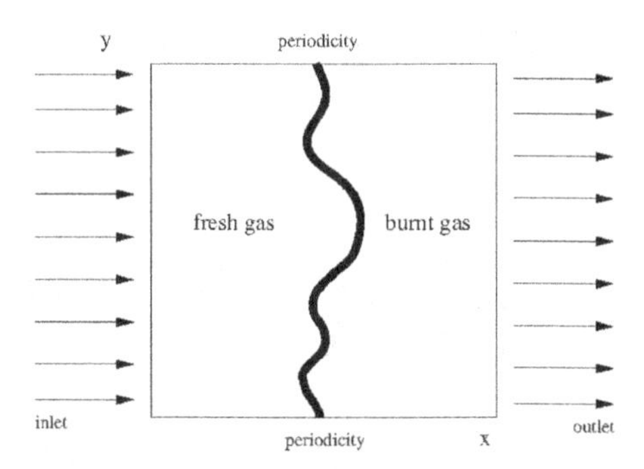

Fig. 7.8 Computational domain and boundary notation of the DNS computation

7.4.2.2 Definition of the thermoacoustic source terms

The source terms, which are examined in this study are briefly summnarized. As was shown in eqn. (7.15), the total time derivative of the density can be decomposed into six parts. Since the density fluctuation includes the substantial time derivative of the pressure, the remaining five terms and the substantial time derivative of the density are examined in this study. Note, all the following sources are included in the RHS of the pressure-density relation (7.21). Plugging in $\alpha/c_p = (\gamma - 1)/c^2$ for an ideal gas [18] the following relations constitute the subsources of the $q_{e,rf}$ term.

effect of unsteady heat release rate:

$$q_{e,rf,1} = -\bar{c}^2 \frac{\bar{\rho}}{\rho} \frac{\gamma - 1}{c^2} \sum_{n=1}^{N} h_n \omega_n \tag{7.52}$$

effect of non-isomolar combustion:

$$q_{e,rf,2} = -\bar{c}^2 \frac{\bar{\rho}}{\rho} \frac{\gamma - 1}{c^2} c_p T \overline{W} \sum_{n=1}^{N} \frac{\omega_n}{W_n} \tag{7.53}$$

effect of species diffusion:

$$q_{e,rf,3} = -\bar{c}^2 \frac{\bar{\rho}}{\rho} \frac{\gamma - 1}{c^2} \left(\sum_{n=1}^{N} h_n \nabla \cdot J_n - c_p T \overline{W} \sum_{n=1}^{N} \frac{\nabla \cdot J_n}{W_n} \right) \tag{7.54}$$

effect of heat diffusion:

$$q_{e,rf,4} = -\bar{c}^2 \frac{\bar{\rho}}{\rho} \frac{\gamma - 1}{c^2} \nabla \cdot (-\lambda \nabla T) \tag{7.55}$$

viscous effects:

$$q_{e,rf,5} = -\bar{c}^2 \frac{\bar{\rho}}{\rho} \frac{\gamma - 1}{c^2} \frac{\partial u_i}{\partial x_j} \tau_{ij} \tag{7.56}$$

7.4.2.3 Results: Acoustic field

The acoustical field is analyzed by directivity patterns and power spectral density plots. Figure 9(d) shows the sound pressure level (SPL) spectrum of three different computations using three different source term compositions of the RHS of the pressure-density relation. The first computation was done using $q_{e,rf,1}$ only, while the second and third simulation were performed using all sources, i.e., $q_{e,rf,1}$ through $q_{e,rf,5}$ and the substantial time derivative of the density, respectively. First of all, it is evident from Fig. 9(d) that the major source term is the unsteady heat release rate,i.e., $q_{e,rf,1}$, since the contribution of the other thermoacoustic sources in Eqs. (7.53)-(7.56) to the radiated acoustic field is on average less than $2dB$.
Using the total time derivative of the density as source term of the pressure-density relation the spectrum distribution is in good agreement in comparison with the other two simulations. Especially in the frequency range close to $10kHz$, the sound pressure level amplitude is underestimated by approx. $5dB$ to the SPL values caused by the source $q_{e,rf,1}$.

7.4.2.4 Characteristic frequencies

Each single source term is analyzed by Fourier transforming the perturbation pressure at a point with a radial distance to the center of the domain of $R/L = 4$. In Fig. 7.9(a) it can be observed that the maximum acoustic energy generated by the unsteady heat release rate $q_{e,rf,1}$, can be found at approximately $1kHz$. From the sound pressure spectrum of the source describing the effect of non-isomolar combustion $q_{e,rf,2}$ in Fig. 7.9(b) no characteristic frequency can be identified. This spectrum

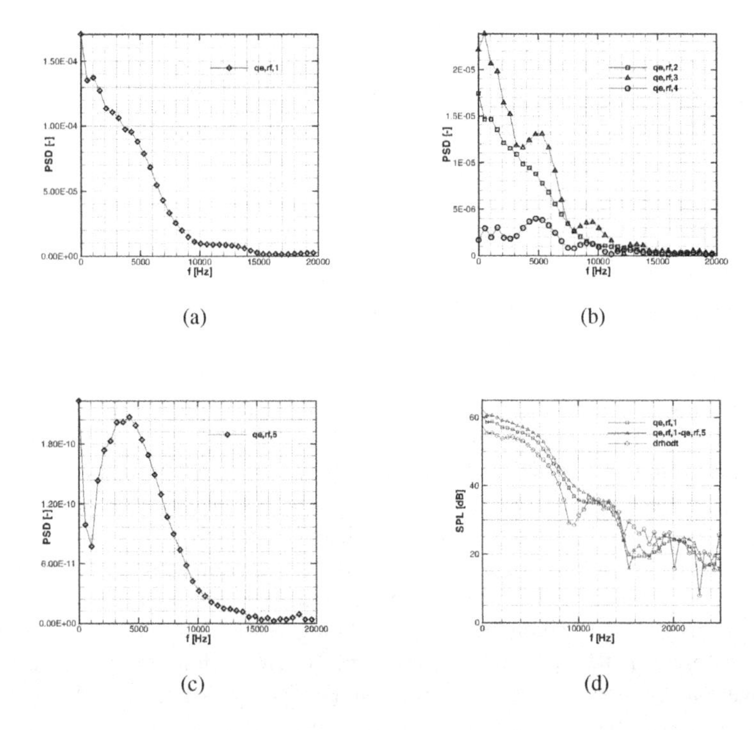

Fig. 7.9 Power spectral density distribution of sources $q_{e,rf,1}$ through $q_{e,rf,5}$ (9(a)-9(c)). Comparison of SPL spectra using different source term formulations (9(d))

shows rather a continuous decay. The remaining three spectra shown in Figs. 7.9(b) and 7.9(c) definitely evidence the existence of characteristic frequencies. The characteristic frequencies of the source $q_{e,rf,3}$ are approx. 550 Hz, 5.3 kHz and 9.5 kHz, while the source terms $q_{e,rf,4}$ and $q_{e,rf,5}$ show a peak value at approx. 5 kHz and 4.3 kHz, respectively. The analysis shows the unsteady heat release to represent the major source term compared to the other source mechanisms, i.e., the effect of non-isomolar combustion, that of species and heat diffusion, and the impact of viscous effects. Moreover, the comparison of the power spectral density of three acoustic simulations using different source terms, shows the total time derivative of the density as major source term as to acoustic radiation of reacting flows in good agreement with the acoustic impact of the unsteady heat release rate or the summation of all sources examined in this study. Note, the simple order-of-magnitude analysis of the source terms does only evidence the significance of the various sources, not the actual generation process of the acoustic sound fields. All source terms but the heat diffusion source term show a monopole character in the low frequency band, whereas the latter possesses a multipole behavior in the frequency range of $1.6kHz < f < 2.6kHz$. Since the time scale of the classical species diffusion is very large compared to the chemical reaction time scale, it can be concluded, that in this case "ordinary" species diffusion is of monopole type while the effect of species creation and species consumption is of multipole type. A detailed analysis about the directivity behavior of the acoustic field caused by each subsource has been published in [11].

7.4.3 DLR-A Flame: A non-premixed open turbulent flame

The numerical applicability of different sound source formulations is investigated in this section. In section 7.2.4, three formulations have been derived based on the inclusion of the major sound generating mechanism, i.e., the unsteady heat release rate. The sound sources of interest are evaluated from a variable density incompressible reactive large-eddy simulation (LES). Among these three options, the results discussed in section 7.4.1 and 7.4.2 show the substantial time derivative of the density being the source formulation of choice. The source expressed via the partial time derivative of the density includes on the one hand most source mechanisms and shows on the other hand the smoothest spatial distribution compared to the other formulations. In other words, for these two reasons, this formulation should yield the best acoustic result, which, however, could not be verified from the numerical findings. Moreover, the spectra obtained from the simulation using the unsteady heat release as the major source term show an unexpected behavior. This source is deemed the major sound source in low MACH number flows, which is not entirely supported by the numerical results. To resolve the uncertainties, about the numerical results of the previous study [8], the sources of the APE-RF system are interpolated onto two CAA grids with different resolutions, i.e., a coarse and a very fine mesh to study the impact of source term interpolation and the numerical applicability of these sound source formulations in terms of numerical accuracy and efficiency. It turned out that the acoustic field is best represented using the partial derivative formulation following the expectations of the previous source analyses. However, the use of this source formulation requires an additional strict constraint, i.e, the convection speed of density inhomogeneities has to be preserved during interpolation to avoid spurious noise due to artificial acceleration caused by interpolation errors, which has been discussed in section 7.3.1.2. A detailed description about the numerical specifications and methods used in the CFD and CAA computations can be found in [14].

7.4.3.1 Results of the DLR-A simulations

To summarize the source mechanisms, included in each source formulation, let us recall the three different simplified sources, which read

$$q_{e,rf,SIA} = -\bar{c}^2 \frac{\bar{\rho}}{\rho} \frac{\gamma - 1}{c^2} \sum_{n=1}^{N} h_n \omega_n \tag{7.57}$$

$$q_{e,rf,D\rho} = -\bar{c}^2 \frac{D\rho}{Dt} \tag{7.58}$$

$$q_{e,rf,\partial\rho} = -\bar{c}^2 \frac{\partial \rho}{\partial t}. \tag{7.59}$$

Table (7.1) lists the source mechanisms inherently contained in the three different source formulations.

Table 7.1 Source mechanisms contained in each simplified formulation of the RHS of the pressure density relation. The effects due to non-isomolar combustion, species diffusion, heat diffusion, and viscosity are denoted by $\sum_{i=2}^{N=5} q_{e,rf,i}$. $\nabla \cdot (u\rho_e)$ represents the effect due to the acceleration of density inhomogeneities.

source	unst. heat release	$\sum_{i=2}^{N=5} q_{e,rf,i}$	$\nabla \cdot (u\rho_e)$	$u \cdot \nabla \bar{\rho}$	pressure fluct.
$q_{e,rf,SIA}$	×	-	-	-	-
$q_{e,rf,D\rho}$	×	×	-	-	-
$q_{e,rf,\partial\rho}$	×	×	×	×	-

Source term analysis

From a source term analysis in wave number space it is evident that the partial time derivative shows the smoothest source distribution, which is characterized by a mainly low wave number content, while the wave number analysis of the source formulation $q_{e,rf,SIA}$, i.e., the unsteady heat release rate, illustrates a very sharp spatial distribution, which is characterized by a wave number content in a wide range. From this analysis it is certain that the interpolated source terms possess the smallest error in case of $q_{e,rf,\partial\rho}$, while the interpolation error for $q_{e,rf,SIA}$ should have a significant impact on the acoustic solution. Additionally, it can be expected from these findings that the interpolation error for the source formulation using the substantial time derivative of the density is lower than for $q_{e,rf,SIA}$. In brief, the formulation with the partial time derivative of the density is expected to reproduce the acoustic field best.

Coarse grid vs. fine grid

In Figs. 7.10, 7.11, and 7.12, experimental and numerical results for different source formulations using a coarse and a fine grid are shown. The acoustic spectra are compared at two axial positions and a radial distance of $r/D = 50$. A significant deviation in the higher frequency range starting from $St > 0.65$ can be observed between the coarse and the fine solution for the source formulations $q_{e,rf,D\rho}$ and $q_{e,rf,SIA}$. This discrepancy has already been presumed and corresponds to a critical maximum wave number of $(kD) = 1.54$ being resolved by the interpolation on the coarse grid. Furthermore, the acoustic solution in Fig. 7.10 excited by $q_{e,rf,SIA}$ on the fine mesh also shows an almost constant deviation of $\Delta L_p = 4dB$ in the lower frequency range compared to the coarse solution. Analyzing the numerical findings of the partial time derivative of the density reveals an unexpected behavior. It has been shown before that this source term exhibits the smoothest source distribution, i.e., the impact of the interpolation error on the acoustic solution should be minimal. The results, however, show that besides the aforementioned deviation in the higher frequency range, which also occurs in the case of $q_{e,rf,\partial\rho}$, there are disparities in the lower frequency range, e.g., in Figs. 7.12 ($0.3 \leq St \leq 0.65$). Those deviations can not be explained by ordinary interpolation errors but rather by a spurious effect, which has been discussed in section 7.3.1.2, i.e., the effect of artificial acceleration due to interpolation. It has been shown in section 7.3.1.2 that the interpolation errors show a behavior, which is similar to a source acceleration. However this artificial acceleration generates spurious sound, which has to be avoided.

Spectral content and directivity behavior

As mentioned before, each source formulation defined in eqs. (7.57,7.58,7.59) includes certain sound generating effects. A statement about the impact of some source mechanisms can be given implicitly when the spectral distributions in Fig. 7.13 are compared. It is evident from Figs. 7.13a, b that on the one hand, the acoustic impact in the lower frequency range ($St \leq 0.45$) is dominated by the effect of unsteady heat release rate. Overall, this is in agreement with theoretical and experimental findings [61, 63]. On the other hand, this source effect appears to be insufficient to reproduce the acoustic field in the higher frequency range ($St > 0.45$).

Recall from table (7.1) that the source formulation with the substantial time derivative ($q_{e,rf,D\rho}$) includes additionally to the unsteady heat release rate, the effect of non-isomolar combustion, heat and species diffusion, and a term describing viscous heating[1]. Comparing the spectra obtained using ($q_{e,rf,D\rho}$) with those using ($q_{e,rf,SIA}$) shows that the additional source mechanisms have an impact especially in the higher frequency range. In the downstream direction, i.e., at $(x/D,\ r/D) = (50,\ 50)$, the spectrum is overpredicted in the frequency range ($St > 0.8$). However, the acoustic field excited by ($q_{e,rf,D\rho}$) shows a good agreement in a wide frequency range at almost all observer locations.

[1] In a hybrid approach, any assumption to the first part, i.e., the equations governing the flow are directly inherited to the subsequent part. Though the APE-RF system is an exact consequence of the conservation equation governing multi-component reacting fluids, the flame has been simulated in this case using a flamelet combustion model, in which viscous heating is neglected.

During the derivation of the simplified source formulation $q_{e,rf,\partial\rho}$ only one simplification has been applied to the RHS of the pressure-density relation, i.e., pressure fluctuations have been neglected. Hence, as also listed in table (7.1), this formulation includes the most sound source effects. Compared with the other formulations, the analysis of Fig. 7.13 shows at each observer location a very good agreement of the numerical results using $q_{e,rf,\partial\rho}$ with the experimental data. The numerical results, especially in Fig. 7.13(a) suggest the acoustic response from the extra sources, expressed through the difference between the two formulations $(q_{e,rf,\partial\rho} - q_{e,rf,D\rho})$, namely the effect of the acceleration of density inhomogeneities and the effect of mean density gradients, to lie in the higher frequency band and to possess a preferred radiation direction of 90° to the flame axis close to the jet exit. A significant contribution of the thermoacoustic sources $\sum_{i=2}^{N=5} q_{e,rf,i}$ to the acoustic field can be observed mainly in the higher frequency range ($St > 0.45$) at each observer position.

Three different simplified formulations of the RHS of the pressure-density relation

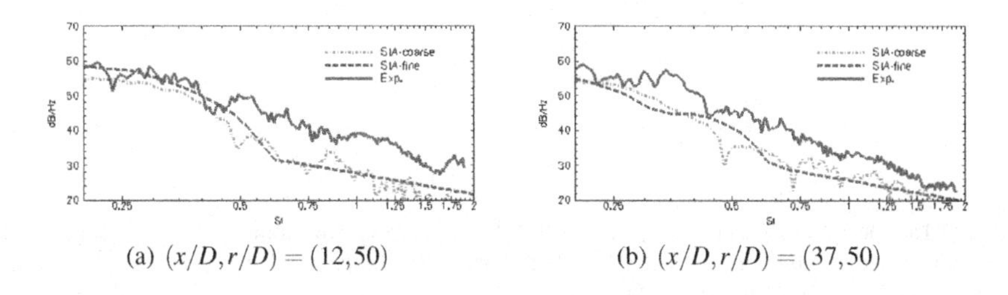

(a) $(x/D, r/D) = (12,50)$ (b) $(x/D, r/D) = (37,50)$

Fig. 7.10 Spectra in the frequency range of $0.2 \leq St \leq 2$ are shown. Comparison of computed sound pressure levels on a coarse and a fine grid at $r/D = 50$ and two different axial x/D locations using $q_{e,SIA}$. The STROUHAL number is based on the jet exit velocity and the nozzle diameter D.

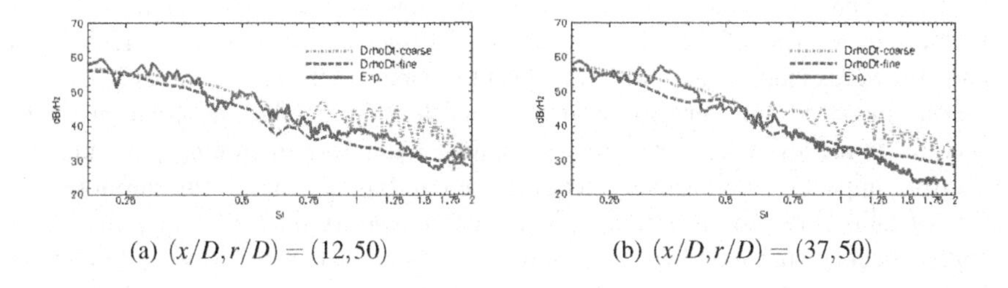

(a) $(x/D, r/D) = (12,50)$ (b) $(x/D, r/D) = (37,50)$

Fig. 7.11 Spectra in the frequency range of $0.2 \leq St \leq 2$ are shown. Comparison of computed sound pressure levels on a coarse and a fine grid at $r/D = 50$ and two different axial x/D locations using q_{e,D_ρ} (right). The STROUHAL number is based on the jet exit velocity and the nozzle diameter D.

have been used to simulate the acoustic field generated by a non-premixed turbulent flame, i.e., the so-called DLR-A flame. The main focus of this study has been

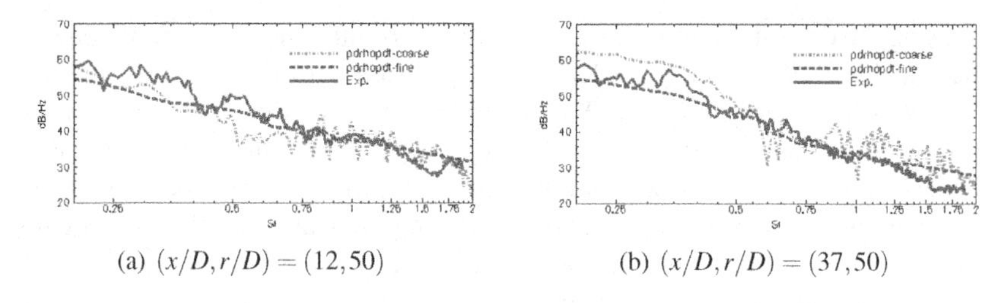

(a) $(x/D, r/D) = (12, 50)$ (b) $(x/D, r/D) = (37, 50)$

Fig. 7.12 Spectra in the frequency range of $0.2 \leq St \leq 2$ are shown. Comparison of computed sound pressure levels on a coarse and a fine grid at $r/D = 50$ and two different axial x/D locations using q_{e,∂_p}. The STROUHAL number is based on the jet exit velocity and the nozzle diameter D.

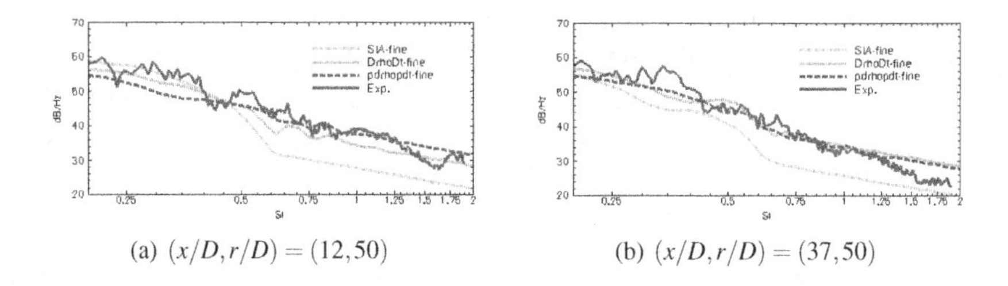

(a) $(x/D, r/D) = (12, 50)$ (b) $(x/D, r/D) = (37, 50)$

Fig. 7.13 Spectra in the frequency range of $0.2 \leq St \leq 2$ are shown. Comparison of measured and calculated sound pressure level at $r/D = 50$ and two different axial x/D locations using different sound source formulations. The STROUHAL number is based on the jet exit velocity and the nozzle diameter D.

the investigation of the numerical applicability of these formulations in terms of numerical accuracy and efficiency. It has been emphasized that the different source formulations include different sound generating mechanisms. To take advantage of the scale separation in aeroacoustics, a hybrid solution approach has been used. That is, the source terms of the APE-RF system have been evaluated on the LES mesh and afterwards mapped onto the CAA grid to propagate the sound waves.

An almost constant energy drop of about $4dB$ due to the interpolation is observable in the lower frequency range when using the formulation $q_{e,rf,SIA}$, i.e., the acoustic impact of the unsteady heat release rate has been used. Furthermore, the CAA results revealed an intricate phenomenon. The formulation containing most source mechanisms and having the smoothest source distribution does not yield the best result on the coarse mesh. On the fine grid, however, the experimental findings are best matched by the acoustic response due to this formulation. One reason for this behavior has been shown to be the effect of "artificial interpolation induced acceleration". The comparison of the source terms shows the substantial time derivative of the density to be a good compromise between numerical accuracy and efficiency. Moreover, this formulation does not suffer from the "artificial interpolation induced acceleration" effect. However, if the convection speed can be preserved during interpolation, the formulation using the partial time derivative of the density is to be preferred.

7.5 Summary and Conclusions

The acoustic perturbation equations for reacting flows (APE-RF) have been derived to simulate combustion generated noise in a hybrid CFD/CAA framework. The APE-RF system is based on the homogeneous APE system by EWERT & SCHRÖDER [21]. The APE system has been chosen as the basis for the APE-RF system due to its special properties concerning the description of acoustic propagation, i.e, the system describes wave propagation in an irrotational non-uniform mean flow. Convection and refraction effects are part of the solution process, which is fundamentally necessary in a hybrid context using incompressible CFD data and a mean flow field with spatially varying mean speed of sound, like in a flame. Moreover, the stability for arbitrary mean flows has been proven for the homogeneous APE system. Unlike aeroacoustics of non-reacting flows, such as jet noise investigations, in which the acoustic excitation is mostly caused at its highest level by vortex sound, combustion generated noise is mainly due to the effect of unsteady heat release rate, which can be found on the RHS of the pressure-density relation. Several other source mechanisms have been identified on the RHS of the pressure-density relation besides the major aforementioned effect, which are the effect due to non-isomolar combustion, heat and species diffusion, viscous heating, non-uniform mean flow ascendancies and the effect due to the acceleration of density inhomogeneities.

To simulate the acoustic field generated by a turbulent flame and to investigate the impact of the different source contributions on the acoustic field, the solution algorithm for the APE-RF system has been implemented in DLR's CAA code PIANO. The acoustic response of a turbulent non-premixed flame, i.e., the H3 flame has been computed using CFD data from an incompressible variable density LES, which has been performed by subproject 3 (Technische Universität Darmstadt [9]). The classical flamelet approach has been used in the LES as combustion model, while the scaled substantial time derivative of the density has been taken as the major source term of the APE-RF system. The comparison of the radial intensity spectrum with the experimental data, conducted by subproject 4 (Technische Fachhochschule Berlin[13]) shows a good agreement in a wide frequency range. A deviation from the experiments in the frequency range of $0.36 < St < 0.45$ has initiated a study about the impact of the thermoacoustic sources, which are implicitly included in the formulation using the material derivative of the density. The acoustic field of a premixed flame front has been simulated in the next step using the APE-RF system in conjunction with a compressible DNS solution. The latter has been performed by subproject 10 (Otto-von-Guericke Universität Magdeburg [10]). DNS input data has been chosen to allow for a most precise source evaluation, since a full DNS up to the chemical time and length scales has been performed. The order of magnitude analysis of the different thermoacoustic sources showed the species diffusion being in the same order as the unsteady heat release rate, while the order of magnitude of the other effects suggests a negligible impact on the acoustic field. Furthermore, from this analysis, the acoustic effect due to viscous heating is negligible, such that neglecting this effect using LES and the flamelet approach is justified.

To maintain the efficient character of the hybrid CFD/CAA method, simplified source formulations have been derived to simulate combustion noise. Besides the scaled unsteady heat release rate, two other formulations involving time derivatives of the density are proposed, which inherently contain the effect of unsteady heat release rate. Moreover, these two formulations are easy to evaluate from a reacting flow CFD simulation, since the unsteady density information is directly available, in contrast to the unsteady heat release rate, which is usually not necessary to be evaluated during the solution process of the flow simulation. The main aim, besides the understanding of the various source mechanisms, is the efficient and accurate prediction of combustion generated noise using a hybrid CFD/APE-RF method. Therefore, these formulations have been investigated in a further step in terms of their numerical applicability. The acoustic field of the DLR-A flame has been numerically investigated using these formulations. Furthermore, to efficiently evaluate the thermoacoustic sources, i.e., the unsteady heat release rate, the effect due to non-isomolar combustion, and the source describing the effect due to species diffusion, they have been directly precompiled into the flamelet library. This case has been simulated on two different CAA meshes with different spatial resolution and with different source formulations. Three main aspects can be summarized from the results. First, the acoustic response in the lower frequency range is dominated by the effect of unsteady heat release rate, which is in agreement with experimental results. However, the results also show in the higher frequency range, that the other source mechanisms, which have been deemed negligible have an significant contribution to the acoustic field. Second, the source formulation expressed through the partial time derivative of the density represents the most complete formulation and hence yields the best agreement with the experimental data. This, however, just holds for the simulation using the very fine mesh. Additional investigations have revealed an artificial effect, which results from the interpolation of the source terms from the LES on the CAA mesh, and leads to spurious noise generation due to the "artificial interpolation induced acceleration" effect of entropy spots, which are just being convected constantly. Third, a good compromise between numerical accuracy and numerical efficiency can be achieved using the source formulation expressed through the substantial time derivative of the density.

The acoustic field from open turbulent non-premixed flames shows an overall directivity pattern of a monopole radiator. Weak deviations from the isotropic behavior, i.e., a preferred acoustic radiation direction between $40^\circ < \alpha < 80^\circ$ has been found from the numerical results, which are in agreement with experimental findings from SMITH & KILHAM [61]. From the detailed source analysis and the numerical experiment to show the influence of the non-uniform mean speed of sound on the acoustic field, it is evident, that the cause for such a directivity behavior is twofold. The refraction effect due to the varying mean speed of sound causes a deviation from the isotropic directivity pattern. Furthermore, this effect is frequency-dependent, i.e., the higher the frequency, the stronger the refraction impact on the directivity. However, since low energy content is transported by the higher frequencies, the impact of the multipole directivity pattern on the overall directivity is fairly small. The source term itself can cause non-isotropic radiation patterns, like the effect of the acceleration of density inhomogeneities, which is of dipole type.

It has been stated before that combustion instabilities are not addressed in this work. However, additionally to the typical combustion instability caused by acoustic induced oscillating fuel feeding, there exists an acoustic wave amplification effect, which does not cause a feedback loop but is also captured by RAYLEIGH's criterion. It has been shown that this criterion can be derived from the APE-RF sources.

7.6 Acknowledgments

The financial support by the German Research Foundation (DFG) through the Research Unit FOR 486 "Combustion Noise" is gratefully acknowledged. The authors also would like to express their gratitude to F. Flemming, R. Piscoya, H. Shalaby, and M. Ihme for helpful discussions, comments, and for providing CFD and experimental data.

References

[1] Arnold JS (1972) Generation of combustion noise. J ASME 52(1):5 – 12
[2] Bailly C, Bogey C (2004) Contributions of computational aeroacoustics to jet noise research and prediction. Int J Comput Fluid Dynamics 18(6):481 – 491
[3] Barlow R (ed) (1996-2004) Proceedings of the TNF Workshops, Sandia National Laboratories, Livermore, CA, www.ca.sandia.gov/TNF
[4] Bogey C, Bailly C (2004) A family of low dispersive and low dissipative explicit schemes for flow and noise computations. J Comput Physics 194:194 – 214
[5] Bragg S (1963) Combustion noise. J Inst of Fuel 36:12–16
[6] Brentner KS, Lyrintzis AS, Koutsavdis EK (1997) Comparison of computational aeroacoustic prediction methods for transsonic rotor noise prediction. J Aircraft 34(4):531–538
[7] Brick H, Piscoya R, Ochmann M, Költzsch P (2005) Prediction of the sound radiation from open flames by coupling a large eddy simulation and a kirchhoff-method. Acta Acustica united with Acustica 91:17–21
[8] Bui TP, Ihme M, Meinke M, Schröder W, Pitsch H (2007) Numerical investigation of combustion noise and sound source mechanisms in a non-premixed flame using LES and APE-RF. In: AIAA Paper 2007-3406

[9] Bui TP, Meinke M, Schröder W, Flemming F, Sadiki A, Janicka J (2005) A hybrid method for combustion noise based on LES and APE. In: AIAA Paper 2005-3014
[10] Bui TP, , Schröder W, Meinke M, Shalaby H, Thévenin D (2006) Source term evaluation of the APE-RF system using dns data. In: In Proc. ECCOMAS CFD, Egmond aan Zee, The Netherlands, 2006, 188/1-188/14
[11] Bui TP, Meinke M, Schröder W (2006) Source term evaluation of the APE-RF system. In: AIAA Paper 2006-2678
[12] Bui TP, Meinke M, Schröder W (2007) Numerical analysis of the acoustic field of reacting flows via acoustic perturbation equations. Comput & Fluids 37(9):1157–1169
[13] Bui TP, Schröder W, Meinke M (2007) Acoustic perturbation equations for reacting flows to compute combustion noise. Int J Aeroacoustics 6(4):335–355
[14] Bui TP, Ihme M, Schröder W, Pitsch H (2009) "analysis of different sound source formulations to simulate combustion generated noise using a hybrid LES/APE-RF method". Int J Aeroacoustics 8(1):95–124
[15] Chiu HH, Summerfield M (1974) Theory of combustion noise. Acta Astronautica 1(7-8):967 – 984
[16] Chiu HH, Plett EG, Summerfield M (1973) Noise generation by ducted combustion systems. Conf. Pap. 73-1024, AIAA, aIAA Aeroacoustics Conf., Seattle, WA, Oct 15-17
[17] Crighton DG (1991) Airframe noise. Tech. Rep. 96-3052
[18] Crighton DG, Dowling AP, Ffowcs Williams JE, Heckl M, Leppington FG (1992) Modern Methods in Analytical Acoustics. Springer
[19] Doak PE (1973) Analysis of internally generated sound in continuous materials: 3. The momentum potential field description of fluctuating fluid motion as a basis for a unified theory of internally generated sound. J Sound Vibr 26(1):91 – 120
[20] Ewert R (2002) A Hybrid Computational Aeroacoustics Method to Simulate Airframe Noise. PhD Thesis, RWTH Aachen University
[21] Ewert R, Schröder W (2003) Acoustic perturbation equations based on flow decomposition via source filtering. J Comput Physics 188:365–398
[22] Ewert R, Schröder W (2004) On the simulation of trailling edge noise with a hybrid LES/APE method. J Sound Vibration 270:509–524
[23] Ewert R, Meinke M, Schröder W (2000) Aeroacoustic source terms for the linearized Euler-equations. Tech. Rep. 2000-2046
[24] Farassat F (1996) The Kirchhoff Formulas for Moving Surfaces in Aeroacoustics - The Subsonic and Supersonic Cases. Tech. Rep. Tech. Mem. 110285
[25] Ffowcs Williams JE (1969) Hydrodynamic noise. Ann Ref Fluid Mech 1:197 – 222
[26] Giammar RD, Putnam AA (1970) Combustion roar of turbulent diffusion flames. J Eng for Power 92(2):157 – 165
[27] Goldstein ME (1974) The low frequency sound from multipole sources in axisymmetric shear flows, with applications to jet noise. Tech. rep.

[28] Gröschel E, Schröder W, Meinke M, Comte P Noise prediction for a turbulent jet using different hybrid methods. Computers & Fluids, to be published in 2007
[29] Hassan H (1974) Scaling of combustion-generated noise. J Fluid Mech 66(3):445 – 453
[30] Hilbert R, Thévenin D (2002) Autoignition of turbulent non-premixed flames investigated using direct numerical simulations. Combust Flame 128:22–37
[31] Hinze JO (1975) Turbulence. 2nd edition, McGraw-Hill
[32] Howe MS (1975) Contributions to the theory of aerodynamic sound, with application to excess jet noise and the theory of the flute. J Fluid Mech 71(4):625 – 673
[33] Hu FQ, Hussaini MY, Manthey JL (1996) Low-dissipation and low-dispersion runge-kutta schemes for computational acoustics. J Comput Physics 124(1):177–191
[34] Hurle IR, Price RB, Sugden TM, Thomas A (1968) Sound emission from open turbulent flames. Proc of the Royal Soc of London, Series A 303:409 – 427
[35] Ihme M, Bodony DJ, Pitsch H (2006) Prediction of combustion-generated noise in non-premixed turbulent jet flames using large-eddy simulation. In: AIAA Paper 2006-2614
[36] Ihme M, Kaltenbacher M, Pitsch H (2006) Numerical simulation of flow- and combustion-induced sound using a hybrid les/caa approach. In: Center of Turbulence Research, Proceedings of the Summer Program 2006
[37] Klein R (1995) Semi-implicit extension of a godunov-type scheme based on low mach number asymptotics i: One-dimensional flow. J Comput Physics 121:213–237
[38] Klein SA, Kok JBW (1999) Sound generation of turbulent non-premixed flames. Combust Sci and Tech 149:267 – 295
[39] Knott PR (1971) Noise generated by turbulent non-premixed flames. Tech. Rep. 71-732, AIAA
[40] Kotake S (1975) On combustion noise related to chemical reactions. J Sound Vibration 42(3):399 – 410
[41] Kotake S, Hatta K (1965) On the noise of diffusion flames. Japan Society of Mechanical Engineers Journal 8(30):211 – 219
[42] Laverdant A, Thévenin D (2003) Interaction of a gaussian acoustic wave with a turbulent premixed flame. Combust Flame 134:11 – 19
[43] Laverdant A, Gouarin L, Thévenin D (2007) Interaction of a gaussian acoustic wave with a turbulent non-premixed flame. Combust Theory Modelling 11(4):585 – 602
[44] Lele SK (1992) Compact finite difference schemes with spectral-like resolution. J Comput Physics 103:16 – 42
[45] Lele SK (1997) Computational aeroacoustics: A review. In: AIAA Paper 97-0018
[46] Lieuwen TC, Neumeier Y, Zinn BT (1999) Determination of unsteady heat release distribution in unstable combustor from acoustic pressure measurements. J Propul Power 15:613–616

[47] Lighthill MJ (1952) On sound generated aerodynamically: I. General theory. Proc R Soc London Ser A 211:564–587
[48] Lighthill MJ (1954) On sound generated aerodynamically: II. Turbulence as a source of sound. Proc R Soc London Ser A 222:1 – 32
[49] McMurtry PA, Jou WH, Riley JJ, Metcalfe RW (1986) Direct numerical simulations of a reacting mixing layer with chemical heat release. AIAA Journal 24(6):962–970
[50] Möhring W (1979) Modelling low Mach number noise. In E.-A. Müller, editor, Mechanics of Sound Generation in Flows. Springer
[51] Press WH, Flannery BP, Teukolsky SA, Vetterling WT (1992) Numerical Recipes, 2nd Edition. Cambridge University Press
[52] Price RB, Hurle IR, Sudgen TM (1968) Optical studies of the generation of noise in turbulent flames. In: Proceedings of the 12th Int. Combustion Symposium, The Combustion Institute, pp 1095–1101
[53] Putnam AA (1968) Noise Measurement of Turbulent Natural-Gas Diffusion Flames of 120,000- to 750,000-Btu/h Output. J ASME 43(4):890 – 891
[54] Ramachandra MK, Straehle WC (1983) Acoustic Signature from Flames as a Combustion Diagnostic Tool. AIAA J 21(8):1107 – 1114
[55] Rayleigh JWS (1945) The Theory of Sound, Vol. II. Dover, New York, NJ, 226
[56] Roberts JP, Leventhall HG (1971) The noise of natural gas burners. Applied Acoustics 4(2):103 – 113
[57] Schröder W, Ewert R (2005) LES-CAA Coupling, LES for Acoustics. Cambridge University Press
[58] Selle L, Lartigue G, Poinsot T, Koch R, Schildmacher KU, Krebs W, Prade B, Kaufmann P, Veynante D (2004) Compressible large eddy simulation of turbulent combustion in complex geometry on unstructured meshes. Combust Flame 137:489–505
[59] Shivashankara BN, Strahle WC, Handley JC (1973) Combustion noise radiation by open turbulent flames. Conf. Pap. 73-1025, AIAA, aIAA Aeroacoustics Conf., Seattle, WA, Oct 15-17
[60] Smith ST (2000) Flap side-edge vortex bursting and airframe noise. Tech. Rep. 2000-1999, AIAA Journal
[61] Smith TJB, Kilham JK (1963) Noise generation by open turbulent flames. J Acoust Soc Am 35(5):715 – 724
[62] Strahle WC (1971) On Combustion Generated Noise. J Fluid Mech 49:399 – 414
[63] Strahle WC (1972) Some results in combustion generated noise. J Sound Vibration 23(1):113 – 125
[64] Strahle WC (1978) Combustion noise. Progress in Energy and Comb Noise 4:157 – 176
[65] Subrahmanyam PB, Sujith RI, Ramakrishna M (2003) Determination of unsteady heat release distribution from acoustic pressure measurements: A reformulation of the inverse problem. J Acoust Soc Am 114:686–696
[66] Tam C, Webb J (1993) Direct computation of nonlinear acoustic pulses using high-order finite difference schemes. Tech. Rep. 93-4325, AIAA Journal

[67] Tam CKW (2004) Computational aeroacoustics: An overview of computational challenges and applications. Int J Comput Fluid Dynamics 18(6):547–567
[68] Tam CKW, Webb JC (1993) Dispersion-relation-preserving finite difference schemes for computational acoustics. J Comput Physics 107(2):262–181
[69] Thévenin D, van Kalmthout E, Candel S (1997) Two-dimensional direct numerical simulations of turbulent diffusion flames using detailed chemistry. In: Direct and Large-Eddy Simulation II: (Chollet, J.P., Voke, P.R. and Kleiser, L., Eds.), Kluwer Academic Publishers, pp 343–354
[70] Thomas A, Williams GT (1966) Flame noise: Sound emission from spark-ignited bubbles of combustible gas. Proc of the Royal Soc of London, Series A 294:449 – 466
[71] Truffaut JM, Searby G, Boyer L (1998) Sound emission by non-isomolar combustion at low mach numbers. Combust Theory Modelling 2:423–428
[72] Yanwen M, Dexun F (1996) Super compact finite difference method (scfdm) with arbitrarily high accuracy. Comput Fluid Dynamics J 5:259–276

Chapter 8
Investigations Regarding the Simulation of Wall Noise Interaction and Noise Propagation in Swirled Combustion Chamber Flows

Christoph Richter, Łukasz Panek, Verina Krause and Frank Thiele

Abstract Applications of a time-explicit high-order method, solving a generalised version of the linearised Euler equations including energy supply by an external heat source in combustion noise are considered. Boundary conditions, including non-reflective and partially-reflective formulations as well as filtering approaches to obtain a stable solution in a swirling reactive transonic flow are presented. The acoustic intensity is used to validate the numerical solution and to locate acoustic sources. Three examples, comparing the numerical solution to experiments are given. It is found that the correct reproduction of the experiment essentially depends on the reflection coefficient of the connected systems outside the computational domain, which is efficiently implemented by a time domain impedance model based on an extended Helmholtz resonator formulation. The results are in a reasonable agreement with the experiments for both indirect and direct combustion noise.

8.1 Introduction

Indirect combustion noise and the related instability mechanism are one of the major difficulties facing the introduction of innovative green technologies such as lean fuel combustion and geared high bypass ratio fans driven by a reduced number of highly-loaded turbine stages. According to Cumpsty and Marble [5] both of the aforementioned techniques lead to a growing importance of the indirect noise generation. Leyko et al [9] predict, using the theory of Marble and Candel [11], that the indirect noise produced in the combustion chamber exit nozzle of a modern aero-engine is one order of magnitude larger than the direct noise of the turbulent flame. Thus, the phenomenon can be considered essential to the understanding of turbo-machinery core noise for current and future aero-engine designs.

Christoph Richter · Luaksz Panek · Verina Krause · Frank Thiele
Technische Universität Berlin, Institute of Fluid Mechanics and Engineering Acoustics, Strasse des 17. Juni 135, 10623 Berlin, e-mail: Christoph.Richter@TU-Berlin.de

A. Schwarz, J. Janicka (eds.), *Combustion Noise*, DOI 10.1007/978-3-642-02038-4_8,

Indirect combustion noise is caused by the acceleration of heated efflux from an unsteady combustion, as may occur in the outlet nozzle of a combustion chamber and the turbine stages. Only a small fraction of the perturbation energy is directly radiating as noise from the flame front. The rest is carried silently by the hydrodynamic modes of perturbation through the combustion chamber. Such hydrodynamic modes of perturbation are non-isentropic perturbations of the fluid state and vortices according to the definition of Chu and Kovásznay [4]. In contrast to the isentropic linear acoustic perturbation, they move with the flow speed as a pattern of "frozen turbulence". In a homogeneous moving fluid the interaction of the hydrodynamic perturbations with each other and with sound waves is of second order according to Chu and Kovásznay [4]. However, the interaction becomes first order, once the gradients of the average flow field reach the order of the average field quantities.[1] Thus, such large flow gradients lead to a relevant energy transfer between acoustic, vorticity and entropy mode waves. The indirect and direct sources of combustion noise are shown in the sketch of Fig. 8.1.

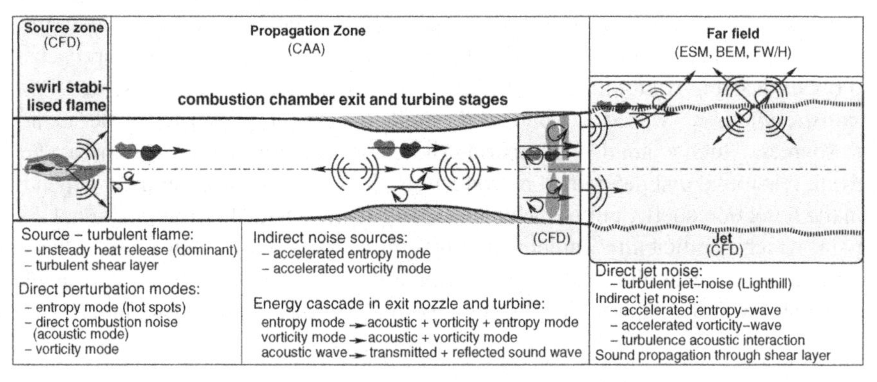

Fig. 8.1 Direct and indirect noise sources in a combustion system, which are correlated due to their common origin from the initial entropy perturbation.

In the case of a combustion system the non-isentropic modes of perturbation which show up as hot and cold spots, which are referred to as entropy modes are thought to carry particularly large amounts of energy. The flow gradient in the contracting outlet nozzle then allows a leakage of energy from the hydrodynamic modes into the acoustic and other hydrodynamic perturbation modes. Sound waves, which travel at the speed of sound, and secondary vortices are generated [4, 11, 7]. The process repeats itself in each turbine stage with the remaining hydrodynamic perturbation and an additional contribution of the secondary vortices to the indirect noise generation as described by Richter and Thiele [13]. The secondary vortices are generated by the interaction of non-isentropic perturbations with the rotor and stator blade flow field. Accordingly, additional indirect noise is radiated from the outlet nozzle of the combustion chamber and each turbine stage. This indirect noise gen-

[1] The speed of sound has to be considered as reference for the velocity to define a large gradient.

eration is sometimes also addressed by the terms entropy noise or excess noise or even as acoustical *bremsstrahlung* by Ffowcs Williams and Howe [7]. As the transformation of perturbation modes by a mean flow gradient can be described as a first order linear interaction of an unsteady perturbation with the constant flow gradient, the result is found to be coherent with the initial perturbation. While theoretically well known, a decomposition of direct and indirect noise in measurements of real combustion systems proves difficult to obtain, as has for instance been described by Bake et al [1]. The difficulty is probably related to the coherence of vortical and acoustic perturbations with the initial entropy mode and acoustic field which is generated directly by the combustion process, such that any spectral decomposition leads to correlations between all signals.

The most accurate way to numerically model the phenomenon would be a direct numerical simulation (DNS) of the whole combustion system, including at least the inflow upstream to the compressor and the outflow through the exit nozzle, which are both usually choked. Such DNS of realistic problems still remains far out of reach for current computing resources. However, there are examples of successful applications of large eddy simulation (LES) to realistic combustion systems described by [3]. LES is nonetheless very expensive in terms of computational resources and LES applications are therefore still limited to one or a few design variations and produce immense computational costs. On the other hand, the available theoretical approaches are limited to low Mach number flows [7] or one-dimensional flow conditions in compact nozzles [11]. Furthermore, there are some successful attempts to adopt an one dimensional theory to practical problems by Cumpsty and Marble [5], but possible three dimensional measures to reduce the indirect noise cannot be assessed by such a theory. The most substantial disadvantage of the compact nozzle assumption in the theory is the limitation to very low frequencies, for which the nozzle appears short compared to the wavelength. This moves turbulent noise far out of reach for a description using the theory for indirect combustion noise. The gap between 1D theory and 3D LES could be closed by a high-order time explicit Computational Aeroacoustics (CAA) method adopted from the description of fan tone propagation, which combines a high cost efficiency in terms of required points per wavelength with a high adaptability to realistic mean flow conditions. Then, a zonal approach, based on the idea of splitting the problem into zones with different physics, as it has been demonstrated for the forward propagation of fan tone noise in [23, 20] becomes obvious.

The combustion zone, featuring nonlinear interaction and chemical reactions in the flow field, could be considered by an LES or unsteady RANS simulation, which would provide the relevant perturbation sources. With an extension to allow the propagation of all hydrodynamic and acoustic perturbations, a classical CAA perturbation approach can be applied to describe indirect combustion noise.

8.2 Theoretical background

This extended mathematical model is presented first in section 8.2.1, following which the CAA method for solving these model equations and the numerical and physical boundary conditions are presented in section 8.2.2. The related analysis methods are shortly described in section 8.2.3.

8.2.1 Mathematical models

The most simple model covering indirect combustion noise, and describing the transport of all possible modes of perturbation in the wake of a flame, are the linearised Euler equations including a non-isentropic energy equation supplemented by a source term to account for an unsteady heat input [19]:

$$\frac{D\rho'}{Dt} + \boldsymbol{u}' \cdot \boldsymbol{\nabla} \rho_0 + \rho_0 \boldsymbol{\nabla} \cdot \boldsymbol{u}' + \rho' \boldsymbol{\nabla} \cdot \boldsymbol{u}_0 = 0, \tag{8.1a}$$

$$\frac{D\boldsymbol{u}'}{Dt} + \frac{\rho'}{\rho_0} \boldsymbol{u}_0 \cdot \boldsymbol{\nabla} \boldsymbol{u}_0 + \boldsymbol{u}' \cdot \boldsymbol{\nabla} \boldsymbol{u}_0 + \frac{1}{\rho_0} \boldsymbol{\nabla} p' = 0 \text{ and} \tag{8.1b}$$

$$\frac{Dp'}{Dt} + \gamma p_0 \boldsymbol{\nabla} \cdot \boldsymbol{u}' + \left(\boldsymbol{u}' - \frac{p'}{p_0} \boldsymbol{u}_0 \right) \cdot \boldsymbol{\nabla} p_0 = [\gamma - 1]\, \dot{q}. \tag{8.1c}$$

The capital D/Dt denotes the substantial derivative with the mean flow as transport velocity:

$$\frac{D\cdot}{Dt} = \frac{\partial \cdot}{\partial t} + \boldsymbol{u}_0 \cdot \boldsymbol{\nabla} \cdot$$

These model equations are then applicable to describe the propagation of hydrodynamic and acoustic perturbations as well as their interaction with a variable compressible mean flow regime. Taking advantage of the axial symmetry of the test set-ups and flow conditions, the efficiency of the numerical method is improved by a 2D axisymmetric approach. The problem is formulated in a cylindrical coordinate system. The boundary conditions, flow field and perturbed flow field are assumed to be constant in the azimuthal direction. Similar to the modal axisymmetric approach for fan tone problems [10], this assumption allows the dimension of the problem to be reduced by one.

8.2.2 Numerical Method

In this section the numerical method including the high-order temporal and spatial discretization, high-order filtering algorithms for stabilizing the finite difference scheme and the boundary conditions are presented.

8.2.2.1 Discretization schemes

The numerical method for solving Eq. (8.1) is based on the fourth order Dispersion Relation Preserving (DRP) scheme [22] for the spatial discretization. At the boundaries such as the walls and open ends of a duct, optimised backward stencils of fourth and third order are used. The time marching is performed by the alternating five/six stage Low Dissipation Low Dispersion Runge–Kutta (LDDRK) method [8] implemented in 2N storage form [21].

8.2.2.2 High order and adaptive filtering

An eleven-point-stencil central Taylor filter of tenth order is used to eliminate parasite waves [15]. At the boundaries, central filters of lower order are applied. Furthermore, a finite difference method could not be applied to describe the abrupt change in a choked flow as grid oscillations from the poorly resolved shock would pollute the computational domain. Any explicit filter, except the second order three point filter, would respond with grid oscillations to such a shock. Therefore, an adaptive reduction of the filtering order is implemented to account for the special conditions in a choked flow. The filter stencil size is reduced as long as the change of the overall Mach number within the filter stencil source is larger than a predefined threshold until the filter stencil size reaches three points. Finally, the mean flow as well as the grid were filtered using the three point, second order filter stencil, too. This is mainly necessary to avoid high wave number oscillations in the source terms and metrics, which would feed point-to-point oscillations in the solution.

8.2.2.3 Radiation and outflow boundary conditions

The principle of the radiation and outflow boundary condition is to formulate the non-reflective boundary condition by modifying the governing equations for the propagation of perturbations in such a way that only outgoing characteristics are described. The direct formulation of a boundary condition in the sense of specifying a boundary value is not necessary in this way. The modified system may be obtained to describe the convective transport and wave propagation in one direction, from inside the computational domain to the outside. The opposite direction is removed from the original PDE system. Two cases have to be considered, which are an inflow/no flow boundary condition, where only acoustic waves are able to leave the computational domain and an outflow boundary condition, which additionally allows all types of hydrodynamic perturbations to leave the computational domain.

Radiation condition

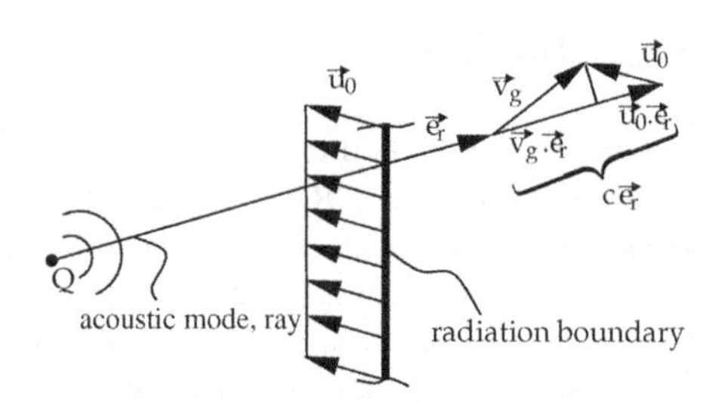

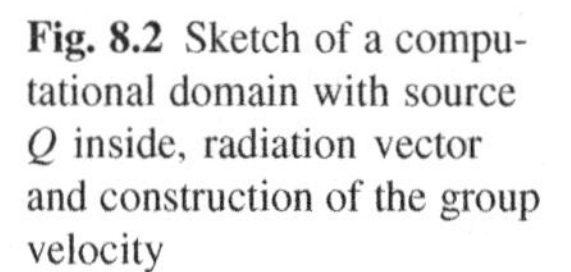
Fig. 8.2 Sketch of a computational domain with source Q inside, radiation vector and construction of the group velocity

The radiation condition, that describes acoustic waves only is presented first, following which the outflow boundary condition is developed as an extension. A split of the governing equations with respect to the propagation direction is difficult in a general varying base flow regime. In addition it requires the source position to be specified in advance by a guess of the solution. To avoid this, the radiation direction is replaced by the normal of the boundary and the distance from the source is assumed to be infinitely large. Under the above assumptions using the normal vector of the radiation boundary, the radiation of acoustic waves can easily be formulated without further assumptions. The resulting modified radiation boundary condition applies exactly for plane waves in a ducted environment and approximately for spherical and cylindrical waves in the far-field of a source. The radiation boundary condition is obtained according to Bogey and Bailly [2] as follows:

$$\frac{\partial}{\partial t}\begin{pmatrix}\rho'\\ \boldsymbol{u}'\\ p'\end{pmatrix} + v_g \frac{\partial}{\partial n}\begin{pmatrix}\rho'\\ \boldsymbol{u}'\\ p'\end{pmatrix} = 0 \tag{8.2a}$$

$$\frac{\partial}{\partial t}\Phi' + v_g \frac{\partial}{\partial n}\Phi' = 0, \tag{8.2b}$$

with the group velocity defined as:

$$v_g = \boldsymbol{u}_0 \cdot \boldsymbol{e}_n + c. \tag{8.3}$$

The geometric construction of the group velocity with respect to the radiation direction is given in Fig. 8.2. The radiation condition is solved instead of the interior PDE system at the outermost three grid lines at the non-reflective boundary. This boundary condition is applied for in-duct problems and in the vicinity of walls. Alternatively, the original formulation of Bogey and Bailly [2] is applied for the radiation into open space, with the centre of the outlet of the duct as the specified source position. Even though the acoustic wave propagation is modelled as a spherical radiation problem and the source position may influence the result significantly, the radiation condition has demonstrated low reflection properties for a wide range of applications.

Outflow boundary condition

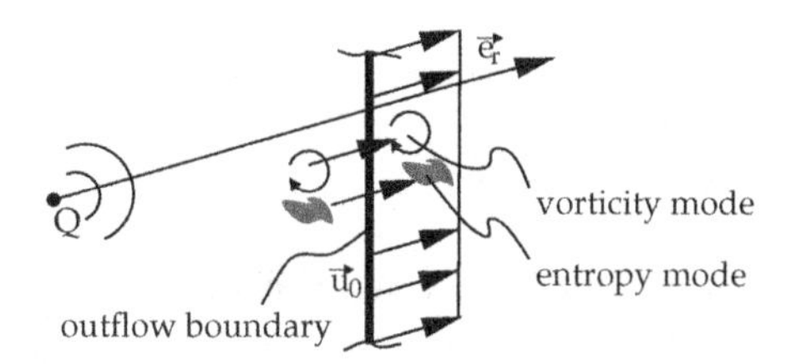

Fig. 8.3 Sketch of a computational domain with source Q inside, outflow of vorticity and entropy mode waves with the flow speed $\boldsymbol{u}_0$ and radiation unit vector $\boldsymbol{e}_n$

The outflow boundary condition is obtained by including the transport of entropy and vorticity perturbations in the system as sketched in Fig. 8.3. The pressure perturbation is purely acoustic, whereas the entropy mode is added to the density perturbation ($\rho' = \rho'_a + \rho'_s$) and the vortical perturbation is superimposed to the acoustic velocity ($\boldsymbol{u}' = \boldsymbol{u}'_a + \boldsymbol{u}'_\omega$). The acoustic density perturbation is given by $\rho_a = \frac{1}{c^2} p'$ and the acoustic velocity is given by $\frac{\partial \boldsymbol{u}'_a}{\partial t} + u_0 \cdot \boldsymbol{\nabla} \boldsymbol{u}'_a = -\frac{1}{\rho_0} \boldsymbol{\nabla} p'$. By using the connection of the acoustic perturbation of the velocity vector and density to p', the outflow condition is obtained as follows (comp. Bogey and Bailly [2]):

$$\frac{\partial \rho'}{\partial t} + \boldsymbol{u}_0 \cdot \nabla \rho' = \frac{1}{c_0^2} \left(\frac{\partial p'}{\partial t} + \boldsymbol{u}_0 \cdot \nabla p' \right) \tag{8.4a}$$

$$\frac{\partial \boldsymbol{u}'}{\partial t} + \boldsymbol{u}_0 \cdot \nabla \boldsymbol{u}' = -\frac{1}{\rho_0} \nabla p' \tag{8.4b}$$

$$\frac{\partial p'}{\partial t} + v_g \frac{\partial}{\partial n} p' = 0. \tag{8.4c}$$

As can easily be seen from Eq. (8.4a) and Eq. (8.4b), variations of the mean flow have been neglected in the construction of the outflow boundary condition. This seem legitimate, as they would describe sources in the boundary region, which are not necessarily part of the numerical solution. However, it is found that strong variations of the mean flow inside the radiation boundary may lead to problems when an instability wave reaches the boundary.

$\dot{p}'$ occurs in Eq. (8.4a) and Eq. (8.4c). Eq. (8.4c) is used to obtain $\dot{p}'$, whereas $\dot{\rho}'$ is obtained from:

$$\frac{\partial \rho'}{\partial t} + \boldsymbol{u}_0 \cdot \nabla \rho' = \frac{1}{c_0^2} \left(-v_g \frac{\partial}{\partial n} p' + \boldsymbol{u}_0 \cdot \nabla p' \right). \tag{8.4d}$$

Automatic detection of outflow and radiation boundary

To automatically detect if a boundary condition is locally an inflow, zero flow or outflow, the normal vector $\boldsymbol{n}$ of the boundary is used. $\boldsymbol{n}$ is defined as the outer normal vector orthogonal to the outermost grid line at the boundary. Then the scalar product with the average velocity vector $\boldsymbol{u}_0$ leads to the following differentiation:

$$\boldsymbol{n} \cdot \boldsymbol{u}_0 = \begin{cases} > 0, \text{ outflow boundary condition} \\ \leq 0, \text{ radition boundary condition.} \end{cases}$$

Based on this criterion the radiation boundary condition or outflow boundary condition are applied.

8.2.2.4 Impedance and wall boundary conditions

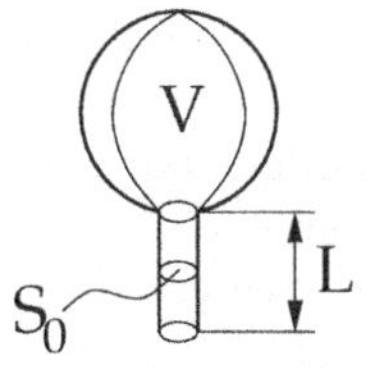

Fig. 8.4 Sketch of the geometrical parameters of a Helmholtz resonator.

The impedance of the connected duct systems is modelled by the extended Helmholtz resonator model [18] (EHR) as a time domain impedance boundary condition. The implementation used here is described in Richter et al [16] and Richter and Thiele [14]. The frequency response of the extended Helmholtz resonator including an extension to high frequencies and for the damping inside the cavity is given by Rienstra [18] as:

$$Z(i\omega) = R + i\omega m - i\beta \cot\left(\frac{1}{2}\omega T_l - i\frac{1}{2}\varepsilon\right). \tag{8.5}$$

An $e^{i\omega t}$ convention is used here. The normal vector $\boldsymbol{n}$ of the impedance surface is defined as positive when pointing into the surface. The cavity parameters ε and β denote the damping of the cavity fluid and the cavity reactance respectively. For zero damping of the cavity fluid ($\varepsilon = 0$) and at the face sheet ($R = 0$), the remaining parameters of the EHR can be identified with the geometric parameters of a low frequency Helmholtz resonator as follows:

$$\frac{T_l}{2} = \frac{2\pi}{c}\sqrt{\frac{VL(1+\Delta_L)}{S_0}}, \tag{8.6a}$$

$$\beta = c\frac{S_0}{V}\frac{T_l}{2} \tag{8.6b}$$

and

$$m = \frac{L(1+\Delta_L)}{c} - \frac{1}{6}\beta T_l. \tag{8.6c}$$

The nomenclature of the Helmholtz resonator dimensions is given in Fig. 8.4. A length correction is applied, which is denoted by Δ_L.

The periodicity of $Z(i\omega)$ is assumed to be T_l such that a z-transformation can be used to obtain the following time domain boundary condition for $\dot{\boldsymbol{u}}'_n$ from Eq. (8.5):

$$\frac{\partial u'_n}{\partial t}(t) = \frac{1}{m}\left[p'(t) - e^{-\varepsilon}p'(t-T_l) - (R+\beta)u'_n(t) + (R-\beta)e^{-\varepsilon}u'_n(t-T_l)\right] + e^{-\varepsilon}\frac{\partial u'_n}{\partial t}(t-T_l). \tag{8.7}$$

8.2.3 Intensity-based analysis of the result

The acoustic intensity is used here to assess the quality of the acoustic solution and locate acoustic sources. In general the conservation of acoustic energy can be expressed in the form:

$$\frac{\partial e_a}{\partial t} + \nabla\cdot\boldsymbol{I} = q, \tag{8.8}$$

where e_a denotes the acoustic energy per unit volume, $\boldsymbol{I}$ is the acoustic intensity, and q is the production rate of acoustic energy per unit volume. Following Morfey [12], the definitions

$$e_a = \frac{p'^2}{2\rho_0 c_0^2} + \frac{\rho_0}{2}\boldsymbol{u}'_a\cdot\boldsymbol{u}'_a + \frac{p'}{c_0^2}\boldsymbol{u}_0\cdot\boldsymbol{u}'_a \tag{8.9a}$$

and

$$\boldsymbol{I} = \left(p' + \rho_0\boldsymbol{u}_0\cdot\boldsymbol{u}'_a\right)\left(\boldsymbol{u}'_a + \frac{p'}{\rho_0 c_0^2}\boldsymbol{u}_0\right) \tag{8.9b}$$

are used. The definitions of Eq. (8.9) have the advantage that q at least vanishes (to second order in fluctuating quantities) in all regions with irrotational flow and uniform specific entropy. Thus, in consideration of the results of Chu and Kovásznay [4] the production of acoustic energy is limited to non-homogeneous flow regions with vorticity or entropy perturbations in an otherwise potential flow field. There, the quantity q represents the production rate at which acoustic energy is generated in the flow.

8.2.3.1 Identification of acoustic sources

Note that q may also be negative, which means that fluctuation energy is annihilated. In the case of a statistically stationary flow the temporal average of equation (8.8) yields

$$\nabla \cdot \langle \boldsymbol{I} \rangle_t = \langle q \rangle_t \,, \tag{8.10}$$

because the mean value of $\partial e_a / \partial t$ vanishes. Equation (8.10) can be used to assess the quality of a numerical solution in the absence of sources for potential flow fields where no analytical solution is available for comparison. Furthermore, based on the identity of Eq. (8.10), acoustic sources are located in the flow by the acoustic energy that they radiate in a time averaged sense. Such sources may for instance be a flow nozzle excited by initial entropy perturbations or the direct noise of an unsteady heat input by a combustion process.

The acoustic source levels span multiple decades. While the artificial sources for instance due to numerical dissipation should be in the range of $< 1\ \%$ of the acoustic intensity, the physical sources can be much larger. To overcome this problem a logarithmic source power level (QPL) in [dB] is introduced, which is based on the logarithmic sound pressure level (SPL) scale. The unit of the acoustic sources is acoustic power per volume. A reference volume is introduced in addition to the reference power, which is analogous to that used to the define the SPL ($P_{\text{ref}} = 1 \times 10^{-12}$ W). The reference volume can be chosen as a typical source volume to obtain a source power level, which corresponds to the observed SPL originating from this source volume. The average source strength can be negative, which means a local annihilation of acoustic energy. Nevertheless, to obtain a compact overview of the sources, two logarithmic scales for positive and negative source strength are combined. The acoustic source power level QPL is defined as:

$$\text{QPL} = \text{sign}\,(P_{in})\ 10\log_{10}\left[\max\left(\frac{|P_{in}|}{10^{-12}W},\ 1\right)\right]. \tag{8.11a}$$

Where the input acoustic power is given by

$$P_{in} = \langle \nabla \cdot I_a \rangle_t \, V_{\text{ref}}. \tag{8.11b}$$

This yields a symmetric logarithmic source power level scale for production and annihilation of acoustic energy, which cuts off sources if the absolute value of the input source power is below the threshold of $10^{-12}\ W/V_{\text{ref}}$.

8.2.3.2 Validation by the conservation of the acoustic energy

Besides analysis, which uses local values of $\langle q \rangle_t$ to asses the quality of a numerical solution, an integral approach presents itself, where the energy balance over a control volume is considered as it was used e. g. by Eversman [6]. The global conservation law is obtained by integrating Eq. (8.10) over a volume V with closed surface S, which leads to

$$\int_S \boldsymbol{n} \cdot \langle \boldsymbol{I} \rangle_t \, \mathrm{d}S = \int_V \langle q \rangle_t \, \mathrm{d}V \,, \tag{8.12a}$$

where $\boldsymbol{n}$ is the outer surface normal, and $\boldsymbol{n} \cdot \langle \boldsymbol{I} \rangle_t$ is the (temporal mean) flux density of acoustic energy across the surface S (compare [6]). Relation (8.12a) is used here to investigate the conservation of energy in a duct with impermeable side walls. Therefore the mean flux $\langle P \rangle_t$ through a cross section of the duct is considered. This flux can be calculated by

$$\langle P(x) \rangle_t = \int_{S_x} \boldsymbol{n} \cdot \langle \boldsymbol{I} \rangle_t \, \mathrm{d}S_x \,, \tag{8.12b}$$

where S_x is the area of the respective lateral cut through the duct and $\boldsymbol{n}$ points in the axial direction. The variation of $\langle P \rangle_t$ with axial position x gives an overview about the energy conservation in the duct. Theoretically the flux of acoustic energy through the impermeable side wall is zero. In the absence of sources, no acoustic energy should be generated or annihilated in the duct. Consequently, because of Eq. (8.12a), the flux $\langle P \rangle_t$ should be identical for all axial positions x. In the presence of sources, the variation of $\langle P \rangle_t$ indicates the global effect of the sources at a given axial position.

8.3 Results and Discussion

In this section the detailed physics of several combustion systems are analysed based on numerical results using different analytical post processing methods. Three different systems are considered, which are an electrically heated flow duct as a model experiment in section 8.3.1, a premixed swirling combustor flow in section 8.3.2 and finally a model combustion system with a single non-premixed swirling flame in section 8.3.3.

8.3.1 The entropy wave generator (EWG) model experiment

The EWG model experiment for indirect combustion noise and the related experimental results are described in chapter 5, whereas an overview on the numerical simulation using CFD tools is provided in chapter 1. The set-up, which features a flow duct with inflow plenum, a heating module in the straight inlet duct, a nozzle and an outflow duct is sketched in Fig. 8.5. The flow in the nozzle ranges from subsonic to transonic conditions. A flat-top peak heating pulse of approximately 100 W over 0.1 s is considered as a perturbation source. The power is equally distributed

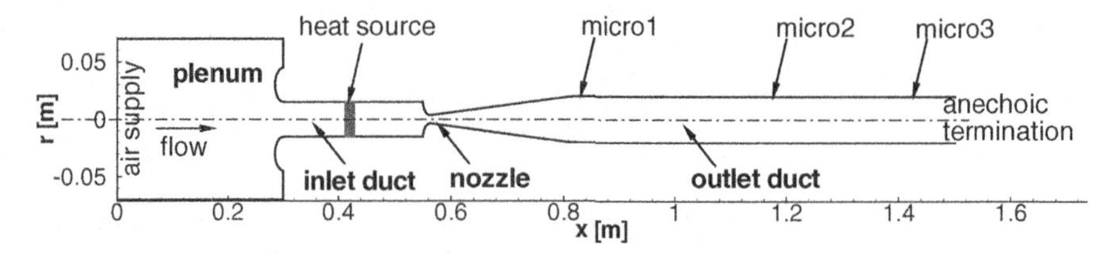

Fig. 8.5 Sketch of the EWG experiment.

over the source region ($x = 0.39 \ldots 0.43$ m) for the axisymmetric simulation. Direct noise from the heating and indirect noise from the acceleration of the hot gas in the nozzle are expected. The CAA method described above is used to simulate the compressible base flow and the unsteady system response to the heating pulse. A nonlinear Euler model as described in Richter et al [15] with adapted boundary conditions, which are given in detail in Richter et al [17] is used to obtain the mean flow conditions.

8.3.1.1 Maximum of the relative peak pressure response

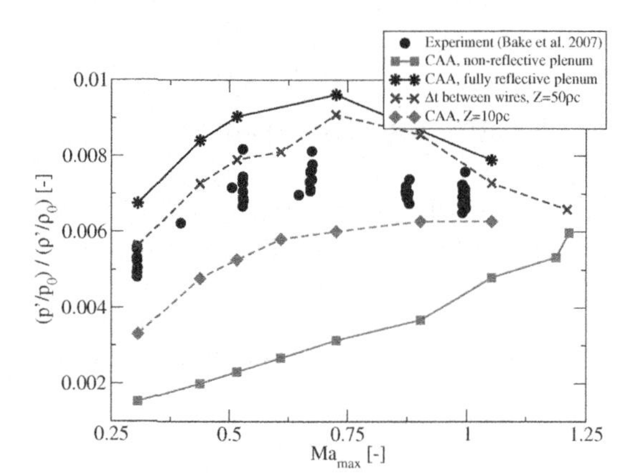

Fig. 8.6 Normalised transmission coefficient between entropy perturbation in the inlet duct and maximum pressure perturbation in the outlet duct.

First, a variation of the maximum nozzle Mach number in the flow duct is carried out following the experiment. All results of the parameter variations are summarised in Fig. 8.6. The perturbation variables are plotted non-dimensionally with the related flow variables. The pressure perturbation is normalised by the pressure in the outlet duct and the non-isentropic density perturbation is normalised by the mean density in the heated duct section. The abscissa of Fig. 8.6 shows a Mach number defined by the maximum flow velocity in the unperturbed state normalised by the local speed of sound.[2]

A fully non-reflective modelling of the plenum produces a large deviation from the variation of the peak pressure response over the nozzle Mach number found in the experiment, with even the trend incorrectly predicted. A hard wall and two impedance models for the plenum inflow are considered. The plenum inflow is made partially reflective by adding a non-infinite real part of the impedance and a negligibly small imaginary part growing with the frequency, respectively. The small imaginary part $m = 1/1000$ is required for the coupling of the EHR model to the time domain CAA method. As can be seen in Fig. 8.6 the plenum reflection leads to a much better reproduction of the experimental data for the peak pressure response. With a reflective plenum a saturation of the peak pressure response is observed around Ma $= 0.7$. The peak pressure level of the anechoic and reflective plenum equalise with the base flow reaching transonic conditions in the nozzle.

8.3.1.2 Quality analysis and source location

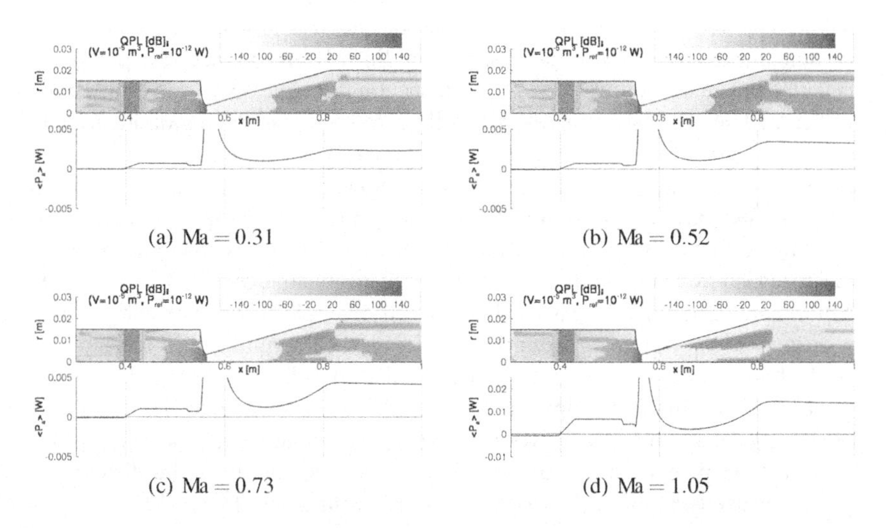

(a) Ma $= 0.31$ (b) Ma $= 0.52$

(c) Ma $= 0.73$ (d) Ma $= 1.05$

Fig. 8.7 Average acoustic source strength QPL (above as contour plot) and axial acoustic power flux P_a (below as lines).

Selected results of the intensity analysis are presented in Fig. 8.7. A partially reflecting plenum is considered. The direct source effect of the heat input is visible on the QPL-scale in the region between $x = 0.39$ m to $x = 0.43$ m. However, the acoustic energy input by the direct source (QPL ≈ 70 dB) is several orders of magnitude below the maximum found in the whole flow duct. The indirect source strength

[2] The experimental data is based on the corrected Mach number in the nozzle throat.

found in the nozzle and diffuser is much larger (QPL $\approx$ 120 dB). The most powerful acoustic sources are found in the narrow cross section of the nozzle. Due to the different sign found in nozzle and diffuser, only a small fraction of the acoustic energy is really radiated into the up- and downstream duct sections. Furthermore, the overall acoustic energy depends on the size of the source. Therefore, the resulting acoustic power is still dependent on the size of the source region. The direct source region forms a volume of one order larger than the strong sources in the nozzle. This means an SPL of one order smaller caused by this source volume. Some additional acoustic sources are located further downstream in the diffuser with the QPL. However, these sources may originate from the simplification of the acoustic velocity perturbation $\boldsymbol{u}_a'$ to the overall velocity perturbation, as a large and strong vortex is generated by the entropy perturbation passing the inhomogeneous flow field in the nozzle. In fact, this vortex is correlated with the acoustic waves in the exhaust duct. Both have the same origin, which is the passage of the initial entropy perturbation through the nozzle. Therefore, the vortical perturbation contributes to the modified intensity in a time averaged sense.

Below the contour plots, the time-averaged overall acoustic power flux $\langle P_a \rangle_t$ is given for each flow speed. While the colour map is the same for all figures, the overall acoustic power shown below has been adjusted to represent the observed levels better. The above observations are supported by $\langle P_a \rangle_t$. In the region of the heat module, the acoustic power increases nearly linearly and the acoustic power flux is negative upstream (left in Fig. 8.7) of the heating module and positive downstream of the module. The linear increase directly corresponds to the region identified as a source region with the QPL. Furthermore, there is an approximately constant acoustic power in the straight duct section between heat module and nozzle. The small decrease and variation in front of the nozzle is due to an interpolation error for the two grid blocks involved in the integration there. This causes an intense increase of the acoustic power flux in the nozzle, which leads to large amplitudes of $\langle P_a \rangle_t$. For a better representation, the peak has been cut off in Fig. 8.7. The intense right-running acoustic waves originating from the nozzle are then annihilated in the diffuser directly following the nozzle throat. The field in the diffuser may include trapped waves, which initially run upstream to be reflected at the nozzle. The net transport of acoustic energy by these waves is zero over the whole pulse. The reflected waves may also be involved in the annihilation of acoustic energy in the diffuser. Altogether the acoustic power flux in the diffuser becomes very low and increases again towards the outlet. The acoustic power flux in the outlet duct becomes constant again, even though there are acoustic sources identified by the QPL.

Taking the information of the QPL together with $\langle P_a \rangle_t$, the direct source mechanism is found to radiate with the full positive source strength, without any perturbation from negative sources into the heated inlet duct section. The sources in the nozzle have a different nature. Strong positive and negative sources are found with the QPL, which cancel on average. The resulting high acoustic power flux is localised at the throat between nozzle and diffuser. The acoustic power flux is then increased by the sources in the exit region of the diffuser again.

8.3.2 Swing-off response of a premixed swirl combustor flow

The premixed swirling flame described in chapter 2 is enclosed by an approximately cylindrical shell with one open surface opposite to the burner as described to obtain an enclosed flame, which features the typical aspiration of hot gases without the indirect noise due to the acceleration in the combustion chamber exit nozzle. The numerical simulation of the problem with CFD methods is also described in chapter 2. A steady RANS simulation and the averaged flow field obtained from an LES simulation, which were both carried out by Zhang et al [24] using CFX, have been made available to the authors. These results are applied as the base flow regime (subscript $_0$) in the perturbation approach described in Eq. (8.1). The resulting mean flow conditions are very similar at first glance as can be seen from Fig. 8.8. However, the flame cone and the recirculation zones differ between the RANS (Fig. 8(a)) and averaged LES results (Fig. 8(b)).

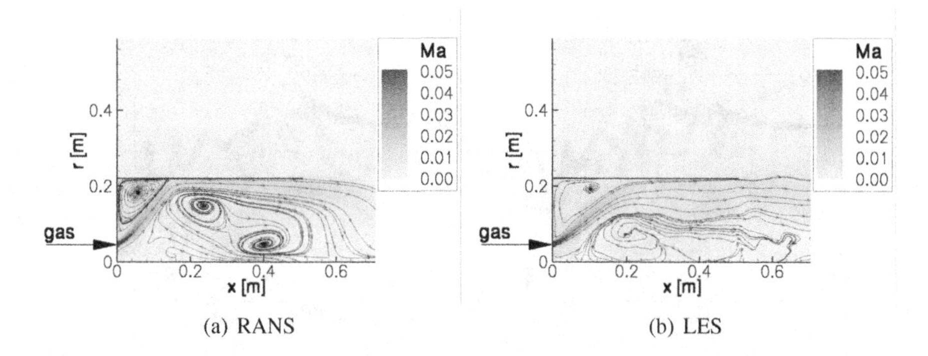

(a) RANS (b) LES

Fig. 8.8 Mach number and stream traces for the average flow fields ($_0$ in Eq. (8.1)) provided by Zhang et al [24].

The CAA method is now used to compute the spectral peak frequencies of the combined geometric and flow system. A swing-off response due to a perturbed initial condition may be considered for this purpose. However, the result is not statistically stationary and therefore, the intensity bases analysis according to section 8.2.3 could not be carried out. Instead of a perturbed initial condition a short heating pulse is therefore used as excitation. The pulse length is chosen such that the frequency band given from the experiment is excited uniformly. The resulting process is statistically stationary after all perturbations have left the computational domain through the non-reflective boundaries.

Even though the combustion chamber is not fully axisymmetric due to the installation of windows and cooling devices, which are necessary for the optical access to the chamber, it can be approximated by the axisymmetric approach very well. However, the swirl nozzle of the burner can not be covered by the modal axisymmetric approach. Therefore, the inflow is modelled as an acoustic impedance using the boundary condition of section 8.2.2.4. The five model parameters of the EHR are calculated based on Eq. (8.6) and the assumption that a sufficiently large cavity damping is caused by the installations inside the plenum of the burner. The geometric parameters are estimated to be $V = 7.6$ l, $S_0 = 56.6$ cm^2 and $L = 14.5$ cm. The length correction is usually found to be in the order of one to two times the diameter of the resonator neck. The neck of the burner is much longer than the diameter, therefore the effect of the length correction remains small. The resonant frequency is obtained as $f_0 \approx 100 \ldots 122$ Hz depending on the length correction. f_0 is the inverse of the halved response time T_l. Consequently the non-dimensional model parameters are obtained as $\beta \approx 0.6$ and $m \approx 10^{-4}$. The parameters for the EHR are chosen as follows in the simulation: $T_l = 1/250$ s, $\beta = 0.6$ and $m = 1/2000$. The latter is necessary to obtain a larger maximum time step size for the simulation and does not significantly change the low frequency impedance. The face sheet resistance is assumed to be very small ($R = 0$) and a considerable cavity damping is obtained by setting $e^{-\varepsilon} = 0.75$.

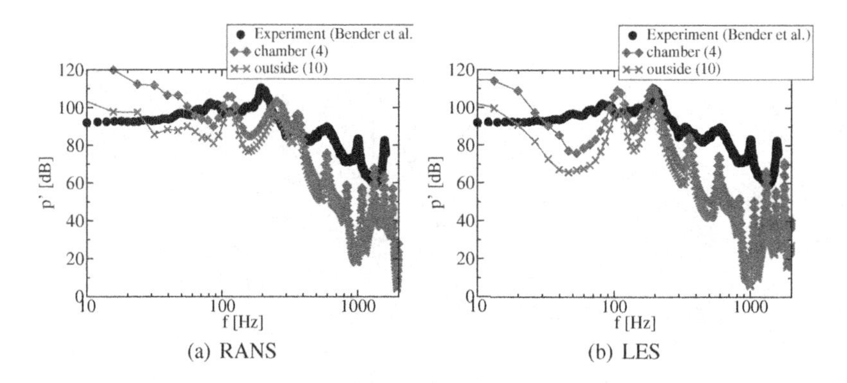

(a) RANS (b) LES

Fig. 8.9 Fourier transform of the swing-off response in the combustion chamber.

For the CAA simulation, only the cylindrical combustion chamber and a cut-out of the exterior are meshed using 0.87×10^5 grid points. The computation of 10^5 time steps takes 33 h on a dual core Athlon 64 X2 CPU with 2.2 GHz using MPI. The resulting swing-off responses for the two different mean flow conditions is given in Fig. 8.9. The geometrical identification of the EHR model parameters is used as boundary condition for the inflow section, whereas the remaining part of the bottom of the model combustion chamber is assumed to be hard walled. The RANS-based mean flow conditions lead to an over prediction of all peak frequencies as shown in Fig. 9(a), whereas, as can be seen from Fig. 9(b), the experimental spectrum is qualitatively and quantitatively well met for the LES-based average flow field. All spectral peak frequencies in the low range ($\lesssim 1$ kHz) are identified. Even though the exact frequency is missed for the lowest peak and the decay is not exactly correct for the higher frequencies because the excitation is not entirely equivalent, the current numerical result can be seen as a good prediction of the experimental observation.

It demonstrates that the combination of average flow field and geometrical boundary conditions including the impedance of the burner fully determines the resulting frequency response. However, it must be noted that the result highly depends on the impedance of the plenum. Furthermore, the RANS-based mean flow leads to the growth of instabilities with the linearised Euler equations. The simulation was stopped before the instability became observable in the result. No such instability is found with the averaged LES flow field as mean flow.

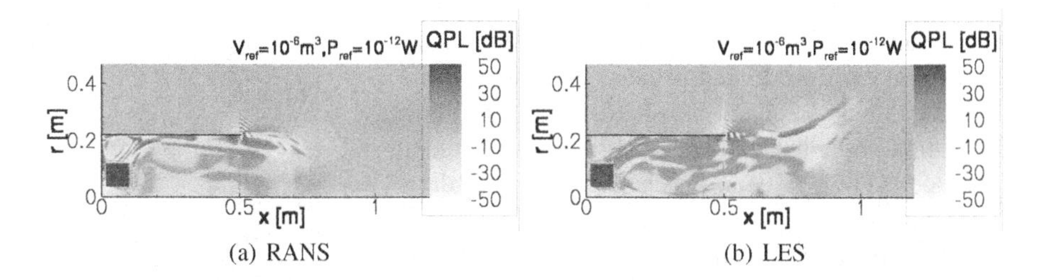

(a) RANS (b) LES

Fig. 8.10 Source location for the swing-off response of the enclosed swirl combustor flow.

Finally, a source location is considered in Fig. 8.10. The QPL clearly identifies the heated region (square around $x = 0.06$ m and $y = 0.08$ m) as the major source of sound. The location for the artificial heating pulse is chosen such that the majority of the reaction zone is covered such that no further information could be obtained from this region. However, an excitation further downstream close to the chamber exit also featured the major source in the heated region. The direct noise clearly dominates the configuration. In addition to this direct source, some indirect sources are found as thin lines of positive and negative source strength following the free shear layers of the swirling flow inside the combustion chamber. The source strength may also be connected to the vorticity generated in the non-isentropic, non-potential flow field. The resulting radiation is probably very small due to the similar level of generation and annihilation of acoustic energy. Furthermore, a vortex shedding from the wall of the chamber becomes visible in the QPL, which cannot be interpreted without a decomposition into acoustic and hydrodynamic velocity fields. The difference between RANS- and LES-based mean flow fields essentially arise due to the different location of the flame front.

8.3.3 10 kW model combustion system with Ø 17 mm exit nozzle

The experimental data for the model combustion chamber is provided by F. Bake, B. Lehmann and I. Röhle, whereas the unsteady RANS simulation, which is used as input for the CAA simulation here was provided by M. Liu, B. Noll and M. Aigner. The results have been published in [15], and represent the oldest results documented here. The case has been included to demonstrate the capability of the approach and to allow an updated interpretation in light of the new insight obtained from the EWG experiment. However, a new simulation has not been carried out, such that neither the EHR model nor the intensity-based analysis could find application here.

8.3.3.1 Numerical case description, base flow and RANS source data

The experimental set-up is sketched in Fig. 8.11. The combustion system consists of a combustion chamber with a swirling, non-premixed flame, Ø 17 mm exit nozzle and exhaust duct. The cylindrical combustion chamber is hard-walled. The low reflection outlet termination used in the experiment is modelled by a fully non-reflective PML boundary condition.

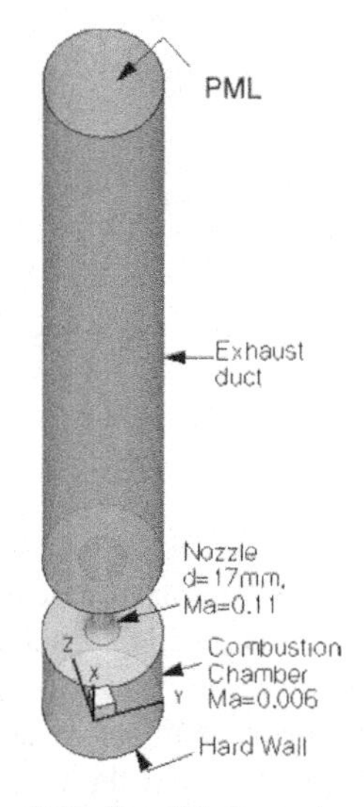

Fig. 8.11 Simplified experimental setup of the 10 kW model combustion system.

The base flow, obtained as a temporal average from a RANS computation, is interpolated by a second order method to the CAA mesh. The maximum flow Mach number reaches from 0.006 to 0.11 in the combustion chamber and the nozzle throat, respectively.

For the CAA-computation the sound propagation into the inlet nozzle and plenum, which are used as fuel-gas and fresh air supply, is neglected. The plenum reflection is assumed to be contained in the compressible U-RANS simulation used as the perturbation source. A zone downstream of the combustion chamber base (100 grid points) is designated for the input of the sound source. The size of this zone is adjusted in a basic investigation of the sound source. Two different concepts for the sound sources were tested, from which the direct prescription of the source as a Dirichlet boundary condition combined with a sponge layer with predefined source as designated input field turned out to be the best method [15]. The results presented here are based on a 2D axisymmetric U-RANS simulation.[3]

The resulting frequency response in shown in Fig. 8.12 in comparison to an experimental spectrum. The result is promising in the sense that several preferred frequencies of the system, even in the high frequency range around 4 kHz, are correctly predicted. However, the major peak is missing. The correct prediction even in the high frequency range shows the general capability of the approach for such problems. The geometrically induced frequency response is related to the correct transport of the low frequency perturbation included in the U-RANS simulation. The missing peak frequency in the low range is probably due to the neglected plenum

[3] The simulation has been provided by M. Liu, B. Noll and M. Aigner from the Institute of Combustion Technology, German Aerospace Center (DLR), Stuttgart.

and termination impedances. As the results in the preceding sections demonstrate, the reflections are an essential part of a correct prediction. Therefore, the application of a measured impedance for the plenum and exhaust duct exit may help to improve the numerical result. However, only the outflow impedance is available at the present time. Furthermore, the application of the combined burner nozzle and plenum impedance would require a different coupling of CFD and CAA to implement the measured impedance.

The observed amplitude of the dominant frequency and the first two harmonics is higher than the amplitude of the corresponding peak in the experiment. This is probably due to the axisymmetric CFD computation, which does not predict 3D structures correctly. Due to the assumed coherence of the perturbation field in the azimuthal direction, sound levels are then over predicted. Furthermore, the vortex interaction leads to growing amplitudes in 2D, which may also cause the over prediction of the noise level.

8.4 Conclusion

Several applications of a time explicit high-order finite difference method in the simulation of direct and indirect combustion noise have been demonstrated. The first is the validation of the method with a model experiment for indirect combustion noise using an electrical heat source to produce the perturbation. Even though the exact pressure response in the exhaust duct of the model experiment has been shown to depend essentially on the unknown reflections from the plenum and probably also from the exhaust duct outflow, the numerical results demonstrate the capability of the method. A source location and validation based on the acoustic intensity has been carried out and the indirect noise source has been identified in the nozzle and diffuser. It is found that there is an interaction of the source contributions. Generation and annihilation of acoustic energy occur at similar levels in the nozzle and diffuser. Therefore, the resulting radiation efficiency of the indirect source is found to be smaller than the theoretical predictions for nozzle and diffuser alone. The second application considers the prediction of the dominant frequencies in an open combustion chamber featuring an enclosed premixed flame. The results are based on the average flow field from a precursor CFD simulation. An estimation of the model parameters for the extended Helmholtz resonator model to describe the plenum impedance is applied, which is based on an identification of the model parameters with a low frequency resonator and its dimensions. It is found that the predicted dominant frequencies fit the experiment very well, if an averaged LES result is considered as mean flow. A RANS-based mean flow cannot reproduce the correct frequencies due to the different location of the flame front. Furthermore, the RANS-based perturbation approach is found to be unstable. The noise generation in the second example is not directly related to indirect combustion noise as the source location shows. Due to the axisymmetric approach and the coarse mesh for the high-order CAA method, the computational time for the swing-off response re-

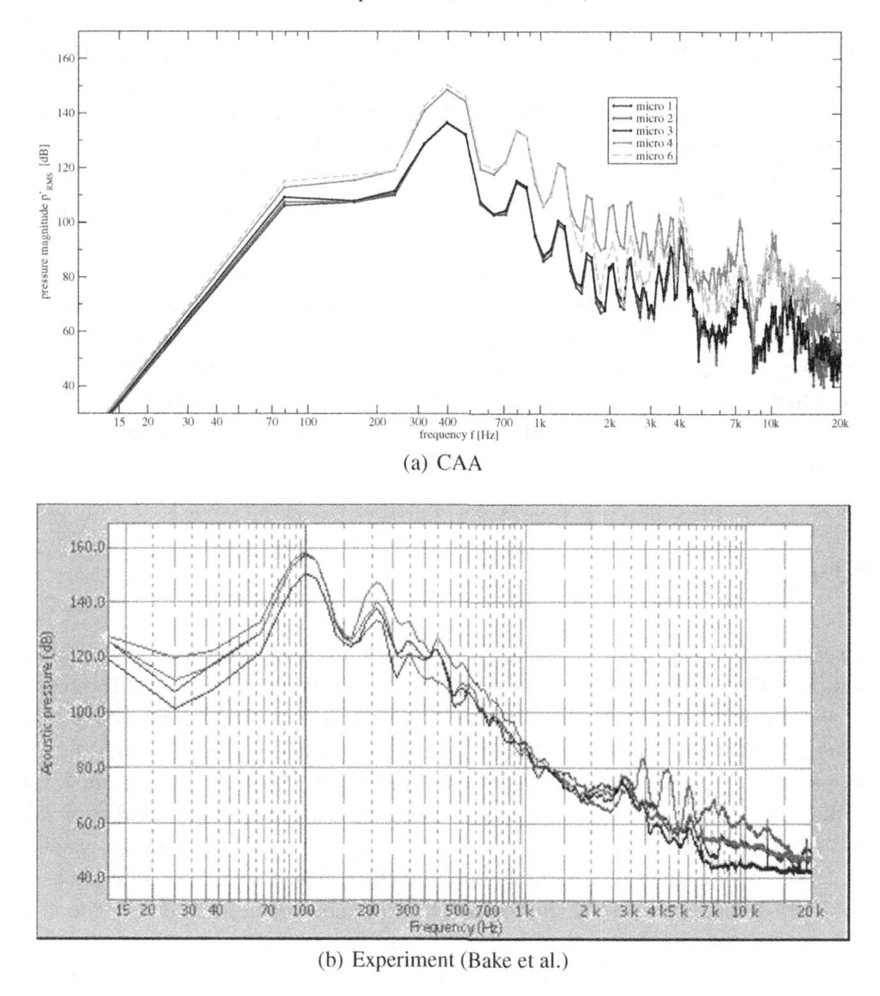

(a) CAA

(b) Experiment (Bake et al.)

Fig. 8.12 Comparison of the experiment and the linear CAA-computation.

mains small. Therefore, the RANS-based result may be seen as a fast prediction with some trade-off effects due to the axisymmetric modelling. Finally, an application of the method as a propagation zone approach in a model combustion system has been considered. The results indicate that a dominant frequency of the combustion system may be missing due to a failure in describing the impedance of the connected systems. On the other hand the qualitatively correct prediction of the dominant frequencies up to 4 kHz shows that the geometrically induced resonances of the system are correctly described. Altogether the results show that the correct modelling of a combustion system requires knowledge about the impedance of connected systems. Furthermore, the encouraging results indicate that dominant frequencies in combustion systems are related above all to the boundary conditions, rather than being related to the highly nonlinear reaction chemistry of the flame.

Acknowledgements The authors gratefully acknowledge the contribution of N. Schönwald and D. Morgenweck, who were involved to the development of the numerical method used here. The comprehensive experimental and numerical data, which helped to validate the method was provided by F. Bake, N. Kings, I. Röhle, B. Mühlbauer, B. Noll, C. Bender and F. Zhang. The special thanks of the authors goes to Klaus Ehrenfried, for the discussions and the contribution of the idea for the intensity-based source location. Finally the authors wish to acknowledge the financial support of the German Research Foundation (DFG).

References

[1] Bake F, Michel U, Röhle I (2007) Investigation of entropy noise in aero-engine combustors. Journal of Engineering for Gas Turbines and Power 129(2):370–376

[2] Bogey C, Bailly C (2002) Three-dimensional non-reflective boundary conditions for acoustic simulations: far field formulation and validation test cases. Acta Acustica united with Acoustica 88:462–471

[3] Boudier G, Lamarque N, Sensieau C, Staffelbach G, Gicquel L, Poisont T, Moureau V (2007) Investigating the thermo-acoustic stability of a full gas turbine combustion chamber using large eddy simulation. In: 11th CEAS-ASC Workshop, Experimental and Numerical Analysis and Prediction of Combustion Noise, CEAS-ASC, Lisbon, Portugal, paper 6

[4] Chu BT, Kovásznay LSG (1958) Non-linear interactions in a viscous heat-conducting compressible gas. Journal of Fluid Mechanics 3(2):494–514

[5] Cumpsty NA, Marble FE (1977) Core noise from gas turbine exhausts. Journal of Sound and Vibration 54(2):297–309

[6] Eversman W (2004) Acoustic power in lined ducts. AIAA Paper 2004-2904

[7] Ffowcs Williams J, Howe M (1975) The generation of sound by density inhomogenities in low Mach number nozzle flows. Journal of Fluid Mechanics 70(3):605–622

[8] Hu FQ, Hussaini MY, Manthey JL (1996) Low-dissipation and Low-dispersion Runge-Kutta Schemes for Computational Acoustics. Journal of Computational Physics 124(1):177–191

[9] Leyko M, Nicoud F, Poinsot T (2007) Comparison of indirect and direct combustion noise in aircraft engines. In: 11th CEAS-ASC Workshop, 27-28. September 2007, Lisbon, Portugal

[10] Li X, Schemel C, Michel U, Thiele F (2002) On the Azimuthal Mode Propagation in Axisymmetric Duct Flows. AIAA Paper 2002–2521

[11] Marble F, Candel S (1977) Acoustic disturbances from gas non-uniformities convected through a nozzle. Journal of Sound Vibration 55(2):225–243

[12] Morfey C (1971) Acoustic energy in non–uniform flows. Journal of Sound Vibration 14(2):159–170

[13] Richter C, Thiele F (2007) Computation of indirect combustion noise by a CAA method. In: Proceedings of the 14th International Congress on Sound and Vibration, ICSV14, Cairns, Australia
[14] Richter C, Thiele F (2007) The stability of time explicit impedance models. AIAA Paper 2007–3538
[15] Richter C, Panek L, Thiele F (2005) On the application of CAA-methods for the simulation of indirect combustion noise. AIAA Paper 2005–2919
[16] Richter C, Thiele F, Li XD, Zhuang M (2007) Comparison of time-domain impedance boundary conditions for lined duct flows. AIAA Journal 45(6):1333–1345
[17] Richter C, Morgenweck D, Thiele F (2009) CAA simulation and intensity based evaluation of a model experiment for indirect combustion noise, to be published in a special issue of Acta Acoustica united with Acoustica on combustion noise
[18] Rienstra SW (2006) Impedance models in time domain, including the extended Helmholtz resonator model. AIAA Paper 2006–2686
[19] Schemel C, Thiele F, Bake F, Lehmann B, Michel U (2004) Sound generation in the outlet section of gas turbine combustion chambers. AIAA Paper 2004–2929
[20] Schönwald N, Panek L, Richter C, Thiele F (2007) Investigation of sound radiation from a scarfed intake by CAA-FWH simulations using overset grids. AIAA Paper 2007–3524
[21] Stanescu D, Habashi W (1998) 2N-storage Low-dissipation and Low-dispersion Runge-Kutta Schemes for Computational Aeroacoustics. Journal of Computational Physics 143(2):674–681
[22] Tam CKW, Webb C (1993) Dispersion-Relation-Preserving Finite Difference Schemes for Computational Aeroacoustics. Journal of Computational Physics 107(2):262–281
[23] Weckmüller C, Guerin S, Richter C (2007) Numerical investigation of geometry and mean flow effects on acoustic radiation from a duct inlet. AIAA Paper 2007–3535
[24] Zhang F, Habisreuter P, Bockhorn H, Büchner H (2008) Les of reactive flow in a strongly swirling combustor system. In: Int. Conf. on Jets, Wakes and Saperated Flows, ICJWSF-2008

Chapter 9
Direct Numerical Simulations of turbulent flames to analyze flame/acoustic interactions

G. Fru, H. Shalaby, A. Laverdant, C. Zistl, G. Janiga and D. Thévenin

Abstract Direct Numerical Simulations (DNS) are becoming increasingly important as a source of quantitative information to understand turbulent reacting flows. For the present project DNS have been mainly used to investigate in a well-defined manner the interaction between turbulent flames and isolated acoustic waves. This is a problem of fundamental interest with practical applications, for example for a better understanding of combustion instabilities. After developing a specific version of the well-known Rayleigh's criterion, allowing to investigate local amplification or damping of an acoustic pulse interacting with a reaction front, extensive investigations have been carried out. The present publication summarizes the main findings of all these studies and describes in detail the underlying numerical and physical models, in particular those used to describe chemical reactions. Post-processing of DNS data in the light of turbulent combustion modeling is also discussed. The results illustrate the complexity of the coupling between reaction fronts and acoustics, since amplification and damping appear mostly side by side, as alternating layers. The influence of individual reactions and species on the damping process can also be quantified in this manner. This publications concludes with perspectives towards higher turbulence levels and effects of differential diffusion.

9.1 Introduction

Turbulent reacting flows are essential for a wealth of practical applications (automobile engines, aircraft turbines, domestic boilers, nano-particle production, electricity

---

G. Fru, H. Shalaby, C. Zistl, G. Janiga, D. Thévenin
Lab. of Fluid Dynamics & Technical Flows, Univ. of Magdeburg "Otto von Guericke", Germany. e-mail: thevenin@ovgu.de

A. Laverdant
Office National d'Etudes et de Recherches Aérospatiales (ONERA), 29, avenue de la Division Leclerc, BP 72, 92322 Châtillon-sous-Bagneux, France. e-mail: laverdan@onera.fr

A. Schwarz, J. Janicka (eds.), *Combustion Noise*, DOI 10.1007/978-3-642-02038-4_9,

generation by a power plant burning coal or oil, etc.). Considering the present environmental issues it is essential to understand and improve combustion processes, in order to reduce fuel consumption and pollutant emissions (including noise, which is indeed one of the most disturbing pollutants in urban areas) as much as possible.

This is indeed a formidable task since turbulent combustion involves many individual, physical processes of high complexity, e.g. chemical kinetics, differential molecular diffusion, radiative heat transfer, phase transition, flame/acoustics interactions, and of course the turbulent flow itself, which is indeed also an unsolved issue. The nonlinear coupling between all these phenomena completely determines the behavior of practical systems. Due to this complex coupling truly predictive numerical models are not available at present for realistic installations, although they are of course necessary to improve existing devices and develop new configurations.

Since the control of combustion instabilities is a fundamental and practical problem of high interest [5], one particularly important phenomenon is the interaction between a turbulent flame and a well-defined, isolated acoustic wave. To understand the birth of combustion instabilities, the interaction of an acoustic wave with a turbulent flame, leading to amplification or damping, is an essential basic element of the coupling process, since acoustics are known to trigger such instabilities in many configurations. This elementary process has therefore been considered in detail during the present project using numerical simulations.

There are three main strategies to simulate turbulent reacting flows in general: the Reynolds-averaged Navier-Stokes formulation (RANS), where all turbulent scales are modeled; Large-Eddy Simulations (LES), in which small scales are modeled while large scales are solved exactly; and as a last alternative, Direct Numerical Simulations (DNS), where all physical scales are resolved on the grid.

RANS require simplified turbulent combustion models and thus cannot be used for an a priori analysis of the coupling process. LES need a subgrid model to describe all physical processes taking place below grid resolution. Both, though highly interesting in particular for solving large scale problems with a complex geometry, are thus associated with many unsolved issues and challenges (see for example [30, 47, 46, 33, 7]). The potential of LES and of Unsteady RANS (URANS) for investigating acoustic emissions in flames is demonstrated in other chapters of this book; it will not be considered further here.

In order to increase the predictive accuracy of numerical simulations, further studies are needed to refine the models required for RANS and LES, or to develop alternatives. These studies can rely either on detailed, quantitative experimental measurements or on DNS computations (sometimes also called "numerical experiments"). The latter will be used in the present work, employing exclusively Direct Numerical Simulations.

The DNS method consists of solving directly and as far as possible exactly the complete, unsteady Navier-Stokes equations. All physical spatial and time scales are resolved accurately by using a suitable discretization in space and time, so that a turbulence modeling is not necessary any more. DNS is the best method with the highest level of accuracy when applicable and its results contain all physical information about the turbulent flow as well as all other variables of interest (concentrations, temperature, density...). Hence, when accurate models are also employed for all other physical processes, DNS qualify indeed as a "numerical experiment", but lead of course to very high requirements in computing time and memory.

Despite the fact that progress in computer technology now sometimes allow direct simulations of turbulent flames relying on complete reaction schemes and realistic, multicomponent transport models, the associated computational cost in terms of computing time and required memory remain tremendous (see for example [9, 10, 11]). Though limited to some simple configurations, corresponding single-processor computing times are expressed at best in months, usually in years! For more realistic configurations (complex three-dimensional geometry), or when systematic studies are required (long physical times, several simulations), such computations remain completely impossible at present. There is therefore an urgent need for suitable model reduction. In particular, it is very important to identify models that will simplify the numerical description of the reacting process, while preserving the coupling between chemistry, heat transfer and fluid dynamics in a quantitative manner. For instance, tabulated chemistry reduction techniques, which are the most recent development in the field of chemical kinetics reduction constitute a vital alternative to the direct use of complex reaction schemes in DNS flow solvers. Tabulated chemistry typically leads to one or two orders of magnitude reduction in computing time compared to a complete reaction scheme. For this reason, the application and validity of tabulated chemistry techniques will be discussed in this paper.

Concerning pressure-velocity coupling, one suitable model reduction for turbulent flames is the low Mach number formalism, leading again to a similar reduction in terms of computing time [3]. Even if this approximation cannot be used directly to investigate acoustics, since the acoustic coupling process has been removed from the corresponding equations, it might still be employed to investigate in detail the structure of turbulent flames. Through a coupling with Computational Aero-Acoustics (CAA) methods, as described in other chapters of this book, the radiated acoustic field can be reconstructed appropriately.

Even with such model reductions, the remaining computing time is still impressive. Therefore, it is essential to employ parallel supercomputers to keep acceptable user waiting times. The corresponding issues are mostly algorithmic and will therefore not be discussed in the present publication.

DNS is the optimal method to foster model analysis and development. For this purpose the huge sets of raw data generated by DNS must be post-processed in an intelligent and efficient manner: generating a bunch of colour videos is by far not sufficient! For instance, the DNS of a 3D hydrogen flame relying on a complete reaction scheme produces typically around 500 MB of raw data *per time-step*. It is obvious that these data have to be post-processed efficiently in order to extract all the available information useful for process understanding and model development. Fast, accurate, flexible and semi-automatic post-processing tools are needed in order to maximize the scientific output from such investigations and thus lead to improved models e.g. for RANS [42] and LES. Such a tool has been developed in our research group and will be described as well.

As already explained, the understanding of turbulent combustion relies on many, strongly coupled and highly complex physical phenomena. Some of the major challenges are:

- *Numerical complexity*: the spatial and temporal dynamics are inherently coupled, with a wide range of relevant scales for the turbulence and chemical reactions, which must be adequately resolved or modeled.
- *Chemical complexity*: complete chemical schemes often involve a large number of species and elementary reactions (for instance hundreds of species, thousands of reactions for gasoline oxidation).
- *Transport complexity*: differential diffusion at molecular scale as well as turbulent transport of heat and species control mixing and heat transfer.
- *Multi-physics complexity*: practical flow configurations often involve multiphase flows (such as in spray combustion, for soot emission, ...), thermal radiation, complex thermodynamics, plasmas, acoustics... Turbulence/chemistry interactions are always essential.
- *Post-processing complexity*: DNS lead nowadays to huge quantities of raw data, from which useful information must be extracted and analyzed as efficiently as possible.

In this work, we will describe some of our contributions in tackling several of the above challenges. After describing the employed numerical tools, methods and models in chapter 9.2, in particular concerning chemistry modeling, the potential of DNS to investigate flame/acoustic interactions is detailed out in chapter 9.3. A recently developed post-processing toolbox is briefly presented in chapter 9.4, before drawing some perspectives concerning higher turbulence levels and effects of differential diffusion.

9.2 Theoretical Background, Numerical methods and procedure

DNS of turbulent combustion problems have been carried out in our research group for a variety of configurations and model problems during the last 20 years. In order to maximize accuracy and numerical efficiency, two different code families have been finally developped. Though all resulting DNS codes are very similar in their building bricks and global structure, they also show noticeable differences depending on the specific problem that must be investigated with the code. The *Parcomb* family considers only complete reaction schemes, while the π^3 family relies on tabulated chemistry to describe reaction processes. In addition, both compressible and low-Mach number formulations are available. The final spezialisation of the different codes is shown in Fig.9.1. Since all codes have been used at some point during this project, the common features and main differences are discussed in what follows.

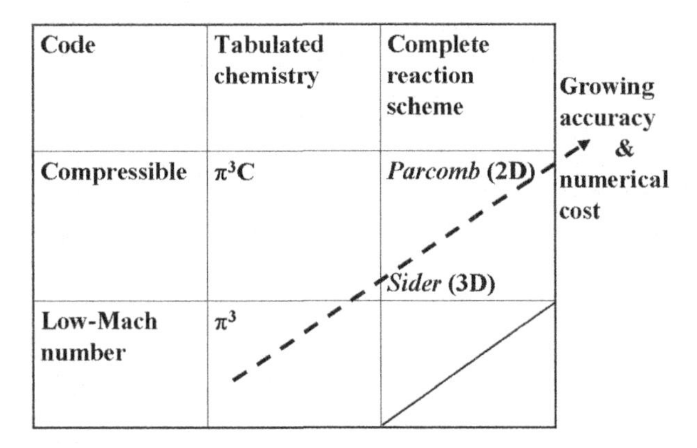

Fig. 9.1 DNS codes employed in this project and main characteristics.

9.2.1 DNS code family Parcomb

Parcomb has been originally developed by Thévenin and coworkers [43, 44]. It is a finite-difference DNS code solving the compressible Navier-Stokes equations for multicomponent reacting flows with accurate physical and chemical models. Since *Parcomb* is the oldest code available in our group (the first version is almost 20 years old), it is limited to two-dimensional computations, corresponding to the available computing power on supercomputers at that time. Derivatives are computed using centered explicit schemes of order six except at boundaries where the order is progressively reduced to four. Temporal integration is realized with a Runge-Kutta algorithm of order four.

Boundary conditions are treated with the help of the Navier-Stokes Characteristic Boundary Condition (NSCBC) technique [32], extended to take into account multicomponent thermodynamic properties [2]. This allows in particular to obtain non-reflecting conditions at the boundaries.

Transport coefficients and chemical kinetics are treated following methods similar to those used in the standard packages CHEMKIN and TRANSPORT [16, 17]. In order to describe chemical reactions a complete reaction scheme involving 9 chemical species (H_2, O_2, O, H, OH, HO_2, H_2O_2, H_2O, N_2) and 37 individual reactions is taken into account for hydrogen combustion. The reaction scheme for syngas contains 13 species (CO, HCO, CH_2O, CO_2, H_2O, O_2, O, H, OH, HO_2, H_2O_2, H_2, N_2) and accounts for 67 individual reactions.

Parcomb has been parallelized and widely used over the last two decades to investigate various properties of turbulent flames (see for example [44, 26, 12]). The numerical domain employed when using *Parcomb* to investigate turbulent flame properties or flame/acoustic interaction is a two-dimensional square box of dimension

L_{box}^2, typically between 1 and 1.5 cm in each direction (Fig.9.2). A fixed mesh is used in each direction, with a typical spatial resolution around 30 μm, necessary to resolve accurately all intermediate chemical radicals as well as the dissipation scale of turbulence. The left-hand boundary condition is usually a subsonic inlet with imposed value of velocity, while the right-hand boundary condition is a non-reflecting subsonic outlet, both implemented using the NSCBC technique. *Parcomb* has been used within this research project to investigate both premixed and non-premixed flames burning hydrogen or syngas. Either isolated acoustic pulses (Gaussian wave) or, recently, realistic acoustic waves involving several periods [40] have been considered.

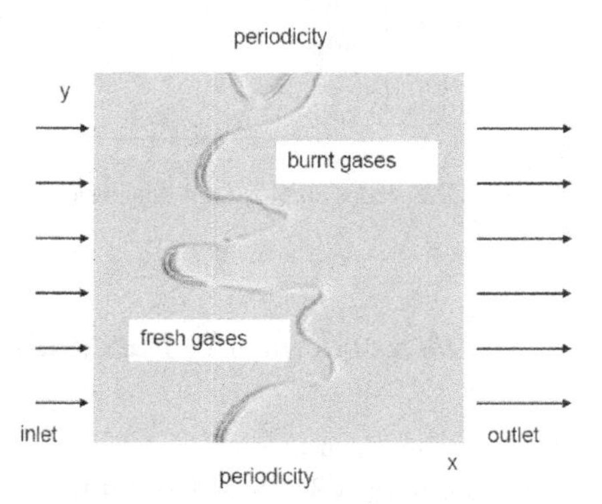

Fig. 9.2 Typical numerical configuration employed for two-dimensional DNS.

Sider (also called *Parcomb*3D) is the three-dimensional version derived from the two-dimensional code *Parcomb*. Simultaneously, the code has been rewritten in FORTRAN 95. This version has been mainly developped and employed at ONERA.

An improved numerical scheme, the skew-symmetric formulation [15] has been implemented in *Sider* for the convective terms, in order to reduce even further numerical dissipation and increase stability. According to this scheme, the derivative of a general convective term can be written as:

$$\frac{\partial(\rho a u_j)}{\partial x_j} = \frac{1}{2}\frac{\partial(\rho a u_j)}{\partial x_j} + \frac{1}{2}\frac{\partial(\rho u_j)}{\partial x_j} + \frac{\rho u_j}{2}\frac{\partial a}{\partial x_j}. \tag{9.1}$$

As a consequence, the transported quantities are globally conserved for a given viscosity approaching zero. Temporal and spatial schemes are unaltered. Further details concerning code structure, optimization on parallel computers, or initialization of the synthetic three-dimensional turbulent velocity field are very similar to that found in *Parcomb* and are described for instance in [12].

In the latest months of the present project, *Sider* has been used to examine flame structure and flame/turbulence coupling for non-premixed hydrogen-air flames in three dimensions. The computational domain (Fig. 9.3) is typically a cube of 1 cm side with a uniform grid spacing of 50 μm (8 million grid points). Non-reflecting boundaries with pressure relaxation are employed in the x−direction while periodicity is assumed along the y- and z-directions.

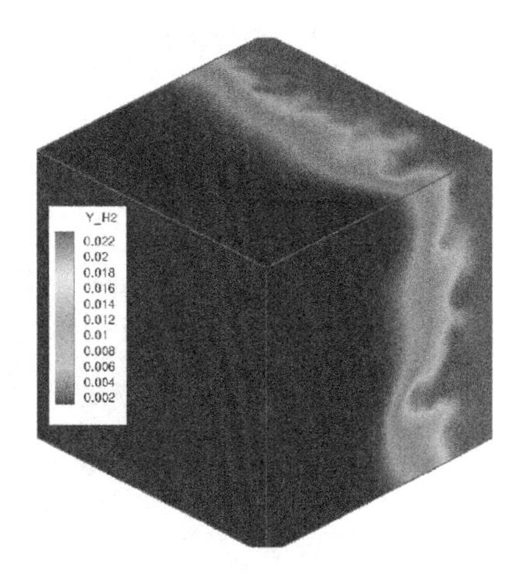

Fig. 9.3 General view of the 3D configuration, showing instantaneous iso-surfaces of the mass fraction of H_2 at $t = 0.94\tau$. The mixing zone is clearly visible, extending around the center of the domain. Non-reflecting boundaries are considered in the x-direction, periodicity is assumed along y and z.

9.2.2 DNS code family π^3

The three-dimensional compressible DNS code family π^3 is also an explicit finite-difference solver, written using FORTRAN 95. Compared to the *Parcomb* family, a considerable reduction of the needed computing times (typically one or two orders of magnitude) is obtained by using an accurate reduction technique for the chemistry, called Flame-Prolongation of Intrinsic Low-Dimensional Manifolds (FPI), described later. The boundaries are again computed using the NSCBC formulation, as in the *Parcomb* family.

As in *Parcomb* the parallelization of π^3 relies on domain-decomposition, but this time only along the y-direction in space, projected onto a corresponding one-dimensional processor topology. Each node thus controls a subdomain of the global computational domain. The integration procedure then consists of two successive steps, repeated until the end of the computation. First, each processor carries out an integration step on its own subdomain independently from the others. Then, all processors communicate the new boundary values to their direct neighbors. The next integration step can afterward be started. The grid is equipartioned among the processors along the y-direction. All communications rely on the Message-Passing Interface (MPI).

The DNS code π^3C (compressible version of π^3) is used in the present work to investigate the interaction of a Gaussian acoustic wave with a turbulent premixed flame in a three-dimensional flow. The flame is initially perfectly spherical, laminar and centered in the middle of the numerical domain (Fig.9.4). Each side of the computational domain is typically 1 cm long. The grid spacing is constant and uniform, again around 30 μm. This leads now to a computational grid with roughly 30 million grid points. A fully premixed methane/air flame at an equivalence ratio $\phi = 1.59$ (rich flame), atmospheric pressure and fresh gas temperature of 300 K is considered for the results shown later.

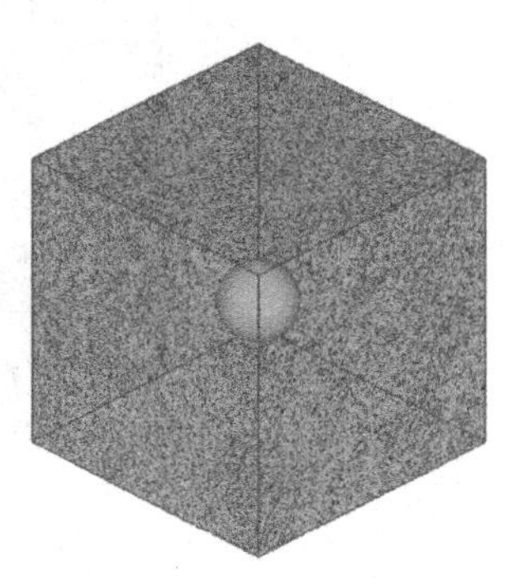

Fig. 9.4 Initially spherical, laminar flame kernel with the superposed turbulent field as initial condition. The turbulence is represented here by isolines of vorticity magnitude along background boundary planes.

In addition, the code π^3 (low-Mach number version) has been employed to analyze turbulent flame structures without any acoustic perturbation. Knowing the resulting flow variables, it is in principle possible to analyze the acoustic emission of such flames by a post-processing relying on Computational Aero-Acoustics (CAA). This issue is considered in other chapters of this book and will therefore not be discussed further here.

9.2.3 Steps involved by the computational procedure

The analysis of turbulent flames and of flame/acoustic interactions using DNS follows always the same procedure. First, the laminar flame configuration is determined, either using a first DNS computation when considering detailed chemistry, or using a simple estimation when relying on tabulated chemistry. Then, the obtained laminar solution is transposed to a two-dimensional (*Parcomb*), respectively to a three-dimensional (*Sider*, π^3 family) configuration, with fuel (resp. fresh gases) on part of the numerical domain, the other part consisting of oxidizer (resp. burnt gases). A field of synthetic homogeneous isotropic turbulence, based on the classical turbulent energy spectrum of Kármán with a Pao correction for near-dissipation scales [44, 12, 14], is then superposed on top of this laminar configuration as exemplified in Figure 9.4. The turbulence properties are imposed by the user in order to obtain the highest possible Reynolds number in the domain, considering its limited size and the really achieved spatial resolution. The turbulence parameters are checked by a separate post-processing of the initial solution. For the results shown afterward in 3D, this leads for example to a Reynolds number based on the integral scale equal to $\mathrm{Re}_t = 256$, a low but typical value for 3D DNS involving realistic chemistry. The integral scale of turbulence is equal to $l_t = 2.06$ mm, with a turbulent fluctuation velocity $u' = 1.56$ m/s and a value of the unburnt mixture viscosity $\nu = 1.76 \times 10^{-5}$ m^2/s. The characteristic time-scale of the large turbulent structures is then $\tau = 1.3$ ms.

When iterating now the DNS computation in time, a turbulent flame develops into the numerical domain. This computation is carried out until reaching equilibrium between chemistry and turbulence (in a statistical sense), which is known to occur at a time around $t \simeq \tau$ from the literature. At that time:

- the structure of the resulting turbulent flame can be analyzed and used for model testing and development;
- or the investigation of the interaction between this turbulent flame and an acoustic perturbation can be started. This time is thus retained as the new origin of time, $t' = 0$. In order to obtain quantitative results concerning the interaction of an acoustic wave with a flame, two DNS computations are now carried out using the same code and the same models, starting with exactly the same initial conditions: the first DNS is simply a restart of the previous DNS computation without adding the acoustic wave (Fig.9.5) ; in the second one, the acoustic wave is added to the obtained configuration at restart (Fig.9.6). This takes place either within the domain for the isolated pulse (Gaussian acoustic wave), by modifying appropriately the fields of velocity, density and pressure; or the corresponding perturbation enters the domain through the inflow boundary, when considering a realistic acoustic wave over several periods. By comparing the results of these two DNS at the same position in time and space, the importance of the acoustic interaction process can be directly quantified using the local Rayleigh's criterion introduced later.

9.2.4 Modeling of chemical reactions

Complete mechanisms constitute the most appropriate model for reacting flows. Here, no simplification is introduced and an equation is solved explicitly for each and every chemical species present in the chemical scheme involved, simultaneously with the Navier-Stokes equations. Though being the most accurate technique, its use is often too demanding, since large reaction mechanisms contain hundreds of species and reactions and rapidly lead to unacceptable computing cost in three dimensions [13]. Generally, calculations with complex chemistry are obtained and used mainly for the validation of simpler or reduced chemical kinetics models, and

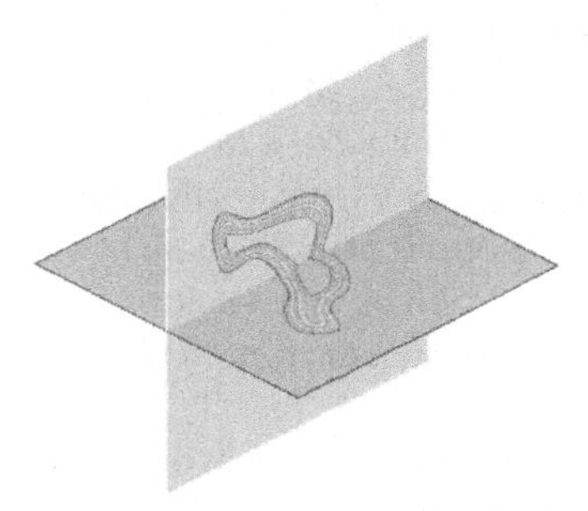

Fig. 9.5 DNS computation without adding an acoustic wave. The flame position is shown using isosurfaces of CO_2 mass fraction.

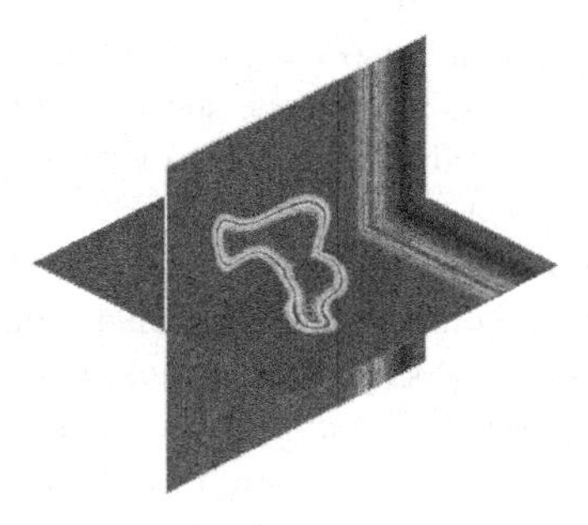

Fig. 9.6 DNS computation when adding a planar, Gaussian acoustic wave coming from the right. The flame position is shown using isosurfaces of CO_2 mass fraction. Compare with Fig.9.5.

also for the understanding of the detailed flame structure. In what follows, complete reaction mechanisms are employed for the simulations relying on the *Parcomb* code family.

Tabulated chemistry reduction techniques offer an attractive and promising alternative to complete reaction schemes and do serve as a smart means of introducing effects of complex chemistry and/or multicomponent transport phenomena in reacting flow computations. A number of prominent complex chemistry tabulation techniques have emerged in the course of the last two decades with a variety of published methods. For instance the In-Situ Adaptive Tabulation (ISAT) technique [34], where a chemical database is constructed from the direct solving of the time evolution of the species compositions and tabulated at run time, restricted to the actually accessed part of the composition space is a very accurate chemistry tabulation technique. Another example is the Intrinsic Low Dimensional Manifold (ILDM) technique [25], which is based on a local study of the characteristic times of the dynamical system associated with the reactive mixture. More recently, the Flamelet Generated Manifold (FGM) [29] and/or the Flame Prolongation of ILDM (FPI) [8] have been developed and lead finally to similar chemistry tabulation techniques. The latter is our method of choice for DNS computations involving the π^3 code family.

The original FPI version is described extensively in other publications (see for a start [8]). The FPI method leads to a tabulation of the chemical reaction terms into a look-up table before the computations [6]. For most cases, flame speed, extinction limits and radical profiles are predicted almost exactly using FPI when compared to the complete reaction scheme, as demonstrated in previous publications. The FPI database used for the results shown afterwards has been already described and employed in [45] and is obtained using a complete reaction scheme taking into account 29 species and 141 reactions to describe in a highly realistic manner methane/air combustion [23]. It has been computed using a unity Lewis number hypothesis for the sake of simplicity. For the conditions retained in the present work, the laminar flame speed obtained using multi-component diffusion velocities is almost identical to that computed with a unity Lewis number hypothesis (relative error below 5%), showing that in this case such a simplification should be acceptable. For the DNS the look-up table is simply parameterized by two coordinates (typically, mass fractions of chosen species, e.g., nitrogen and carbon dioxide, or linear combinations of such mass fractions). Standard transport equations are solved for these coordinates. All other mass fractions and thermodynamic quantities can be obtained by reading the look-up table.

The standard FPI technique shows some drawbacks and has therefore been reconsidered during the course of this project, leading to two major modifications. First, the classical FPI table is built on an unstructured triangular/tetrahedral grid and information localization and retrieval from within the database rely on a complex tree search and multi-linear interpolation. This procedure, though reliable and efficient in terms of storage, is computationally not very efficient, so that a considerable computing time is spent for information retrieval out of the database. Therefore, the presently employed FPI look-up table can rely as an alternative on a regular, cartesian description of the coordinate space. After computing the original flame structures employed to define the look-up table (for the present project, laminar premixed flames), a triangular/tetrahedral grid is first generated and then projected onto a regular cartesian grid (Fig,9.7). This procedure leads to increased storage requirements, but allows a considerably faster information retrieval. For all tested configurations, the computing time requested for reading the look-up table during real DNS computations is reduced by a factor around 4 when using the cartesian grid, which is very attractive in practice.

As a second major improvement, a FPI version allowing computations with multicomponent transport models has been recently derived. The original FPI technique, though not fully constrained to unity Lewis numbers (see for example the develop-

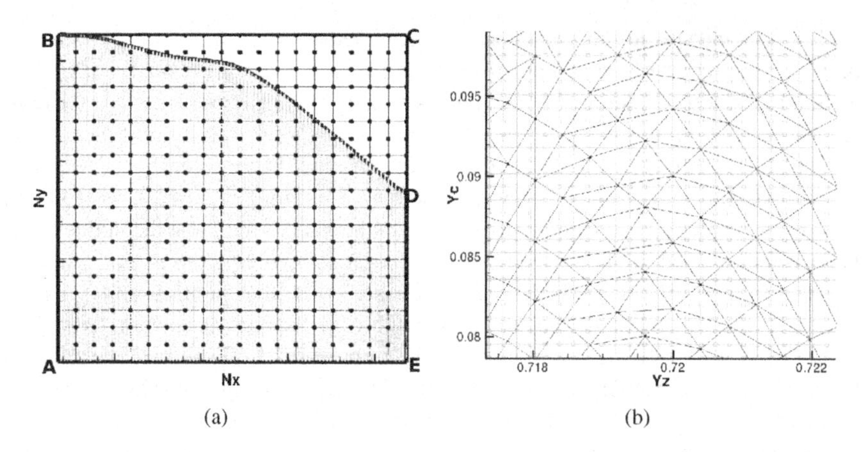

Fig. 9.7 7(a) Sketch of the triangular-to-rectangular grid mapping. The shaded region, *ABDEA*, defines the actual flame domain. 7(b) Zoom on the triangulation with the regular rectangular grid in the background

ments in [8]) has been in practice almost always used for cases relying on such an approximation. Since it is not sure that differential diffusion effects will always be negligible, in particular when considering hydrogen flames, introducing realistic molecular diffusion models into the FPI method is an important issue.

In DNS the standard FPI look-up table is simply parameterized by two coordinates (typically, mass fractions of chosen species, e.g., nitrogen and carbon dioxide, or linear combinations of such mass fractions) in order to describe premixed or partially-premixed flames. The use of a combination of one or more species mass fractions to reconstruct an appropriate progress variable Y_c is a measure taken to ensure that the variation of Y_c across the flame front is monotonic. For hydrocarbon fuels for example, Fiorina *et al.* [6] proposed the use of a linear combination of Y_{CO_2} and Y_{CO} for Y_c, an approach that is also used for the results presented later.

When using the standard look-up table, standard transport equations are directly solved for the coordinates of the table and all other mass fractions and thermodynamic quantities can then be obtained by data retrieval from the table.

On the other hand, if a detailed transport model is used to generate the look-up table, appropriate modifications must be made in the solver to ensure that, first of all, global mass is conserved and secondly, that the linear combination strategy with which the progress variable is reconstructed is appropriate for all the tabulated flame variables associated with it. Corresponding improvements have been successfully implemented within our research group. Of course, the generation of the look-up table then relies on laminar flame computations involving as well realistic diffusion models (Fig.9.8).

The improved FPI procedure has been validated in two steps. First, an a priori test has proved that the new mapping does not lead to any loss of accuracy, even for minor radicals. Figure 9.9 shows a comparison between the results obtained by solving directly the exact flame structure using the complete reaction scheme

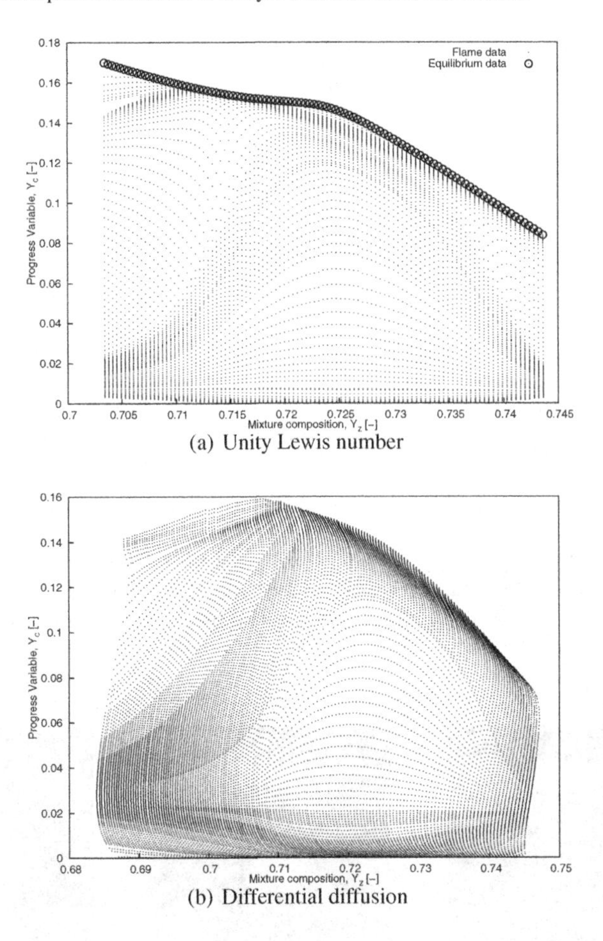

(a) Unity Lewis number

(b) Differential diffusion

Fig. 9.8 Projection of a series of 1-D freely propagating laminar premixed methane/air flames on the (Y_z, Y_c) subspace for $0.50 \leq \phi \leq 1.97$; $0.688 \leq Y_z(\phi) \leq 0.745$ and a stepping of $\Delta\phi = 0.01$. Each vertical set of points (dots) constitute a single flame, covering the entire Y_c space from the fresh gas mixture ($Y_{c,min}$) to chemical equilibrium ($Y_{c,max}$) within the given Y_z flamability limit. The $\odot$ symbols are the corresponding equilibrium values. When using the assumption Le=1, each flame is aligned vertically (top), since differential diffusion is not allowed. On the other hand, each flame has a complex structure when taking into account a realistic diffusion model (bottom).

(symbols) compared to the FPI result obtained with the interpolated (rectangular) look-up table for the mass fractions of CH_2O, HCO and HO_2 radicals in a freely propagating laminar premixed methane/air flame. The agreement is perfect.

Then, a direct comparison has been carried out for an expanding premixed CO/H_2/air flame configuration under the influence of a field of synthetic homogeneous isotropic turbulence. Results obtained with a complete reaction scheme and a multicomponent diffusion model are directly compared to the results obtained with the advanced FPI formulation. Three different time instants are shown for such an

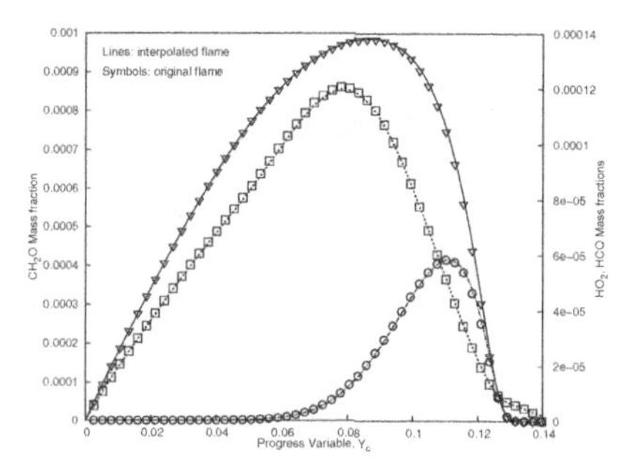

Fig. 9.9 Comparison of the results obtained with complex chemistry with the FPI results obtained using a rectangular parameterization of the look-up table. Shown are the mass fractions of HO_2 (□, dashed line, right axis), HCO (∘, dash dotted line, right axis) and CH_2O (▽, solid line, left axis) radicals, for a freely propagating laminar premixed methane/air flame projected on the progress variable Y_c.

expanding turbulent flame in two dimensions in Fig. 9.10, where the iso-surface of the HCO radical is plotted. An excellent agreement is obtained, demonstrating the accuracy of the extended FPI approach, that will be used from now on for all further investigations.

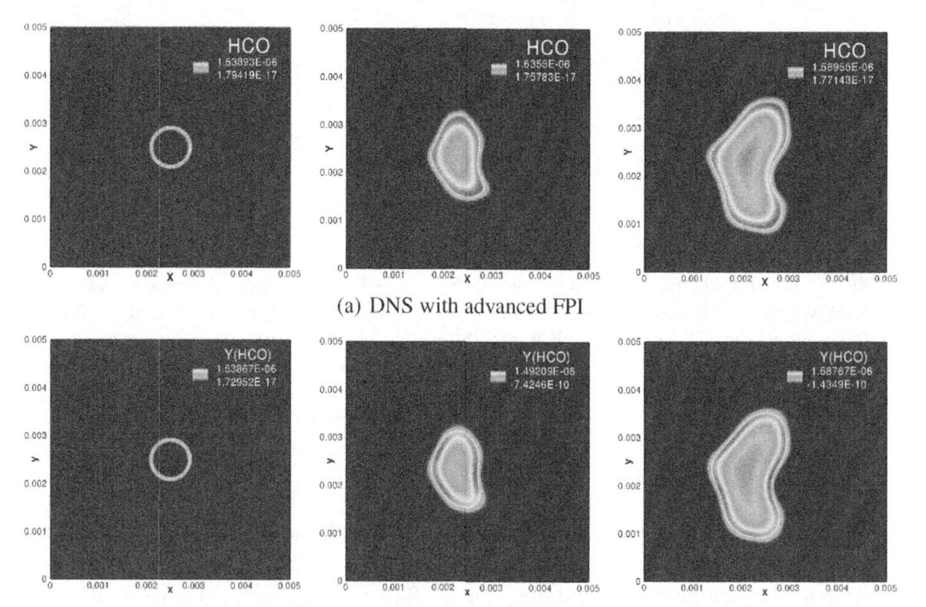

(a) DNS with advanced FPI

(b) DNS with complete chemistry and transport

Fig. 9.10 Validation of the advanced FPI procedure showing the iso-levels of HCO mass fraction at three times (from left to right 0.0, 0.25 and 0.5 ms): top, advanced FPI results; bottom, complex chemistry and transport models.

9.3 Flame/acoustics interactions investigated with DNS

To understand the origin of combustion instabilities, the interaction of an isolated acoustic wave with a turbulent flame, leading to amplification or damping, is an important basic element of the coupling process, since acoustics is known to trigger combustion instabilities in many configurations. This coupling phenomenon has first been explained qualitatively by Lord Rayleigh [36, 37] and has been the subject of many later investigations (e.g. [27, 22]), relying mostly on theoretical considerations or experimental measurements. This issue of flame/acoustic coupling has been investigated at depth in our research group during the last ten years, using only Direct Numerical Simulations and theoretical considerations. Only a summary of the most important findings is proposed here. Complementary information and further references can be found in particular in [41].

To the knowledge of the authors, no other research groups have been specifically using Direct Numerical Simulations to investigate the interaction between turbulent flames and acoustic perturbations. On the other hand, closely related aspects are the subject of an abundant and very interesting literature. For a first review, the recent book by Poinsot and Veynante is perhaps the best starting point [33]. Early references are particularly important to understand the most important coupling processes in laminar configurations. For example, wave propagation and scattering has been considered analytically in [4]. Prasad [35] was the first to carry out a corresponding investigation relying on a detailed reaction scheme. McIntosh and co-workers have examined at depth the details of the interaction process for laminar flames (e.g. [24, 1]).

In spite of all these investigations, many questions remain, which cannot be solved analytically and are extremely difficult to consider using experimental techniques. As a consequence, after developing a specific version of the well-known Rayleigh's criterion to investigate local amplification or damping of an acoustic pulse interacting with a reaction front [18], extensive investigations of flame/acoustic interactions have been carried out in our group. Premixed [19] as well as non-premixed [20] flames have been considered, using different fuels (for example hydrogen [20] or synthetic gas [39]). The influence of the direction of propagation of the wave on the observed amplification process has been checked [38]. All these studies have already delivered essential information in order to understand the local coupling between chemical reactions and acoustic waves and are now used to develop and test Computational Aero-Acoustics (CAA) models [31]. The main findings will be summarized in what follows, with an emphasis on premixed flames.

Our previous investigations have always considered a 2-dimensional flow field, using the DNS code *Parcomb*. This code has been used in the present project to investigate the structure of hydrogen and syngas (CO/H_2/air) turbulent premixed and non-premixed flames. Since turbulence is in general a 3-dimensional phenomenon [21], some of its aspects cannot be captured with such simulations. It is therefore a priori questionable how general the obtained results can be for real turbulence. As a consequence, the second DNS code family (π^3, see Fig.9.1) has been developed and used for three-dimensional computations. As explained in the previous section, while *Parcomb* considers complete reaction schemes, the π^3 codes use the FPI chemistry tabulation procedure.

9.3.1 Local Rayleigh's criterion

The developments first presented in [18] lead back to the classical stability condition first proposed by Lord Rayleigh, and based on a product between pressure fluctuation and heat release fluctuation. At the difference of the original formulation, the criterion proposed in what follows considers now *local* values of the important quantities in space and time. We restrict ourselves to a brief summary, in which only the source term due to combustion is considered. Readers interested in all details of the derivation are referred to [40, 18].

Using standard assumptions the balance equation for the acoustic energy E finally takes the form:

$$\frac{\partial E}{\partial t} + \nabla \cdot F = \frac{\gamma - 1}{\gamma p} \Pi' Q' \tag{9.2}$$

where $\Pi' = (1/\gamma)\log(1 + p'/p_0)$ [28], with$(\cdot)_0$ denoting the unperturbed values and $(\cdot)'$ the acoustic perturbation. The variable Q' is the heat release fluctuation, E and F are respectively the acoustic energy and the acoustic flux. These variables are computed as:

$$E = \frac{1}{2}\left(\frac{\mathrm{v}'^2}{c^2} + \Pi'^2\right) \tag{9.3}$$

$$\mathrm{F} = \mathrm{v}' \Pi' \tag{9.4}$$

$$Q' = -\sum_{i=1}^{N} h_i \dot{w}_i' \tag{9.5}$$

In these equations, c is the speed of sound, v is the flow velocity and thus v' is the acoustic velocity fluctuation N is the total number of chemical species, h_i the enthalpy of species number i and $\dot{w}_i$ its total mass production rate due to all reactions. In all equations the stoichiometric coefficients of a total number of R elementary reactions (complete reaction mechanism) are taken into account, along with the molar production rates $\dot{\omega}_{ik}$ of species i in the elementary reaction number k. Since, if the right side of Eq.(9.2) is negative (resp. positive), the wave is attenuated (resp. amplified), one obtains finally that the acoustic wave is amplified (locally) for $\Pi' Q' > 0$ and damped for $\Pi' Q' < 0$.

The term Q' is the fluctuation of the heat source term in the temperature equation. It can be directly computed as a function of the fluctuations of the mass production rate of any species, $\dot{w}_i$ (species number i from a total of N, see Eq.9.6); or at the level of each chemical reaction, considering the reaction production rate $\dot{\omega}_{ik}$ (reaction number k from a total of R, see Eq.9.7):

$$Q' = -\sum_{i=1}^{N} h_i \dot{w}'_i \tag{9.6}$$

$$= \sum_{k=1}^{R} \left(-\sum_{i=1}^{N} (\nu^r_{ik} - \nu^f_{ik}) W_i h_i \dot{\omega}'_{ik} \right) \tag{9.7}$$

As a consequence, the resulting local stability criterion $\Pi' Q' < 0$ can now be used to investigate the influence of any individual reaction (choosing a constant value for k in Eq.9.7) or of any individual species (keeping a constant value for i in Eq.9.6).

Since flame propagation is always much slower than acoustic propagation in all our studies (low flame Mach number), it can be shown that it is not necessary to account for flame movement in the local Rayleigh's criterion [40].

9.3.2 Examples of results and discussion

9.3.2.1 Interaction with an isolated pulse in two dimensions

This first example considers a premixed flame burning syngas, interacting with a Gaussian acoustic wave coming from the burnt gas side. Since the propagation of this isolated pulse is very fast (speed of sound), the reaction zone does not move noticeably during the interaction.

The pressure perturbation Π' induced by the interaction is shown in Figs.9.11(a) and (b) at $t' = 10\ \mu$s and $15\ \mu$s respectively. These times correspond respectively to a wave crossing the reaction zone (a) and after arriving in the fresh gases (b). At time (a), the wave has already crossed most of the reaction zone and is now considerably wrinkled compared to the initial, planar structure. At time (b), the wave has finished crossing the reaction zone. The wave profile is complex and reveals multiple fronts.

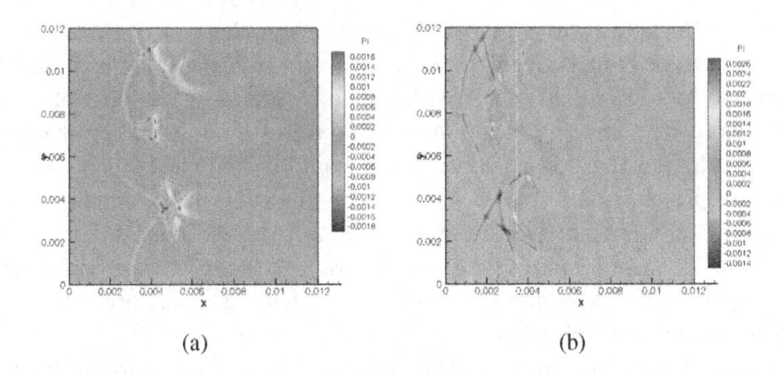

Fig. 9.11 Pressure perturbation Π' induced by the interaction at $t' = 10\ \mu$s (a) and $15\ \mu$s (b)

The corresponding heat release fluctuation Q' induced by the interaction with the Gaussian wave is shown in Fig.9.12 at 10 μs (see also Fig.9.11a). These fluctuations are limited to a sharp front inside the reaction zone and locally show large positive and negative values. It is interesting to note that the fluctuations are completely negligible for some regions. Note that the peak heat release in the unperturbed flame is equal to 4.2 10^9 J/(m^2s). As a consequence, the relative perturbation induced by the interaction with the acoustic wave is indeed very small, less than 0.3%.

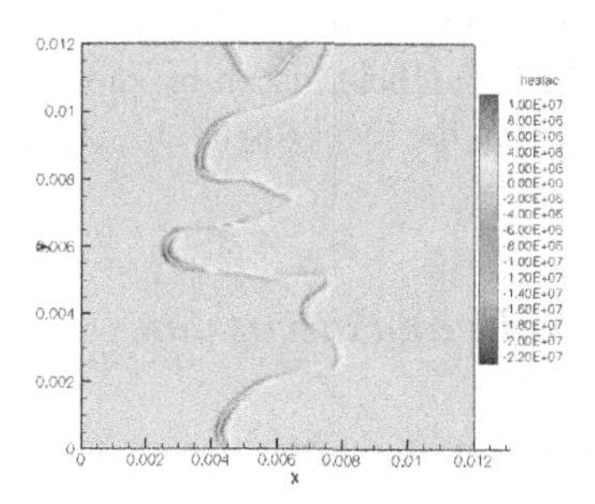

Fig. 9.12 Heat release fluctuation Q' induced by the interaction at $t' = 10$ μs. See also Fig.9.11a.

The local Rayleigh's criterion introduced previously (right side of Eq.9.2) is shown in Fig.9.13 at $t' = 10$ μs (a) and 15 μs (b). In principle, Fig.9.13a is the "product" of Figure 9.11a with Figure 9.12. It can be observed that only very few, well-localized amplification (positive values) and damping zones (negative values) appear. Focusing effects and wave scattering are observed when the acoustic wave crosses regions associated with large gradients of acoustic impedance, i.e. in the flame zone.

It can be observed for example in Fig.9.13 that amplification and damping regions appear very often close to each other, usually as double layers corresponding to opposite effects. It is therefore useful to make sure that this is not an artefact resulting from some numerical error leading to a very small shift in the flame position between the two DNS. This can best be done by checking the possible influence of the direction of wave propagation. In the case of a numerical error one would expect that the positive and negative layers would be switched when changing the direction of propagation of the wave. To check this issue, DNS simulations involving an identical Gaussian acoustic wave coming from the right direction (starting in the burnt gas zone), or from the left direction (starting in fresh gas zone) have been compared [38]. A detailed analysis reveals that the spatial organization of the layers associated to amplification and damping remains mostly unchanged in both simulations, at least qualitatively. Therefore, this double-layer arrangement is not the result of a numerical artefact. When examining these sandwich layers in more details, it can be observed that they are a direct consequence of the similar, layered

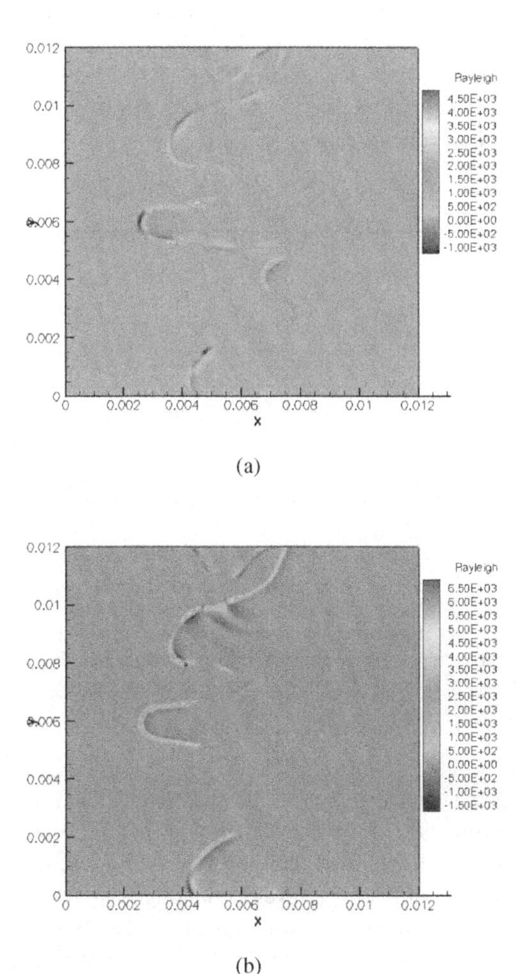

Fig. 9.13 Local Rayleigh's criterion at 10 μs (a) and 15 μs (b)

structure observed in most cases for Q' (see for example Fig.9.12). This can again be unterstood when considering the influence of the wave on the chemistry [40].

9.3.2.2 Influence of individual species

The local Rayleigh's criterion introduced previously can also be used to investigate the isolated influence of each species on wave modification. This can easily be done by considering Eq.(9.2) and keeping a constant value for index i (species index) in Eq.(9.6) in order to compute Q'. In this manner, the influence of each species on amplification or damping is quantified as a function of space and time. In what follows the resulting influence is averaged over the complete numerical domain in order to get the global influence of each species (Fig.9.14). The present fresh gas conditions correspond to a stoichiometric $CO:O_2$ ratio with slight hydrogen enrichment, leading to a global equivalence ratio $\phi = 1.12$.

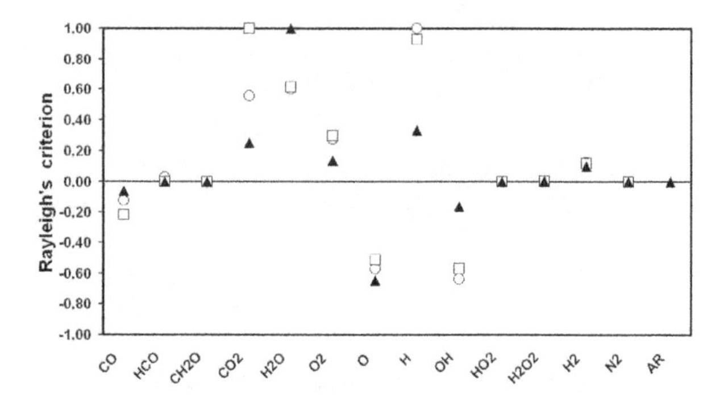

Fig. 9.14 Normalized contribution of each species to Rayleigh's criterion, averaged over the full numerical domain. Wave coming from the fresh gas, i.e. the left side (squares) or from the burnt gas, i.e. the right side of the numerical domain (circles), with both using the chemical scheme of [25]. Filled triangles: wave coming from the fresh gas using the chemical scheme of [48].

Three different results are presented in this figure, in which the influence of the most important species has always been normalized to 1 in order to facilitate comparisons. First, the direction of wave propagation is not important when determining if a species leads to amplification or damping, supporting the original, global analysis of Lord Rayleigh. Both results (wave originating from the fresh gas: squares; or from the burnt gas: circles) are almost identical, confirming again the previous results [38]. The only noticeable difference is observed for species CO_2, for which the qualitative effect is unchanged (amplification), but the efficiency varies by roughly 50% depending on the direction of propagation. We have not been able to explain this observation yet. Note that, in a relative manner, the same is true for species CO, but with only a small damping effect.

Second, the employed kinetic scheme does not modify these findings. In order to check this issue, the same computation (wave starting in the fresh gas) has been repeated twice using two different chemical schemes ([25]: squares; [48]: filled triangles). Once again, the two schemes lead qualitatively to the same findings and show only limited quantitative differences. Finally, the influence of the different species on amplification and damping is clear for all conditions: species CO_2, H and H_2O mostly control wave amplification, while species O, OH and CO dominate damping. As expected, short-lived radicals like HCO, CH_2O, HO_2, H_2O_2 and neutral species (N_2, Ar) have a negligible influence on the process.

9.3.2.3 Three-dimensional results

All the results obtained up to now rely on two-dimensional simulations. As a consequence, turbulence evolves in a different manner compared to real, three-dimensional turbulence properties. In order to check this possible issue, the three-dimensional DNS code π^3C has been developed and is used in this section to compute the interaction between a Gaussian acoustic wave and a realistic, three-dimensional turbulent premixed flame. By comparing the interaction process in the present 3D simulations with the previous observations, it is possible to check if differences are observed, at least qualitatively. The flame considered now corresponds to an initially spherical flame kernel freely expanding in a fresh gas mixture associated with homogeneous isotropic turbulent conditions (Fig.9.4). More information on this configuration can be found for instance in [45].

The acoustically-induced pressure fluctuation Π' is plotted at $t' = 12$ μs and $t' = 18$ μs in Figs.9.15 and 9.16. This corresponds respectively to an acoustic wave in the middle of the reaction zone, and soon leaving the flame. For both figures, the initially planar structure of the Gaussian wave remains visible outside of the flame region, near the boundaries of the computational domain. Since the wave propagates much faster than the flame (corresponding flame Mach number based on the laminar flame speed $M_f = S_l/c$ of the order of 0.02), the flame appears to be frozen during wave propagation.

A low-amplitude reflected wave on the right side of the reaction zone and a more intense transmitted wave within the burnt gases are visible. The expansion of the reflected wave is easy to see in Figs.9.15 and 9.16, while the transmitted wave appears to be attenuated compared to the unperturbed, surrounding Gaussian wave. This is mostly the result of the broadening of this wave induced by the interaction with the gradients of acoustic impedance associated with the reaction zone. The pressure fluctuation becomes broader in space but at a lower level.

As such, these figures do not allow a direct investigation of amplification or damping. For that purpose, the local Rayleigh's criterion introduced previously must again be employed.

The heat release fluctuation induced by the acoustic interaction is plotted at $t' = 18$ μs in Fig.9.17. As expected, this fluctuation is restricted to the active reaction zone. In agreement with previous observations relying on two-dimensional DNS results [18, 39], alternating layers associated with positive and negative fluctuations are visible in this figure. They are a consequence of a slight shift in space between different intermediate radicals, that react in an opposite manner to a local change in flow state conditions. Note, however, that the induced heat release fluctuation is extremely small compared to the mean value of heat release in the flame.

The resulting field of Q' is completely different from the associated field of pressure fluctuation Π' (compare Fig.9.17 to Fig.9.16). This illustrates the complexity of the coupling process. As a matter of fact, amplification or damping is conditioned by the "product" of Fig.9.17 with Fig.9.16, leading again to very small active regions in space, since the overlap between both fluctuations is very limited.

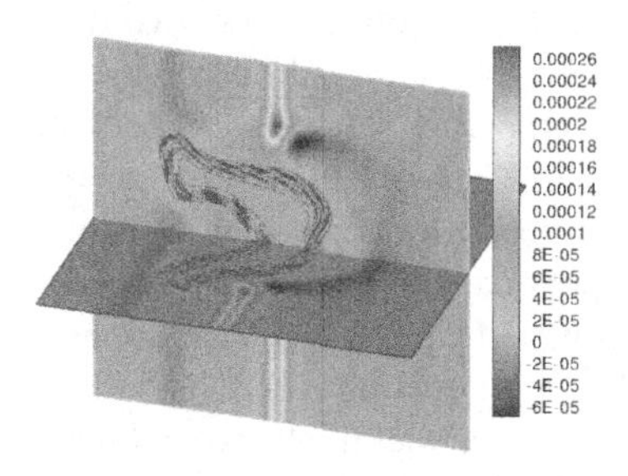

Fig. 9.15 Acoustically-induced pressure fluctuation Π' at $t' = 12\ \mu$s (color field). The flame position is shown using isosurfaces of CO_2 mass fraction (lines).

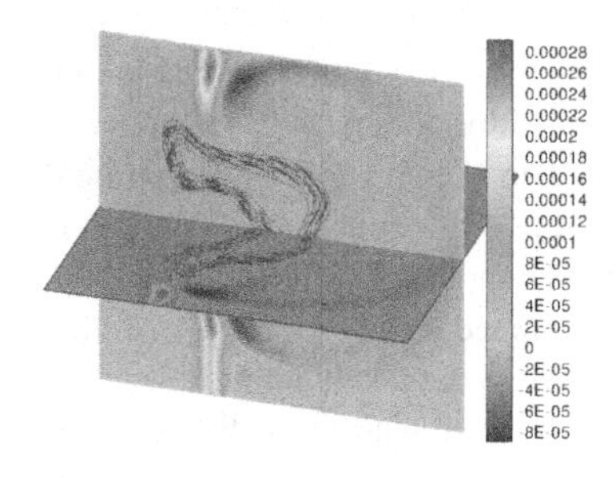

Fig. 9.16 Acoustically-induced pressure fluctuation Π' at $t' = 18\ \mu$s (color field). The flame position is shown using isosurfaces of CO_2 mass fraction (lines). See also Fig.9.17 for the associated heat release fluctuation.

The local Rayleigh's criterion (right side of Eq.9.2) is plotted in Fig.9.18 at $t' = 18\ \mu$s. The modification of the wave due to the coupling with the flame is restricted to very small regions in space. Similarly to the observation concerning heat release fluctuation, amplification and damping mostly take place in neighboring, sandwich layers, illustrating the complexity of the local process. This corroborates previous observations based on two-dimensional DNS.

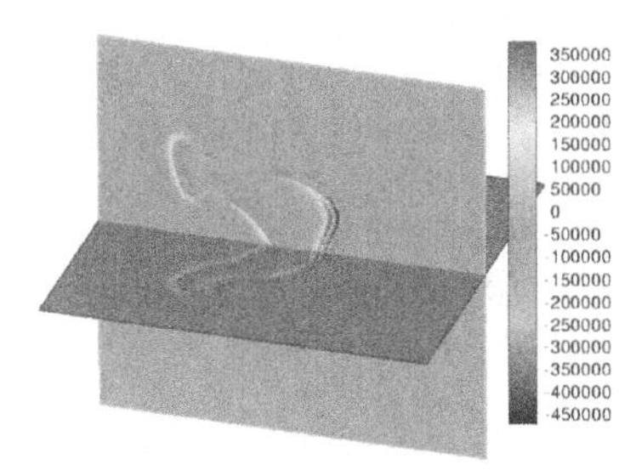

Fig. 9.17 Acoustically-induced heat release fluctuation Q' at $t' = 18\ \mu s$. See also Fig.9.16 for the associated pressure fluctuation.

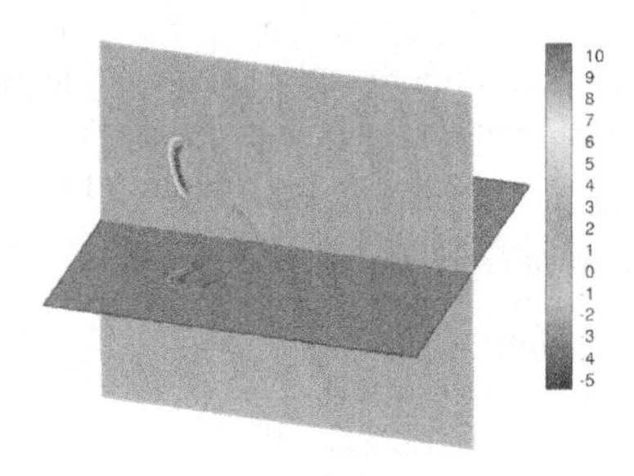

Fig. 9.18 Local Rayleigh's criterion (right side of Eq.9.2) at $t' = 18\ \mu s$. See also Figs.9.16 and 9.17 for the associated fields of Π' and Q'.

9.4 Post-processing challenge: AnaFlame

9.4.1 Introduction

In order to improve combustion processes, two complementary means are classically used: experimental investigations and numerical simulations. Both means are often three-dimensional in space, unsteady and involve a large number of chemical species beyond classical flow variables, leading to huge quantities of raw data. As such, post-processing becomes a real and essential challenge. In other words, the user time needed for an intelligent post-processing become comparable to the time requested for the DNS simulations itself. For this purpose, a dedicated library called ANAFLAME [49], containing essential post-processing methods has been developed as a MATLAB toolbox: it contains numerous tools to analyze and visualize 2D and 3D flames and flow fields; to investigate the geometry and structure of flames both locally and globally; to quantify the interaction between flow fields and flames; and to determine statistics and correlations of all variables that are essential for model development.

Available tools are briefly described below. Throughout this publication, the toolbox has been used to analyze and possibly visualize the obtained DNS results.

9.4.2 Content of the post-processing toolbox AnaFlame

In many cases the most suitable way to analyze the wealth of raw data produced by DNS or obtained experimentally is through graphical visualization relying on complex plots. MATLAB provides many such basic graphical functions. The development of ANAFLAME is intended for post-processing data produced on a structured, orthogonal grid. Values obtained on another grid type must first be interpolated on such a grid before continuing post-process. The developed toolbox contains presently five different categories of complementary tools:

1. *Geometrical analysis tools:* containing scripts to
   - define and locate a flame front.
   - compute the flame length (in 2D),
   - compute the flame surface area (in 3D),
   - compute the flame thickness,
   - compute the flame curvature and
   - compute the flame shape factor (in 3D).
2. *Flame structure analysis tools*, containing tools to
   - extract desired flame quantities along the flame front and along any prescribed isolevel or isosurface of an existing variable.
   - compute linear cuts through the flame front and any variable of interest can be extracted, integrated or correlated along these cuts.
   - compute nonlinear cuts, following the direction of the steepest gradient of a user-chosen variable,
3. *Flow analysis tools*, to analyze some flame/turbulent flow field interaction phenomena, e.g.
   - the vorticity,
   - the strain rate,

- the streamlines

4. *Statistical tools*, used for example

   - to compute the moments of a distribution,
   - to represent graphically such distributions using for instance histograms, and
   - to determine Probability Density Functions.

5. *Quantification of turbulent flow field properties*

   - turbulent energy distribution,
   - integral scale(s) and
   - velocity fluctuations.

For the interested reader, more details (documentation and tutorials) about ANAFLAME can be obtained from [49] and on the ANAFLAME website, where the corresponding library can be freely downloaded (http://www.uni-magdeburg.de/isut/LSS).

9.5 Conclusions and perspectives

Before concluding, we would like to illustrate briefly two aspects that will play a major role for DNS studies of turbulent flames in the near future.

First, it is clear that higher values of the turbulent Reynolds number (Re) must be accessed. Practical applications involve Re values much higher than those currently accessible when considering realistic chemistry models. For this purpose, the most powerful existing super-computers must be employed in an efficient manner to carry out corresponding DNS simulations. This is the purpose of a project presently realized in our research group in the framework of the DEISA Extreme Computing Initiative (DECI), supported by the European Union. Turbulent flame structures associated with a value of Re increased by at least an order of magnitude will be obtained at the end of this project. This is already exemplified in Fig.9.19, where a premixed flame with a higher turbulence level is shown during expansion in a turbulent velocity field. This result has been obtained on our own, local PC-cluster using 540 cores. Even more interesting DNS computations at higher Re values should be possible using several thousands high-capacity nodes on dedicated supercomputers.

Second, the influence of the employed transport model on the resulting turbulent flame structure must be carefully revisited. For the rich premixed flame considered previously, using the assumption Le=1 is acceptable, since it corresponds closely to the real value. Nevertheless, this is not true in general. For example, the advanced FPI approach described previously has been employed to compare the time-dependent growth of the flame surface area for an initially spherical flame expanding in a turbulent flow field (Fig.9.20). For stoichiometric conditions, major differences appear between the computation relying on differential diffusion and that associated with Le=1. These discrepancies must be analyzed in more details in the future.

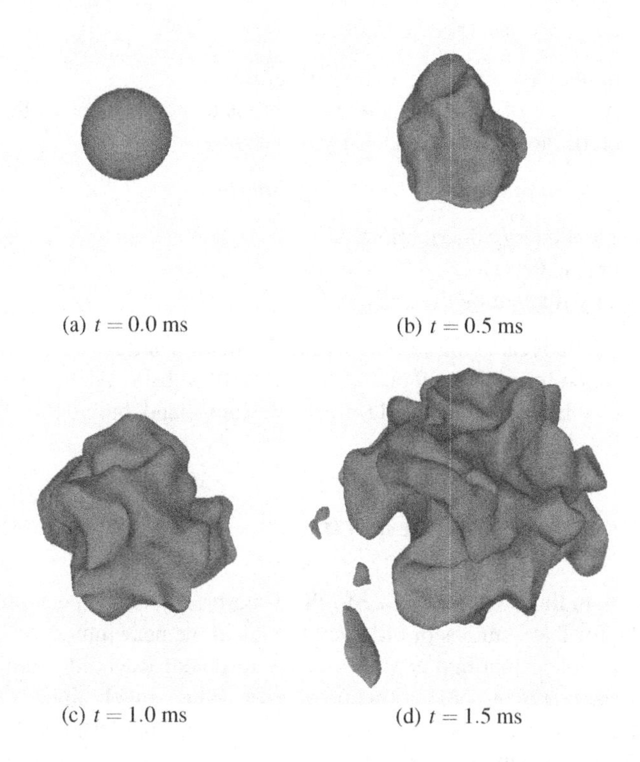

Fig. 9.19 Time evolution of a premixed methane/air turbulent flame kernel at high turbulence level

Coming now to the conclusions, this paper has shown that Direct Numerical Simulations are able to deliver extremely interesting information for strongly coupled physical problems controlling turbulent reacting flows. In the present project, flame/acoustics interactions have been investigated extensively using DNS, for premixed and for non-premixed configurations and for different fuels. As a whole, this comprehensive study can be briefly summarized as follows:

- The structure of the initially planar acoustic wave is always strongly modified during interaction with the flame. At the end of the interaction process, the wave is highly wrinkled, leading to a non-planar geometry, related to that of the flame front. Large focusing effects are observed in connection with gradients of acoustic impedance.
- Regions associated with a noticeable amplification and damping of the flame are very localized and mostly located near to each other in a double-layer structure. This is a result of the similar structure found for the heat release fluctuation Q'.
- Two mechanisms control wave modifications: 1) the influence of the gradients of acoustic impedance, which is not symmetric with respect to the fresh or burnt gas but will globally play a minor role since both curvatures are found with a simi-

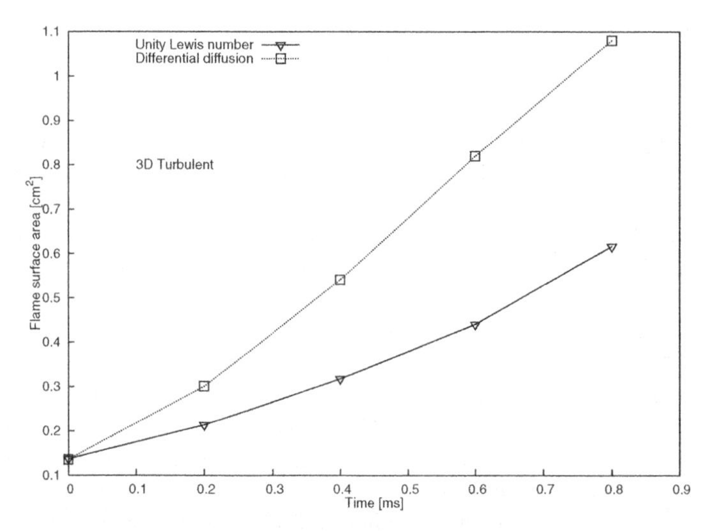

Fig. 9.20 Time evolution of the surface area for a premixed stoichiometric methane/air flame computed using differential diffusion (□) and unity Lewis number (▽) FPI tables.

lar probability in highly turbulent flames, leading to large cancelling effects; 2) coupling with the chemical reactions, as described originally by Lord Rayleigh. This last effect will play a dominant role in explaining wave amplification and damping, since it is almost independent of flame orientation or wave propagation direction.

- The developed Rayleigh's criterion can be used to quantify the influence of each individual chemical species. For syngas combustion CO_2, H and H_2O mostly control wave amplification, while species O, OH and CO dominate damping. Short-lived radicals are independent from the employed reaction mechanism. This suggests new possibilities to enhance flame stability, either by modifying slightly the burning regime in order to shift the local composition towards more favorable conditions; or by injecting additional species, leading to damping.

Presently, DNS at much higher and thus more realistic values of the turbulent Reynolds number are being carried out. Together with companion studies pertaining to the possible influence of differential diffusion, these simulations will shed additional light on the processes controlling turbulent combustion.

Acknowledgements Most of this work has been financially supported by the Deutsche Forschungsgemeinschaft (DFG) in the frame of the Research Unit # 486 "Combustion Noise". Part of the DNS computations have been carried out thanks to the support of the Leibniz Supercomputing Center in Munich (Project h1121).

References

[1] Batley G, McIntosh A, Brindley J, Falle S (1994) A numerical study of the vorticity field generated by the baroclinic effect due to the propagation of a planar pressure wave through a cylindrical premixed laminar flame. J Fluid Mech 435:289–303

[2] Baum M, Poinsot T, Thévenin D (1994) Accurate boundary conditions for multicomponent reactive flows. J Comput Phys 116:247–261

[3] Bell J, Day M, Grcar J, Lijewski M, Driscoll J, Filatyev S (2007) Numerical simulation of a laboratory-scale turbulent slot flame. Proc Combust Inst 31:1299–1307

[4] Candel S (1979) Numerical solution of wave scattering problems in the parabolic approximation. J Fluid Mech 90:465–507

[5] Candel S (2002) Combustion dynamics and control: Progress and challenges. Proc Combust Inst 29:1–28

[6] Fiorina B, Baron R, Gicquel O, Thévenin D, Carpentier S, Darabiha N (2003) Modelling non-adiabatic partially premixed flames using flame-prolongation of ILDM. Combust Theory Modelling 7:449–470

[7] Fox R (2003) Computational Models for Turbulent Reacting Flows. Cambridge University Press

[8] Gicquel O, Darabiha N, Thévenin D (2000) Laminar premixed hydrogen/air counterflow flame simulations using flame prolongation of ILDM with differential diffusion. Proc Combust Inst 28:1901–1908

[9] Hawkes E, Chen J (2004) Direct numerical simulation of hydrogen-enriched lean premixed methane-air flames. Combust Flame 138(3):242–258

[10] Hawkes E, Sankaran R, Sutherland J, Chen J (2005) Direct numerical simulation of turbulent combustion: fundamental insights towards predictive models. J Phys Conf Ser 16:65–79

[11] Hawkes E, Sankaran R, Sutherland J, Chen J (2007) Scalar mixing in direct numerical simulations of temporally evolving plane jet flames with skeletal CO/H_2 kinetics. Proc Combust Inst 31:1633–1640

[12] Hilbert R, Thévenin D (2002) Autoignition of turbulent non-premixed flames investigated using direct numerical simulations. Combust Flame 128(1-2):22–37

[13] Hilbert R, Tap F, El-Rabii H, Thévenin D (2004) Impact of detailed chemistry and transport models on turbulent combustion simulations. Prog Energy Combust Sci 30:61–117

[14] Hinze JO (1975) Turbulence. McGraw-Hill

[15] Honein A, Moin P (2004) Higher entropy conservation and numerical stability of compressible turbulence simulations. J Comput Phys 201:531–545

[16] Kee R, Miller J, Jefferson T (1980) Chemkin, a general purpose problem-independent transportable fortran chemical kinetics code package. Sandia National Laboratories Report SAND 80-8003

[17] Kee R, Warnatz J, Miller J (1983) A fortran computer code package for the evaluation of gas-phase viscosities, conductivities, and diffusion coefficients. Sandia National Laboratories Report, SAND83-8209
[18] Laverdant A, Thévenin D (2003) Interaction of a gaussian acoustic wave with a turbulent premixed flame. Combust Flame 134(1-2):11–19
[19] Laverdant A, Thévenin D (2005) Direct numerical simulation of a gaussian acoustic wave interaction with a turbulent premixed flame. Comptes Rendus de l'Académie des Sciences-Mécanique 333:29–37
[20] Laverdant A, Gouarin L, Thévenin D (2007) Interaction of a gaussian acoustic wave with a turbulent non-premixed flame. Combust Theory Modelling 11(4):585–602
[21] Lesieur M (1997) Turbulence in Fluids. Kluwer Academic Publishers
[22] Lieuwen T (2003) Modeling premixed combustion-acoustic wave interactions: a review. J Propu Power 19(5):765–781
[23] Lindstedt P (1998) Modeling of the chemical complexities of flames. Proc Combust Inst 27:269–285
[24] Liu F, McIntosh A, Brindley J (1993) A numerical investigation of Rayleigh-Taylor effects in pressure wave-premixed flame interactions. Combust Sci Tech 91:373–386
[25] Maas U, Pope SB (1992) Simplifying chemical kinetics: intrinsic low-dimensional manifolds in composition space. Combust Flame 88:239–264
[26] Maas U, Thévenin D (1998) Correlation analysis of direct numerical simulation data of turbulent non-premixed flames. Proc Combust Inst 27:1183–1189
[27] Markstein G (1964) Nonsteady Flame Propagation. Pergamon Press, Paris
[28] Monin A, Yaglom A (1979) Statistical Fluid Mechanics: Mechanics of Turbulence. MIT Press, Cambridge, MA
[29] van Oijen J, Lammers F, de Goey L (2001) Modeling of complex premixed burner systems by using flamelet-generated manifolds. Combust Flame 127:2124–2134
[30] Peters N (2000) Turbulent Combustion. Cambridge University Press
[31] Phong Bui T, Schröder W, Meinke M, Shalaby H, Thévenin D (2007) Source term evaluation of the APE-RF system using DNS data. Proc ECCOMAS CFD, (Wesseling, P, Oñate, E and Périaux, J, Eds) 188:1–14
[32] Poinsot T, Lele S (1992) Boundary conditions for direct simulations of compressible viscous flow. J Comput Phys 101:104–129
[33] Poinsot T, Veynante D (2005) Theoretical and Numerical Combustion. Edwards Publishing, PA, 2nd Edition
[34] Pope S (1997) Computationally efficient implementation of combustion chemistry using in situ adaptive tabulation. Combust Theory Modelling 1:41–63
[35] Prasad K (1997) Interaction of pressure perturbations with premixed flames. Combust Flame 97:173–200
[36] Rayleigh J (1878) Nature 18(319)
[37] Rayleigh J (1945) The Theory of Sound, vol II. NJ, Dover, New York

[38] Shalaby H, Thévenin D (2006) Influence of the propagation direction for an acoustic wave interacting with a turbulent premixed flame. Proc Appl Math Mech 6:545–546

[39] Shalaby H, Laverdant A, Thévenin D (2005) Interaction of an acoustic wave with a turbulent premixed syngas flame. In: Twelfth International Congress on Sound and Vibration, Lisbon, Portugal, pp 635/1–635/8

[40] Shalaby H, Laverdant A, Thévenin D (2009) Direct numerical simulation of realistic acoustic wave interacting with a premixed flame. Proc Combust Inst 32:in press

[41] Shalaby H, Laverdant A, Thévenin D (2009) Potential of direct numerical simulations to investigate flame/acoustic interactions. Acta Acust in press

[42] Tap F, Hilbert R, Thévenin D, Veynante D (2004) A generalized flame surface density modelling approach for the auto-ignition of a turbulent non-premixed system. Combust Theory Modelling 8:165–193

[43] Thévenin D, Behrendt F, Maas U, Przywara B, Warnatz J (1996) Development of a parallel direct simulation code to investigate reactive flows. Comput Fluids 25(5):485–496

[44] Thévenin D, van Kalmthout E, Candel S (1997) Two-dimensional direct numerical simulations of turbulent diffusion flames using detailed chemistry. In: Direct and Large-Eddy Simulation II, (Chollet, J.P., Voke, P.R. and Kleiser, L., Eds.), Kluwer Academic Publishers, Amsterdam, pp 343–354

[45] Thévenin D, Gicquel O, de Charentenay J, Hilbert R, Veynante D (2002) Two-versus three-dimensional direct simulations of turbulent methane flame kernels using realistic chemistry. Proc Combust Inst 29:2031–2038

[46] Veynante D, Vervisch L (2002) Turbulent combustion modeling. Prog Energy Combust Sci 28(3):193–266

[47] Warnatz J, Maas U, Dibble R (2001) Combustion. Springer, 3d Ed.

[48] Yetter U, Dryer F, Rabitz H (1991) A comprehensive reaction mechanism for carbon monoxide/hydrogen/oxygen kinetics. Combust Sci Tech 79:97–128

[49] Zistl C, Hilbert R, Janiga G, Thévenin D (2009) Increasing the efficiency of post-processing for turbulent reacting flows. Comput Visual Sci in press

Chapter 10
Localization of Sound Sources in Combustion Chambers

Christian Pfeifer, Jonas P. Moeck, C. Oliver Paschereit and Lars Enghardt

Abstract In this project, an algorithm for the determination of the positions and the strengths of sound sources in closed combustion chambers by the evaluation of wall-flush-mounted microphones is presented. The theoretical background of the reconstruction of sound sources is the nearfield acoustic holography. In the theory of nearfield acoustic holography, evanescent modes of the sound pressure field have to be taken into consideration. The choice of these modes is described in detail. Since the problem is ill-conditioned, different regularization methods are used for the solution of the inverse acoustic problem. The algorithm is applied to optimize the arrangement of a sensor array with 48 microphones and the dependence on contaminating noise is investigated. The focus of this project is the localization of real sound sources not exactly located on the assumed source distribution. For this case, a scanning technique is introduced and its applicability is investigated.

10.1 Introduction

One of the main goals in modern gas turbine development for power generation and aeroengine applications is the reduction of pollutant and noise emissions. Flow disturbances in the combustion chamber cause fluctuations of the flame which, in turn, generate acoustic waves due to the unsteady expansion across the region of heat release [4]. Significant sound pressure levels can be observed if this mechanism excites acoustic resonances of the combustion chamber. Even worse, if the time scales of the acoustic-heat release interactions match those of the wave resonance period, positive feedback may occur. If the damping in the combustion chamber is not suffi-

Christian Pfeifer, Jonas P. Moeck, C. Oliver Paschereit
Institute of Fluid Dynamics and Engineering Acoustics, TU Berlin
e-mail: christian.pfeifer@pi.tu-berlin.de

Lars Enghardt
Engine Acoustics Department, DLR Berlin

A. Schwarz, J. Janicka (eds.), *Combustion Noise*, DOI 10.1007/978-3-642-02038-4_10,

ciently high, perturbations grow in amplitude until limited by nonlinear effects [9]. This phenomenon, commonly referred to as thermoacoustic instability or combustion oscillations, may lead to exceptionally high pressure pulsations with amplitudes of up to several percent of the mean static pressure in the combustion chamber [13]. These self-excited oscillations are an unacceptable noise source, and furthermore, they reduce the durability of the combustion chamber significantly. Unfortunately, the lean premixed combustion mode, which was introduced by the gas turbine industry as the key step to lower NO_x emissions, turned out to be particularly susceptible to these kind of instabilities [16].

In contrast to, e.g., jet associated noise, the sound generation in flames is much less understood. As a result, modeling tools for the direct prediction of combustion noise are rarely available. For an enhanced understanding of the physical processes associated with sound generation in flames, as well as for modeling and control of combustors, experimental access to the acoustic sources in premixed flames is necessary. In an anechoic environment, measuring the acoustic sources due to combustion is fairly well developed [20, 15]. In case of an enclosed configuration, experimental access to the acoustic sources of the flame is not straightforward. The pressure that is measured at the combustor wall is strongly influenced by wave reflections from the boundaries.

Reduced order acoustic models in combination with semi-empirical relations for the flame response to velocity fluctuations and the spectral distribution of the global acoustic sources have been used to predict combustion chamber pressure spectra as well as thermoacoustically unstable operating regimes (see, e.g., Refs. [14, 18, 8]). This type of models is fairly mature and has been successfully used to reduce combustion induced noise by active and passive means [3, 2]. Due to the reduced complexity, however, these models do not allow to draw any conclusions about the local physical phenomena responsible for the sound generation inside the flame. To gain more insight into the complex interaction mechanisms between the sound field and the flame, detailed information on the acoustic source distribution in the combustion zone is desirable. Due to sensor robustness limitations, direct acoustic measurements inside the flame are obviously not possible.

The acoustic source distribution in the flame can, however, be reconstructed from wall-mounted pressure sensors by solving an inverse problem based on the acoustic near-field. The focus of this project is the localization of flame associated noise sources in enclosed combustion chambers. This can be achieved using the acoustic nearfield information obtained from pressure measurements with wall-flush mounted sensors. The generic combustion chamber considered in this work is simply a hard-walled flow duct with a circular cross section.

A schematic representation of the reconstruction procedure is displayed in Figure 10.1. The combustion chamber with several wall-flush-mounted microphones and the flame are shown. In a first step, the position and the shape (i.e., the region of heat release) of the enclosed flame is determined by chemiluminescence imaging techniques (see, e.g., [5]). With this knowledge, the domain where the sound sources can be expected is discretized with a set of N (say) acoustic monopoles with unknown amplitudes. Using a Green's function representation for the acoustic field in the duct, the contribution of every individual monopole source to the pressure field on the duct wall can be computed. For a finite number M of sensor locations, we obtain a linear mapping from N source amplitudes to M pressure sensors. Given now the M complex pressure values, formal inversion of the linear system yields the unknown source amplitudes. The discretization of a continuous source domain by a set of monopole sound sources with unknown amplitudes was earlier used by Kim et al. [6, 7] and Lowis et al. [10] in the investigation of the estimation of acoustic source strengths within a cylindrical duct and the determination of the strength of rotating broadband sources in ducts, respectively.

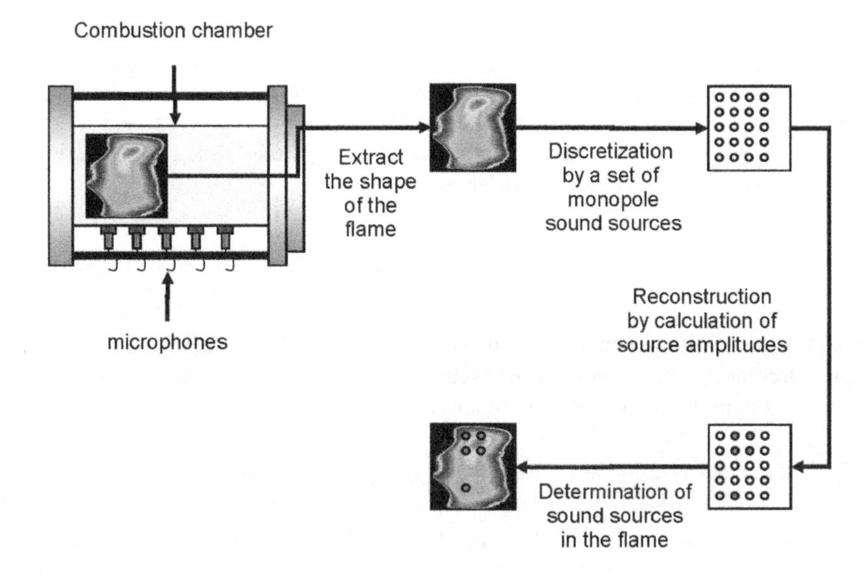

Fig. 10.1 Main ideas of reconstruction algorithm

10.2 Theoretical Background

10.2.1 Theory of Nearfield Acoustic Holography

The theoretical background for the identification of sound sources in compact regions is the theory of nearfield acoustic holography developed in [11]. The main difference of nearfield acoustic holography to ordinary acoustic holography is that evanescent modes are taken into account in the former. Evanescent modes can only be observed in the acoustic nearfield, since they decay exponentially along the duct axis. The nearfield of a sound source can be defined as follows: the distance between the observer and the sound source r is less than the wavelength λ of the signal $r < \lambda$.

10.2.2 Sound Pressure Field in the Combustion Chamber without Mean Flow

The starting point of the investigation is to assume the combustion chamber as an infinite long cylindrical flow duct. The sound sources are represented by a set of monopole sources. The acoustic pressure field caused by N monopoles can be described by the Helmholtz equation in the frequency domain

$$\nabla^2 p + k^2 p = \sum_{\lambda=1}^{N} q_\lambda \delta(\boldsymbol{x} - \boldsymbol{x}_\lambda) \ . \tag{10.1}$$

In Eq. (10.1), p is the complex pressure field and $k = \omega R/c$, the dimensionless frequency (or Helmholtz number), where R denotes the duct radius and c the speed of sound. q_λ is the (complex) amplitude of the monopole source located at $\boldsymbol{x}_\lambda$, $\boldsymbol{x} = [x, \vartheta, r]$ is the cylindrical coordinate vector and δ is the Dirac delta function. The solution of Eq. (10.1) can be written as

$$p(\boldsymbol{x}) = \sum_{\lambda=1}^{N} G(\boldsymbol{x}|\boldsymbol{x}_\lambda) q_\lambda \ , \tag{10.2}$$

where $G(\boldsymbol{x}|\boldsymbol{x}_0)$ is the Green's function satisfying the Helmholtz equation with one point source at $\boldsymbol{x}_0$ and homogeneous Neumann conditions at the duct wall. For a hard-walled cylindrical duct of infinite length, G can be written as [17]

$$G(\boldsymbol{x}|\boldsymbol{x}_0) = \sum_{m=-\infty}^{\infty} \sum_{n=1}^{\infty} \frac{\mathrm{i}}{2\pi} \frac{J_m\left(\sigma_{mn}\frac{r}{R}\right) J_m\left(\sigma_{mn}\frac{r_0}{R}\right)}{\kappa_{mn}\left[1 - \left(\frac{m}{\sigma_{mn}}\right)^2\right] J_m^2(\sigma_{mn})} \times \exp[-\mathrm{i}m(\vartheta - \vartheta_0)] \exp[-\mathrm{i}\kappa_{mn}(x - x_0)] \ . \tag{10.3}$$

The part of the solution associated with a certain pair (m,n) is called mode. In Eq. (10.3), J_m is the m-th order Bessel function of the first kind and σ_{mn} is the the n-th zero of $\mathrm{d}J_m/\mathrm{d}r$. The axial wave number κ_{mn} is given by

$$\kappa_{mn} = \sqrt{k^2 - \left(\frac{\sigma_{mn}}{R}\right)^2} \ , \tag{10.4}$$

where the square root is defined such that $\mathrm{sgn}(x - x_0)\mathrm{Re}(\kappa_{mn}) < 0$ and $\mathrm{sgn}(x - x_0)\Im(\kappa_{mn}) < 0$. The axial wavenumber κ_{mn} is either purely real or purely imaginary, depending on the sign of $k^2 - (\sigma_{mn}/R)^2$. In the latter case, i.e., $(\sigma_{mn}/R)^2 > k^2$, the amplitude of the mode (m,n) decays exponentially along the duct axis with decay rate $-\Im(\kappa_{mn})$. Uniquely associated with each mode (m,n) is its cut-on frequency $\omega_{c,mn}$ at which $\kappa_{mn} = 0$. For frequencies larger than ω_c, the mode is called propagating, otherwise it is called evanescent (or cut-off).

Based on Eq. (10.2), we can now define a linear system, which maps the N source amplitudes to the M pressure signals at the duct wall, as follows

$$\boldsymbol{p} = \boldsymbol{G}\boldsymbol{q}\,, \tag{10.5}$$

where $\boldsymbol{p} = [p_1, \ldots, p_M]^T$, $\boldsymbol{q} = [q_1, \ldots, q_N]^T$ and the elements of the transition matrix $\boldsymbol{G}$ are given by $[\boldsymbol{G}]_{m,n} = G(\boldsymbol{x}_m|\boldsymbol{x}_n)$. For the computation of the elements of $\boldsymbol{G}$, only a finite number of modes (m,n) in the double series in Eq. (10.3) can be taken into account. The proper choice of the modes to be included in the computation of $\boldsymbol{G}$ is discussed in the next section.

10.2.3 Modal Composition of G

An intuitive and straightforward choice for the modal composition of the Green's function G is to truncate the double series in Eq. (10.3) at upper bounds m_0 and n_0 and only use modes with $|m| \leq m_0$ and $n \leq n_0$ for the computation of G. In this case, the number of modes taken into account is given by $(2m_0+1)\,n_0$, or, equivalently, by presetting the number of modes, m_0 is a function of n_0 and vice versa. This scheme for choosing the modes to be included is, however, not the best method for the following reason. As explained in the last section, the amplitude of the evanescent modes, carrying the nearfield information, decay exponentially in axial direction with

$$decay_{mn}(\Delta x) = \exp\left[-i\kappa_{mn}\Delta x\right]\,. \tag{10.6}$$

Using the decay of mode (m,n), a damping rate δ_{mn} can be defined as follows

$$\delta_{mn} = 20\log_{10}\left(\frac{|decay_{mn}(\Delta x)|}{|decay_{mn}(\Delta x = 0)|}\right)\,. \tag{10.7}$$

Therefore, a more meaningful choice is to include all modes in the computation of G, which have a damping rate greater than some given bound δ^0_{mn} (for fixed Helmholtz number k). From the distribution of the zeros σ_{mn} of $\mathrm{d}J_m/\mathrm{d}r$, it is easy to see that the two compositions are not identical.

In Table 10.1, the damping rates δ_{mn} for the evanescent modes associated with the lowest mode numbers are shown for the axial distance $\Delta x = R$ from the source plane and a dimensionless frequency $k = 1$. The cells bounded by the red frame correspond to the damping coefficients for the modes which are chosen according to the first scheme with $m_0 = 2$ and $n_0 = 5$ (overall 25 modes). Obviously, e.g., the damping coefficient of mode $(2,5)$ is much higher than the damping coefficient of mode $(7,1)$.

The modal composition of G according to the second scheme is shown in Table 10.1 for the case $\delta_{mn} \geq -75$ as the union of the cells highlighted in grey. This corresponds to the same total number of modes as in the first case, but the pattern is different. In Table 10.1, the damping rate for the only propagating mode ($(0,1)$) is printed as

well. It can be seen that δ_{01} is equal to 0, which is due to the fact that mode $(0,1)$ is a propagating mode.
In Fig. 10.2, the damping rates of some modes (m,n) are plotted over the axial distance Δx from the source plane. The axial distance Δx is varied from 0 to R. It can be seen that in the axial distance of $\Delta x = R$, the amplitudes of the modes $(0,5)$ and $(2,5)$ are strongly damped. In comparison to that, the modes $(3,1)$ and $(7,1)$ are moderately damped. So, the amplitude contribution of the modes $(3,1)$ and $(7,1)$ to the overall sound pressure field is higher than the contribution of the modes $(0,5)$ and $(2,5)$. For this reason, the choice for the modal composition of the Green's function G for the reconstruction of the sound field is made by defining a damping rate threshold $\delta^0_{mn} = -75$.

Table 10.1 Damping coefficients δ_{mn} according to mode (m,n) choosing modes fulfilling condition a) $(-2 \leq m \leq 2) \wedge (1 \leq m \leq 5)$ (red frame), b) $\delta^0_{mn} = -75.0$ (gray shaded boxes)

$m \backslash n$	1	2	3	4	5	...
⋮			⋮			
−7	−74.0	−112.0	−143.3	−173.0	−201.9	
−6	−64.6	−101.6	−132.3	−161.6	−190.55	
−5	−55.0	−90.9	−121.2	−150.1	−178.5	
−4	−45.4	−80.2	−109.8	−138.4	−166.5	
−3	−35.4	−69.1	−98.2	−126.4	−154.3	
−2	−25.1	−57.6	−86.2	−114.1	−141.7	
−1	−13.4	−45.5	−73.6	−101.3	−128.8	
0	0	−32.1	−60.6	−87.9	−115.4	...
1	−13.4	−45.5	−73.6	−101.3	−128.8	
2	−25.1	−57.6	−86.2	−114.1	−141.7	
3	−35.4	−69.1	−98.2	−126.4	−154.3	
4	−45.4	−80.2	−109.8	−138.4	−166.5	
5	−55.0	−90.9	−121.2	−150.1	−178.5	
6	−64.6	−101.6	−132.3	−161.6	−190.55	
7	−74.0	−112.0	−143.3	−173.0	−201.9	
⋮			⋮			

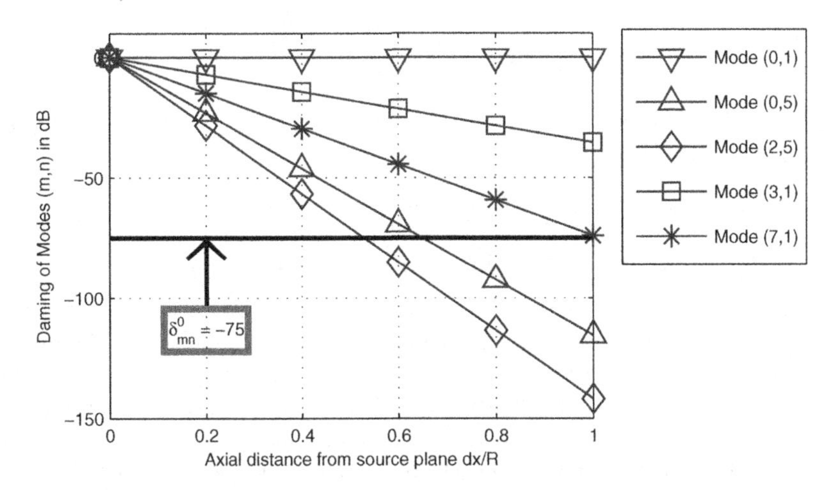

Fig. 10.2 Damping of modes

10.2.4 Sound Pressure Field in the Combustion Chamber with Mean Flow

In the previous sections, we assumed no mean flow inside the infinitely long flow duct. In this section, we extend the flow field by a mean flow with Mach number $M =$ const. in axial direction (plug flow). The acoustic pressure field can be described in the frequency domain by the following equation

$$\nabla^2 p - \left(\mathrm{i}k + M\frac{\partial}{\partial x}\right)^2 p = \sum_{\lambda=1}^{N} q_\lambda \delta\left(\boldsymbol{x} - \boldsymbol{x}_\lambda\right) . \tag{10.8}$$

For the solution of Eq. (10.8), in the sense of superposition of the contribution of each sound source, an expression for the Green's function satisfying Eq. (10.8) with a point source at $\boldsymbol{x}_0$ and homogeneous Neumann conditions at the duct wall is required. The necessary Green's function is [17]

$$G(\boldsymbol{x}|\boldsymbol{x}_0) = \sum_{m=-\infty}^{\infty}\sum_{n=1}^{\infty} \frac{\mathrm{i}}{2\pi} \frac{J_m\left(\sigma_{mn}\frac{r}{R}\right) J_m\left(\sigma_{mn}\frac{r_0}{R}\right)}{Q_{mn} J_m^2\left(\sigma_{mn}\right)} \times \exp\left[-\mathrm{i}m\left(\vartheta - \vartheta_0\right)\right] \exp\left[-\mathrm{i}\kappa_{mn}\left(x - x_0\right)\right] , \tag{10.9}$$

with

$$Q_{mn} = \left[\kappa_{mn}\left(1 - M^2 + kM\right)\right]\left[1 - \left(\frac{m}{\sigma_{mn}}\right)^2\right] , \tag{10.10}$$

and

$$\kappa_{mn} = \frac{-kM + \sqrt{k^2 - (1-M^2)\left(\frac{\sigma_{mn}}{R}\right)^2}}{1-M^2}, \quad (10.11)$$

with the same convention for the square root as in Sec. 10.2.2, i.e. $\operatorname{sgn}(x-x_0)\operatorname{Re}(\kappa_{mn}) < 0$ and $\operatorname{sgn}(x-x_0)\Im(\kappa_{mn}) < 0$.

10.2.5 Reflection at the Combustion Chamber Outlet

In the previous sections, we approximated the combustion chamber as an infinitely long flow duct with circular cross section. In this section, we consider a geometry resembling that of a typical combustion chamber (see Fig. 10.3). We approximate the combustion chamber as a finite flow duct with circular cross section connected to two flow ducts having smaller radii. The two smaller flow ducts represent the inlet (e.g. fuel and air supply) and outlet (e.g. exhaust device). Using this configuration,

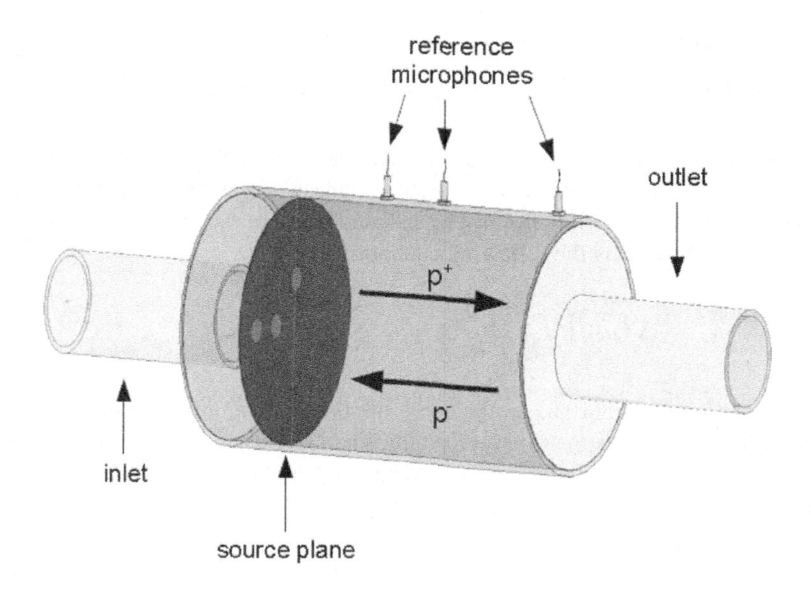

Fig. 10.3 Combustion chamber assumed as finite flow duct with circular cross section with fuel/air supply and exhaust oulet, plane wave p^+ and reflected wave p^-

the sound pressure field is reflected at the ends of the finite flow duct. As the non-propagating mode amplitudes are damped exponentially with respect to the axial distance from the source plane (see Sec. 10.2.3), just the reflection of the plane wave is to be considered. Here, only the reflection of the plane wave at the downstream side of the combustion chamber (outlet) was investigated. The same procedure has to be applied for the reflection of the plane wave at the upstream side of the combustion chamber (inlet).

The reflection of the plane wave at the downstream end of the flow duct can be calculated by the evaluation of at least two axially distributed microphones located in the far field of the pressure field (see Fig. 10.3). The simplified representation (10.12) for the pressure field is

$$p(\boldsymbol{x}) = p^{+} + p^{-} , \tag{10.12}$$

where $p^{+} = A\exp(-ikx)$ is the plane wave caused by the sound sources at the source plane and $p^{-} = B\exp(ikx)$ is the reflected part of the plane wave p^{+}. Here, we use the sound pressure field acquired by three microphones. The pressure data can be written as

$$\begin{bmatrix} p_{\mathrm{ref}_1} \\ p_{\mathrm{ref}_2} \\ p_{\mathrm{ref}_3} \end{bmatrix} = \underbrace{\begin{bmatrix} \exp(-ikx_1) & \exp(ikx_1) \\ \exp(-ikx_2) & \exp(ikx_2) \\ \exp(-ikx_3) & \exp(ikx_3) \end{bmatrix}}_{=:\boldsymbol{M}} \begin{bmatrix} A \\ B \end{bmatrix} . \tag{10.13}$$

From this representation, the reflection can be calculated by inverting the system of equations

$$\begin{bmatrix} A_{\mathrm{calc}} \\ B_{\mathrm{calc}} \end{bmatrix} = \left(M^{H} M\right)^{-1} M^{H} \boldsymbol{p} . \tag{10.14}$$

The subscript H denotes the Hermitian transpose of a matrix or a vector. Hence, by discriminating the reflected and transmitted plane wave components, the combustion chamber can be treated as an infinitely long flow duct with circular cross section. The measured pressure field p_{finite} for the finite combustion chamber has to be corrected subtracting the reflected part of the plane wave p_{refl}:

$$p_{\mathrm{infinite}} = p_{\mathrm{finite}} - p_{\mathrm{refl}} \quad \text{with} \quad p_{\mathrm{refl}} = B_{\mathrm{calc}} \exp(ikx) . \tag{10.15}$$

In this way, the reconstruction algorithm for the infinite flow duct can be applied to the finite flow duct case, as well.

10.2.6 Reconstruction of Sound Sources

In this section, the reconstruction algorithm using a measured sound pressure field is discussed in detail. Here, we assume an infinite flow duct with circular cross section.

10.2.6.1 Reconstruction Strategy

The problem can be stated as follows: Find an estimate of the source distribution inside the duct for a given pressure field at discrete points on the duct boundary. In this approach, a continuous region, where the sound sources are expected, is approximated by a finite number of discrete sources, so that the positions of the distributed sound sources $\boldsymbol{x}_0$ are known a priori. Only the amplitudes of the sources are treated to be unknown [12, 10]. Assuming that the generated pressure field can be represented in the frequency domain as a weighted superposition of the signals of each sound source, a linear system of equations for the source amplitudes can be established:

$$\hat{\boldsymbol{p}} = \boldsymbol{G}\boldsymbol{q} + \boldsymbol{e}\,, \tag{10.16}$$

where $\boldsymbol{e}$ is a vector of contaminating noise.
Now, the goal is to determine the vector of optimal discretized source amplitudes, associated with the estimated source positions, such that the corresponding pressure field matches the pressure measurements at the duct wall in a least squares sense. This is well known as an *inverse acoustic problem*. The optimal vector of source amplitudes can be formally obtained by pseudo-inversion of Eq. (10.16). A major difficulty is, however, the inherently ill-conditioned system of equations, resulting from the small singular values of the transfer matrix $\boldsymbol{G}$. This undesirable property is a well-known characteristic of inverse acoustic problems [12] and has negative effects on stability and accuracy of the reconstruction algorithm. One way to deal with ill-conditioned problems are so-called regularization methods [1, 7]. Some techniques for regularization, like *Tikhonov* regularization and the method of *Moore-Penrose* are presented in the next section.

10.2.6.2 Solution of the Inverse Acoustic Problem

The intuitive solution of the problem (10.16) is the *least-mean-squares*-solution. In this approach, the sum of the squared errors, called *residuals*, between the measured microphone output $\hat{\boldsymbol{p}}$ and the modeled microphone output $\boldsymbol{p}$ will be minimized:

$$\min_{\boldsymbol{q}} ||\boldsymbol{e}||_2^2 = \min_{\boldsymbol{q}} ||\boldsymbol{G}\boldsymbol{q} - \hat{\boldsymbol{p}}||_2^2\,. \tag{10.17}$$

The solution of Eq. (10.17) can be calculated by the determination of the pseudoinverse $\boldsymbol{G}^+_{\mathrm{LMS}}$ of the system matrix $\boldsymbol{G}$:

$$\boldsymbol{q}_{\mathrm{LMS}} = \boldsymbol{G}^+_{\mathrm{LMS}}\hat{\boldsymbol{p}} = \left[\boldsymbol{G}^H\boldsymbol{G}\right]^{-1}\boldsymbol{G}^H\hat{\boldsymbol{p}}\,. \tag{10.18}$$

The solution $\boldsymbol{q}_{\mathrm{LMS}}$ can also be calculated by making use of the singular value decomposition of $\boldsymbol{G}$

$$\boldsymbol{q}_{\mathrm{LMS}} = \boldsymbol{V}\boldsymbol{\Sigma}^+_{\mathrm{LMS}}\boldsymbol{U}^H\hat{\boldsymbol{p}} \tag{10.19}$$

with $\boldsymbol{G} = \boldsymbol{U\Sigma V}^H$. The entries of the matrix Σ^+_{LMS} are the inverses of the singular values σ_i of the matrix $\boldsymbol{G}$

$$\left(\Sigma^+_{\text{LMS}}\right)_{ij} = \left(\Sigma^{-1}\right)_{ij} = \begin{cases} \dfrac{1}{\sigma_i} & \text{for } i = j\,, \\ 0 & \text{else}\,. \end{cases} \tag{10.20}$$

The right hand side of Eq. (10.20) emphasizes the problem of small singular values. In the case of small (or zero) singular values, the reciprocal on the right hand side of Eq. (10.20) is very large. The problem of small singular values of a matrix $\boldsymbol{G}$ can be characterized by its condition number defined as

$$\kappa(\boldsymbol{G}) = ||\boldsymbol{G}||_2 \, ||\boldsymbol{G}^+||_2\,, \tag{10.21}$$

where $\boldsymbol{G}^+$ is a pseudoinverse of $\boldsymbol{G}$. The condition number can also be calculated from

$$\kappa(\boldsymbol{G}) = \frac{\sigma_{\max}}{\sigma_n}\,, \tag{10.22}$$

where $\sigma_{\max}$ is the largest and σ_n is the smallest non-zero singular value of $\boldsymbol{G}$. Detailed investigations on the characteristics of the condition number of $\boldsymbol{G}$ can be found in [6, 12, 19, 21].

This well-known property of ill-posedness of the problem, caused by vanishing singular values of the system matrix, results in an unstable behavior of the solution when using the method of *least-mean-squares* [6, 7, 12, 21]. For this reason, regularization methods have to be applied.

Tikhonov-regularization

A widely spread method is the regularization according to *Tikhonov*. Here, a (regularization) parameter β is introduced to control the necessary degree of regularization to mitigate the numerical problems. The solution of the problem (10.16) can be written as

$$\boldsymbol{q}_{\text{Tikh}} = \boldsymbol{G}^+_{\text{Tikh}}\hat{\boldsymbol{p}} = \left[\boldsymbol{G}^H\boldsymbol{G} + \beta I\right]^{-1}\boldsymbol{G}^H\hat{\boldsymbol{p}}\,. \tag{10.23}$$

Using the singular value decomposition of $\boldsymbol{G}$, the solution can be calculated from

$$\boldsymbol{q}_{\text{Tikh}} = \boldsymbol{V}\Sigma^+_{\text{Tikh}}\boldsymbol{U}^H\hat{\boldsymbol{p}}\,, \tag{10.24}$$

but now with the matrix

$$\left(\Sigma^+_{\text{Tikh}}\right)_{ij} = \begin{cases} \dfrac{\sigma_i}{\sigma_i^2 + \beta} & \text{for } i = j\,, \\ 0 & \text{else}\,. \end{cases} \tag{10.25}$$

In contrast to Eq. (10.20), the problem of small singular values is avoided by introducing the regularization parameter $\beta > 0$. This parameter can be determined, for example, by the *Generalized-Cross-Validation-* or the *L-Curve*-method. Detailed information about the calculation of the regularization parameter β can be found in [6, 1, 12, 21]. In the following, the *Tikhonov*-regularization is used with β parameters determined by means of the *L-Curve*-method.

Moore-Penrose-regularization

Another method is the calculation of the *Moore-Penrose*-pseudoinverse of the matrix $\boldsymbol{G}$. The matrix $\boldsymbol{G}_{\mathrm{MP}}$ is called the *Moore-Penrose*-pseudoinverse of the matrix $\boldsymbol{G}$, if $\boldsymbol{G}_{\mathrm{MP}}$ satisfies the following properties

$$\boldsymbol{G}_{\mathrm{MP}}\boldsymbol{G}\boldsymbol{G}_{\mathrm{MP}} = \boldsymbol{G}_{\mathrm{MP}}\,, \quad \boldsymbol{G}\boldsymbol{G}_{\mathrm{MP}}\boldsymbol{G} = \boldsymbol{G}\,,$$

$$(\boldsymbol{G}\boldsymbol{G}_{\mathrm{MP}})^H = \boldsymbol{G}\boldsymbol{G}_{\mathrm{MP}}\,, \ (\boldsymbol{G}_{\mathrm{MP}}\boldsymbol{G})^H = \boldsymbol{G}_{\mathrm{MP}}\boldsymbol{G}\,.$$

The *Moore-Penrose*-pseudoinverse $\boldsymbol{G}_{\mathrm{MP}}$ can be calculated from the method of *truncated singular-value-decomposition*

$$\boldsymbol{G}_{\mathrm{MP}} = \boldsymbol{V}\Sigma_{\mathrm{MP}}^{+}\boldsymbol{U}^H\,, \tag{10.26}$$

where

$$\left(\Sigma_{\mathrm{MP}}^{+}\right)_{ij} = \begin{cases} \dfrac{1}{\sigma_i} & \text{if } i = j \wedge \sigma_i > \text{tol.} \\ 0 & \text{else}\,, \end{cases} \tag{10.27}$$

where tol. $\approx O\left(10^{-12}\right)$ (depending on machine accuracy and system matrix $\boldsymbol{G}$). The common goal of *Tikhonov-* and *Moore-Penrose*-regularization is the improvement of solving ill-posed problems with small singular values. *Tikhonov*-regularization increases all singular values of the system matrix $\boldsymbol{G}$, whereas *Moore-Penrose*-regularization is applied only to small singular values and sets them to zero. In the next section, the described methods are applied to different model configurations.

10.3 Results and Analysis

In this section, the reconstruction algorithm is applied to a model configuration. The dependence of the accuracy of the reconstructed source positions and source amplitudes is studied by varying the sensor arrangement. Also, the effect of contaminating

noise is investigated. Finally, a search algorithm is presented to locate the positions of the sound sources not coinciding with the assumed source distribution.
In order to avoid an underdetermined system of equations, the number of sensors have to be at least equal to the number of model sources (compare Eq. (10.5)). The two different methods to invert the linear system of equations (described in Sec. 10.2.6) are applied and evaluated separately.

In Fig. 10.4, the source plane at axial position $x_0 = 0$ and the assumed distribution of modeled sound sources (21) are plotted (blue squares). Three sound sources with non-zero source amplitudes (red open circles), which are to be reconstructed by the algorithm, are plotted as well. The positions and amplitudes of the non-trivial sources chosen for this case are:

$$\begin{aligned} \boldsymbol{x}_{01} &= [0.0,\ \pi,\ 2/3]\,, \quad q_{01} = 1.0\,, \\ \boldsymbol{x}_{02} &= [0.0,\ \pi,\ 1/3]\,, \quad q_{02} = 0.8\,, \\ \boldsymbol{x}_{03} &= [0.0,\ \pi/2,\ 1/3]\,, \ q_{03} = 0.5\,. \end{aligned} \tag{10.28}$$

Firstly, the dependence of the accuracy of the reconstruction algorithm on the arrangement of the microphone array is investigated.

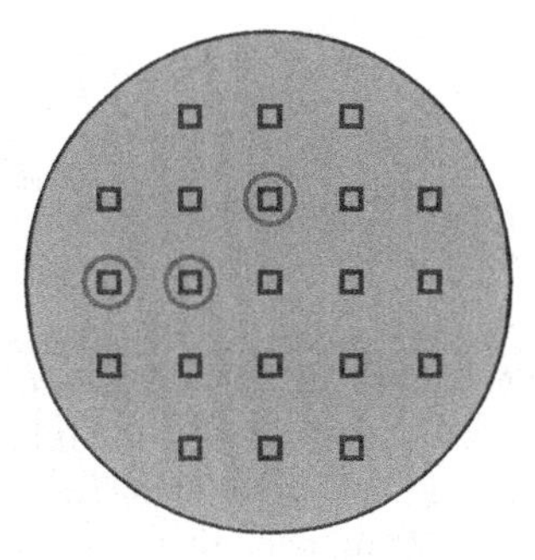

Fig. 10.4 Assumed source distribution (blue squares) in the source plane $x_0 = 0$, red circles: sources with non-zero amplitudes

10.3.1 Optimization of the Sensor Arrangement

In this section, the dependence of the reconstruction error on the sensor arrangement for 48 microphones in total is investigated. The number of axial sensor rings N_x and the number of azimuthal sensors per ring N_ϑ is varied while keeping the overall sensor count $N_\vartheta \cdot N_x = 48$ constant. In Table 10.2, the numbers N_ϑ and N_x for the different investigated microphone arrays are given. The dimensionless frequency of the sound sources is $k = 1$. For this investigation, a superposition of 0.1 % white gaussian noise is assumed. The dependence of the reconstruction accuracy on contaminating noise is investigated in Sec. 10.3.2. The evanescent modes taken

Table 10.2 Arrangement of 48 microphones

arrangement number	number of microphones azimuthal N_ϑ	axial N_x
1	1	48
2	2	24
3	3	16
4	4	12
5	6	8
6	8	6
7	12	4
8	16	3
9	24	2
10	48	1

into account for modeling the sound field (see Sec. 10.2.6) are chosen using the a maximum decay rate of $\delta^0_{m\mu} = -75$ (see Sec. 10.2.3).
To assess the accuracy of the reconstruction results, the relative error of the reconstructed to the given source amplitudes is defined as

$$\varepsilon = \frac{||\boldsymbol{q}_{\text{reconstr.}} - \boldsymbol{q}_{\text{given}}||_2}{||\boldsymbol{q}_{\text{given}}||_2} \,. \tag{10.29}$$

Table 10.3 lists and Fig. 10.5 shows the relative errors ε for the ten sensor arrangements and the two different inversion techniques. It is found that the accuracy of the reconstruction results is increasing with increasing number of azimuthal microphones. Using the *Moore-Penrose*-inversion, for configurations with no more than eight azimuthal sensors, the relative error between the reconstructed and the given amplitudes is greater than 1 and can be considered as unsatisfactory. By increasing the number of azimuthal sensors N_ϑ, the relative error decreases rapidly and the reconstruction can be considered as exact. With *Tikhonov*-inversion, the relative error is less sensitive to the sensor geometry. However, the same trend can be identified. The relative error decreases with increasing number of azimuthal sensors N_ϑ, albeit slower than using *Moore-Penrose*-inversion. The best reconstruction results can be achieved with *Moore-Penrose*-inversion and a microphone array where 48 or 24 sensors are located at one or two axial position (arrangement # 9 and # 10). In Figs. 10.6 and 10.7, the results for the reconstruction of the three specified sound sources using the two inversion techniques with arrangement # 9 is shown. The assumed source distribution (blue squares), the given (red open circles) and the reconstructed source amplitudes (black solid circles) are plotted over the source plane.

As expected from Table 10.3, the reconstruction with *Moore-Penrose*-inversion is exact. The three given sound sources are reconstructed exactly and no artifical sound sources appear. Figure 10.7 shows that the reconstruction with *Tikhonov*-regularization is also satisfactory. The positions of the three sound sources are reconstructed exactly. However, the source amplitudes are not reconstructed exactly, but the result is still reasonable. In this case, a few small artifical sound sources

Table 10.3 Relative error of reconstruction using *Moore-Penrose* or *Tikhonov* with *L-Curve* for different microphone arrangements calculated with $\delta^0_{m\mu} = -25$

arrange-ment number	relative error ε of reconstruction using *Moore–Penrose*	*Tikhonov–L – Curve*
1	$3.11e+08$	0.82
2	$2.04e+06$	0.535
3	447	0.584
4	255	0.465
5	5.46	0.477
6	0.6	0.345
7	0.0456	0.389
8	0.047	0.141
9	0.0309	0.115
10	0.0163	0.0687

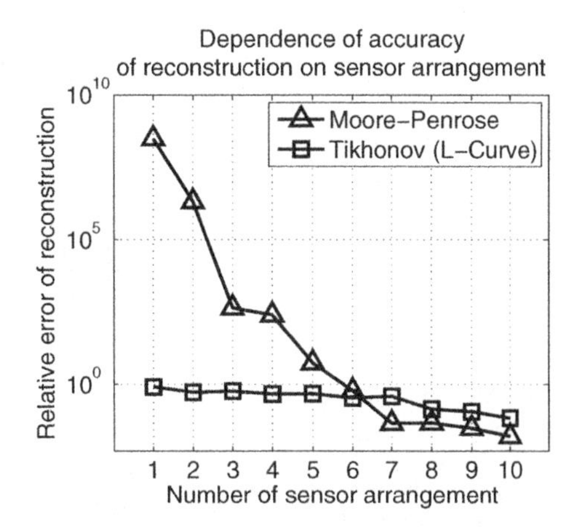

Fig. 10.5 Dependence of relative error of reconstruction on sensor arrangement using *Moore-Penrose*-inversion (triangles) and *Tikhonov*-regularization (squares)

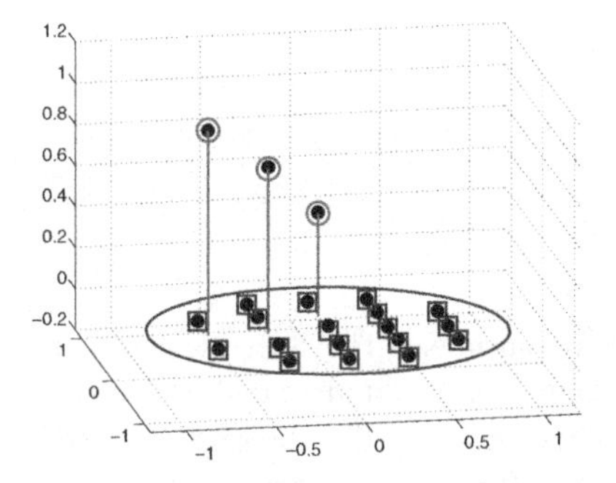

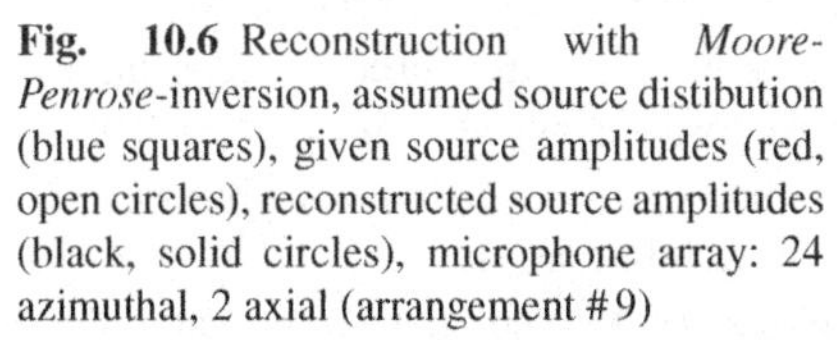

Fig. 10.6 Reconstruction with *Moore-Penrose*-inversion, assumed source distibution (blue squares), given source amplitudes (red, open circles), reconstructed source amplitudes (black, solid circles), microphone array: 24 azimuthal, 2 axial (arrangement #9)

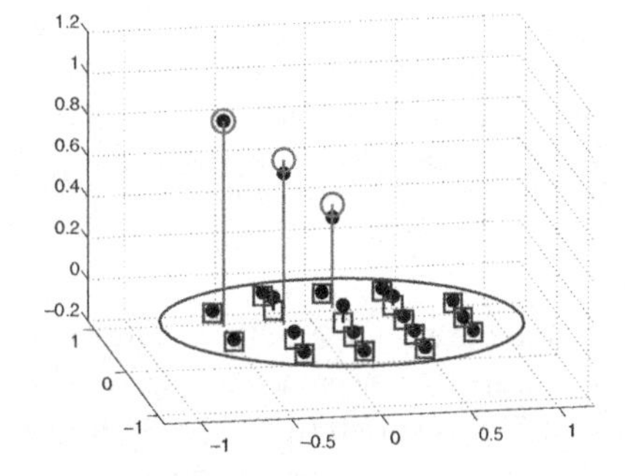

Fig. 10.7 Reconstruction with *Tikhonov (L-Curve)*-inversion, assumed source distibution (blue squares), given source amplitudes (red, open circles), reconstructed source amplitudes (black, solid circles), microphone array: 24 azimuthal, 2 axial (arrangement #9)

are introduced during the reconstruction. In the following, a microphone array of two microphone rings with 24 sensors per ring was used for the reconstruction. The arrangement of the array is displayed in Fig. 10.8.

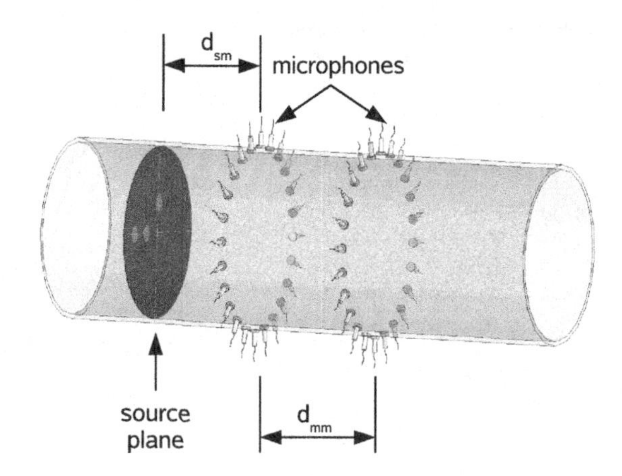

Fig. 10.8 Microphone array: two axial microphone rings each with 24 microphones and the source plane, $d_{sm} = 0.02R$, $d_{mm} = 0.02R$

10.3.2 Effect of Noise on the Reconstruction Accuracy

In this section, the dependence of the reconstruction results on contaminating noise is investigated. White Gaussian noise is added to the vector of pressure measurements $\boldsymbol{p}$. The noisy input data $\hat{\boldsymbol{p}}$ is given by

$$\hat{\boldsymbol{p}} = \boldsymbol{p} + \boldsymbol{e}, \tag{10.30}$$

where the vector $\boldsymbol{e}$ is the vector of white gaussian noise. The dependence of the reconstruction accuracy on the Signal-Noise-Ratio (*SNR*) is studied in the following. The *SNR* is varied from 5 to 40, followed by an average of 1000 measured pressure data. Sensor arrangement #9 was used (see Sec. 10.3.1). One source plane at axial position $x_0 = 0$ and 21 model sources are assumed. Again, three sources having non-trivial source amplitudes are presetted. According to the result of Sec. 10.3.1, one can expect that the reconstruction with the *Moore-Penrose*-algorithm is more accurate than the reconstruction with *Tikhonov*-regularization.
In Fig. 10.9, for both configurations the relative error between the reconstructed and the given source amplitudes according to Eq. (10.29) is plotted. Here, the importance of the application of regularization methods becomes evident. For small *SNR*, the reconstruction with *Tikhonov*-regularization is more accurate than the other method. For high *SNR* (> 20) the reconstruction with the *Moore-Penrose*-algorithm is more accurate, as it was expected from the simuation results of Sec. 10.3.1.

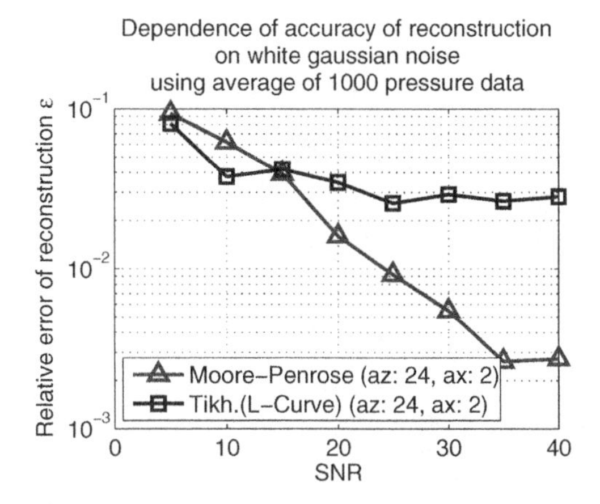

Fig. 10.9 Relative error of reconstruction for different *SNR* using *Moore-Penrose* (blue triangles)- and *Tikhonov (L-Curve)* (black squares)-regularization with modes with $\delta_{m\mu} \geq \delta^0_{m\mu} = -75$

10.3.3 Reconstruction of Sound Sources not located on assumed Source Distribution

In the preceding sections, it was assumed that the location of the actual sources coincided with the position of some model sources on the reconstruction grid. Mostly, this will not be the case in common applications. For this reason, a *scanning-technique* is introduced to estimate the positions of the sound sources without knowing them a priori. The reconstruction algorithm using the *scanning technique* should find the presetted sound source in an approximate way.

The idea of the *scanning-technique* is to subdivide the source plane into smaller partitions (here: division of the circular source plane in four parts). The first part is discretized by a number of assumed monopole sources, and the reconstruction algorithm is applied. The positions of the sources having amplitudes above a certain threshold are stored. Subsequently, the second part is discretized and the new source distribution is the union of these new model sources and those stored from the reconstruction in the first part. After having applied the reconstruction to all parts and in-line stored all positions of sources with large amplitudes, the reonstruction algorithm is applied again to the resulting source distribution comprising all dominant source locations from the individual partitions.

Now, the *scanning technique*, as described above, is applied to three cases. The configurations are shown in Figs. 10.10a–10.10c. The source plane at axial position $x_0 = 0$ and the assumed distribution of modeled sound sources (21) are plotted (blue squares) together with the presetted sound sources providing non-zero source amplitudes (red open circles), which are to be reconstructed by the algorithm.
In case 1, only one real source was presetted near the assumed grid. In case 2, two real sources not matching the assumed grid are presetted in the same quadrant. Case 3 is an extension of case 2: Here the two real sources are presetted in different quadrants. The reconstruction results based on the proposed *scanning technique* and *Tikhonov*-inversion for the 3 model configurations are presented in Figs. 10.11a–

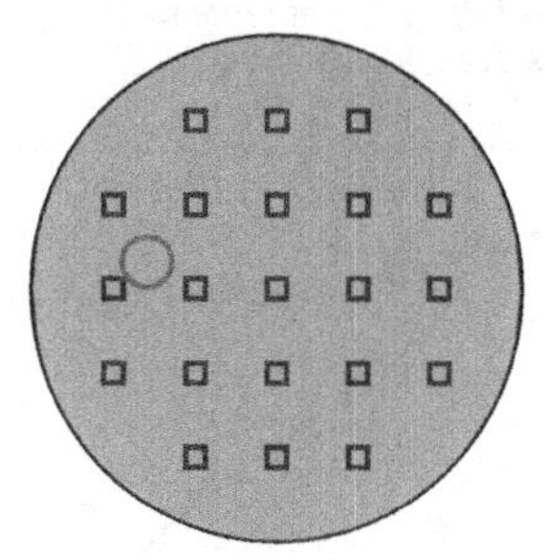

Fig. 10.10a Assumed source distribution (blue squares) in the source plane $x_0 = 0$, red circles: sources with non-zero amplitudes, case 1

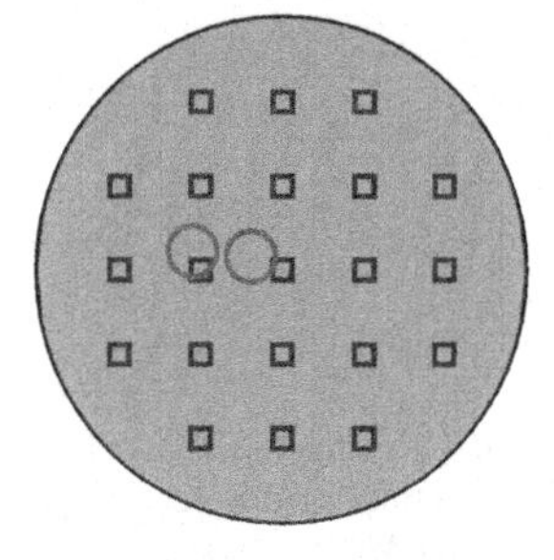

Fig. 10.10b Assumed source distribution (blue squares) in the source plane $x_0 = 0$, red circles: sources with non-zero amplitudes, case 2

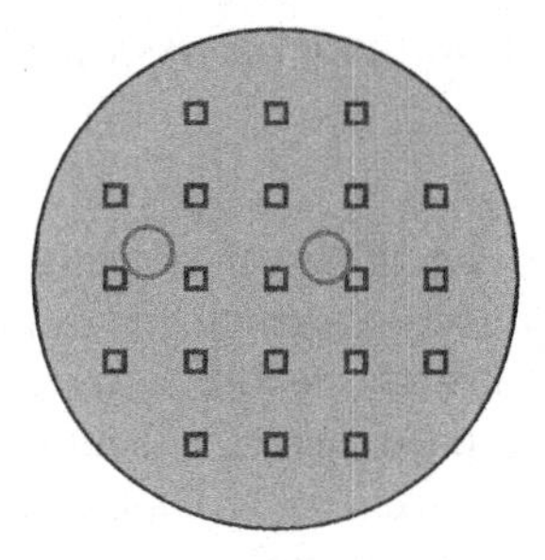

Fig. 10.10c Assumed source distribution (blue squares) in the source plane $x_0 = 0$, red circles: sources with non-zero amplitudes, case 3

10.11c as follows: the amplitudes of the presetted sound sources are plotted as red open circles and the amplitudes of the reconstructed sound sources are plotted as black solid circles over the source plane. For clarity, the assumed source grid is not shown.

The reconstruction result of the first model configuration, plotted in Fig. 10.11a, exhibits some inaccuracy in the position and amplitude of the presetted sound source. A second artifical sound source is introduced, as well, which is an undesired event. In case 2, the algorithm is able to reconstruct the two given sources. The positions and the amplitudes of the reconstructed sound sources are approximated with a high level of accuracy. In case 3, two sources in different quadrants are presetted. Again, the algorithm is able to reconstruct the two given sources. The positions and the amplitudes of the reconstructed sound sources are reconstructed with reasonable accuracy. It can be noted that in the cases 2 and 3, no artifical sound sources

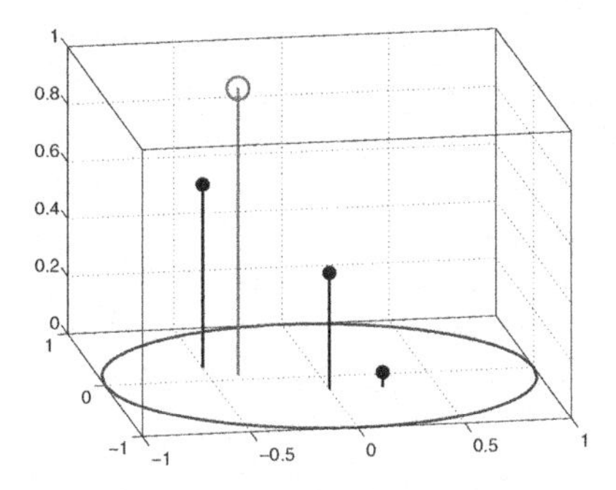

Fig. 10.11a Reconstruction of case 1 with *Tikhonov*-inverting and *scanning technique*, given source amplitudes (red open circles), reconstructed source amplitudes (black solid circles)

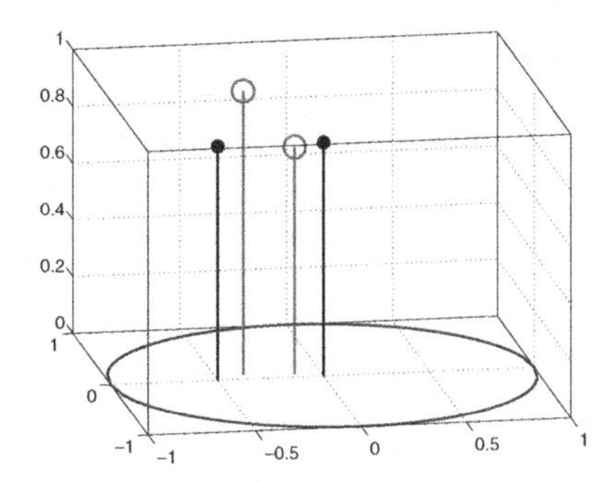

Fig. 10.11b Reconstruction of case 2 with *Tikhonov*-inverting and *scanning technique*, given source amplitudes (red open circles), reconstructed source amplitudes (black solid circles)

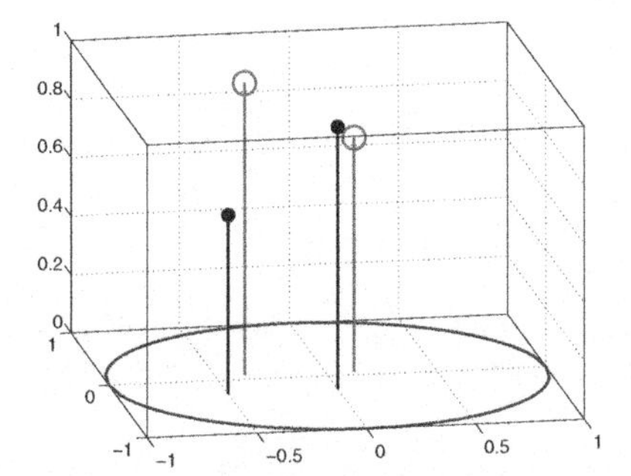

Fig. 10.11c Reconstruction of case 3 with *Tikhonov*-inverting and *scanning technique*, given source amplitudes (red open circles), reconstructed source amplitudes (black solid circles)

are introduced by the reconstruction algorithm. It was shown that the reconstruction of sound sources having a mismatch between the assumed source distribution and real source position by application of a *scanning technique* is successful with at least reasonable accuracy for all investigated cases.

10.3.4 Effect of Reflection at the Combustion Chambers Outlet

In this section, we investigate the dependence of the accuracy of the reconstructed source positions and amplitudes on the reflection of the plane wave at the combustion chamber outlet of a finite combustion chamber. We assume a reflection of 95% of the plane wave at the downstream end of the combustion chamber. We preset one sound source having non-zero amplitude, which is to be reconstructed. In Fig. 10.12a, the reconstruction result using *Moore-Penrose*-regularization with reflection and no correction is plotted. It can be seen that the presetted source is reconstructed exactly in position and amplitude. However, an artifical source is introduced by the reconstruction. This is an undesired result.

If the sound pressure field in the finite combustion chamber is corrected from the reflection of the plane wave at the combustion chamber outlet (see Sec. 10.2.5), the reconstruction is exact and no artifical sound source appears. The result is plotted in Fig. 10.12b.

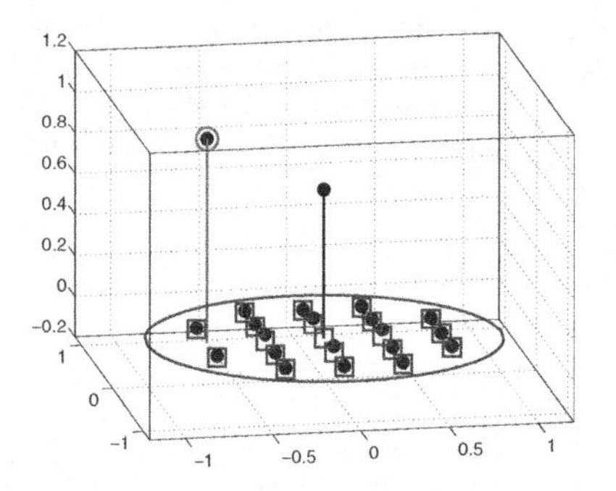

Fig. 10.12a Reconstruction result of one sound source using *Moore-Penrose*-regularization with reflection of 95% of plane wave at downstream end of combustion chamber without correction of the pressure field

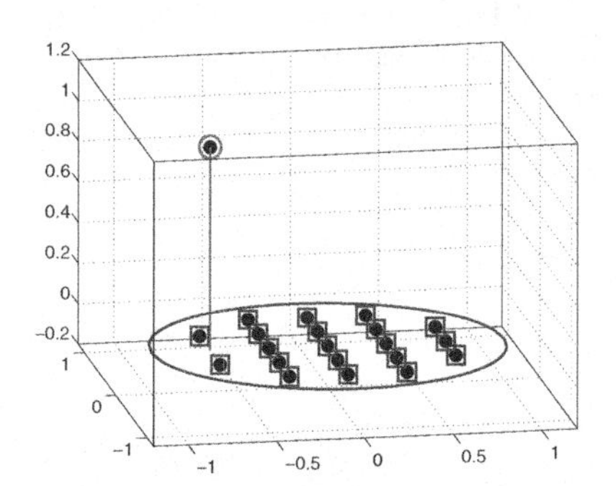

Fig. 10.12b Reconstruction result of one sound source using *Moore-Penrose*-regularization with reflection of 95% of plane wave at downstream end of combustion chamber with correction of the pressure field

10.3.5 Effect of Mean Flow

In this section, we investigate the dependence of the accuracy of the reconstructed source positions and amplitudes from the mean flow. Here, we use the configuration described at the beginning of this section. We preset three sound sources in one axial source plane. Additionally, we assume a constant mean flow in axial direction with Mach number $M = 0.2$.

In the first case, the mean flow will be ignored by the reconstruction algorithm: the Green's function for a stationary gas is used for the reconstruction (see Eq. (10.3)). Of course, this modelling error results in an incorrect reconstruction of the three sound sources, which can be seen in Fig. 10.13a. Using the correct Green's function (Eq. (10.9)), the reconstruction is exact (see Fig. 10.13b).

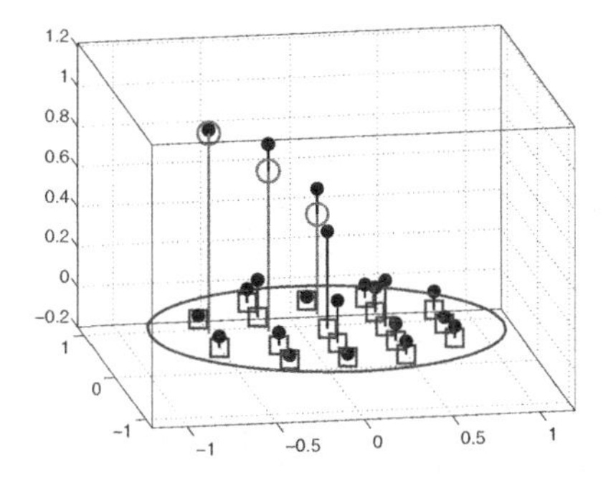

Fig. 10.13a Reconstruction result of three sound sources using *Moore-Penrose*-regularization and mean flow with $M = 0.2$, mean flow not considered

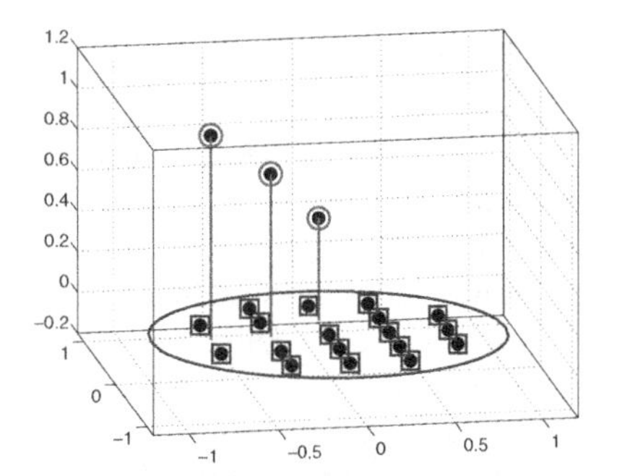

Fig. 10.13b Reconstruction result of three sound sources using *Moore-Penrose*-regularization and mean flow with $M = 0.2$, mean flow considered

10.4 Conclusion

In this project, an algorithm for the determination of the locations and the strengths of sound sources in closed combustion chambers using measured data of wall-flush-mounted microphones is presented. The generic procedure of sound source reconstruction was introduced based on the discretization of the region of interest by a set of monopole sources. The unknown source amplitudes are determined by applying different regularization methods.
In simulations, the arrangement of a microphone array consisting of 48 sensors was optimized with almost non-existing background noise. Here, the algorithm of *Moore-Penrose* was found to be the best reconstruction tool. On the other hand, the reconstruction with *Moore-Penrose*-inversion is highly sensitive to contaminating noise. In this case, the reconstruction with *Tikhonov*-regularization using *L-Curve*-based parameter estimation turned out to be a successful strategy. Afterwards, the applicability of a *scanning technique* for the reconstruction of sound sources not exactly located on the assumed source distribution was investigated with positive result. Subsequently, the influence of the reflection of the plane wave at the downstream end of the combustion chamber was investigated. Here, the importance of correcting the measured sound pressure field from the reflected plane wave was shown. Finally, the reconstruction algorithm was applied to a pressure field with constant mean flow. An extended Green's function yields to good reconstruction results.

In future research, it is planned to determine the axial position of the source plane, especially for the case of sources located in more than one axial plane. Another point of continuing investigation is the extension of the reconstruction algorithm to cases with a temperature gradient in axial direction. It is planned to validate the reconstruction algorithm experimentally in a cold environment and later-on in a real combustion chamber.

References

[1] Ahmadian H, Mottershead JE, Friswell MI (1998) Regularisation methods for finite element model updating. Mechanical Systems and Signal Processing **12**(1): 47–64

[2] Bellucci V, Schuermans B, Nowak D, Flohr P, Paschereit CO (2005) Thermoacoustic Modeling of a Gas Turbine Combustor Equipped with Acoustic Dampers. Journal of Turbomachinery **127**(2): 372–379

[3] Dowling AP, Morgans AS (2005) Feedback Control of Combustion Oscillations. Annual Review of Fluid Mechanics **27**(2): 151–182

[4] Ducruix S, Schuller T, Durox D, Candel S (2003) Combustion Dynamics and Instabilities: Elementary Coupling and Driving Mechanisms. Journal of Propulsion and Power **19**(5): 722–734

[5] Güthe F, Lachner R, Schuermans B, Biagioli F, Geng W, Inauen A, Schenker S, Bombach R, Hubschmidt W (2006) Flame imaging on the ALSTOM EV-burner: thermo acoustic pulsations and CFD-validation. AIAA paper 2006-437

[6] Kim Y, Nelson PA (2003) Estimation of acoustic source strength within a cylindrical duct by inverse methods. Journal of Sound and Vibration **275**: 391–413

[7] Kim Y, Nelson PA (2003) Optimal regularisation for acoustic source reconstruction by inverse methods. Journal of Sound and Vibration **275**: 463–487

[8] Krebs W, Flohr P, Prade B, Hoffmann S (2002) Thermoacoustic Stability Chart for High-Intensity Gas Turbine Combustion Systems. Combustion Science and Technology **174**(7): 99–128

[9] Lieuwen TC, Yang V eds. (2005) Combustion Instabilities in Gas Turbine Engines. Progress in Astronautics and Aeronautics **210**. AIAA, Inc.

[10] Lowis CR, Joseph P (2005) Inversion Technique for Determining the Strength of Rotating Broadband Sources in Ducts. 11th AIAA/CEAS Aeroacoustics Conference, Monterey, CA

[11] Maynard JD, Williams EG, Lee Y (1985) Nearfield acoustic holography: I. Theory of generalized holography and the development of NAH. Journal of the Acoustical Society of America **78**: 1395–1412

[12] Nelson PA, Yoon SH (2000) Estimation of acoustic source strength by inverse methods: Part I, Conditioning of the inverse problem. Journal of Sound and Vibration **233**(4): 643–668

[13] Paschereit CO, Gutmark E, Weisenstein W (1998) Structure and control of thermoacoustic instabilities in a gas-turbine combustor. **138**: 213–232

[14] Paschereit CO, Schuermans B, Bellucci V, Flohr P (2005) Implementation of Instability Prediction in Design: ALSTOM Approaches. In: Combustion Instabilities in Gas Turbine Engines, Lieuwen TC, Yang V eds Progress in Astronautics and Aeronautics **210**: 445–480. AIAA, Inc.

[15] Rajaram R, Lieuwen T (2003) Parametric Studies of Acoustic Radiation from Premixed Flames. Combustion Science and Technology **175**(12): 2269–2298

[16] Richards GA, Thornton JD, Robey EH, Arellano L (2007) Open-Loop Active Control of Combustion Dynamics on a Gas Turbine Engine. Journal of Engineering for Gas Turbines and Power **129**(1): 38-48

[17] Rienstra SW, Tester BJ (2006) An Analytic Green's Function for a Lines Circular Duct Containing Uniform Mean Flow. 11th AIAA/CEAS Aeroacoustics Conference, Monterey, CA

[18] Schuermans BBH, Polifke W, Paschereit CO, van der Linden JH (2000) Prediction of Acoustic Pressure Spectra in Combustion Systems Using Swirl Stabilized Gas Turbine Burners. ASME paper 2000-GT-0105

[19] Tapken U, Enghardt L (2006) Optimization of Sensor Arrays for Radial Mode Analysis in Flow Ducts. 12th AIAA/CEAS Aeroacoustics Conference, Cambridge, MA

[20] Wäsle J, Winkler A, Sattelmayer T (2005) Influence of the Combustion Mode on Acoustic Spectra of Open Turbulent Swirl Flames. Proceedings of the 12th International Congress on Sound and Vibration, Lisbon, Portugal

[21] Yoon SH, Nelson PA (2000) Estimation of acoustic source strength by inverse methods: Part II, Experimental investigation of methods for choosing regularization parameters. Journal of Sound and Vibration **233**(4): 669–705

Made in United States
North Haven, CT
02 August 2025